Interfacial Wave Theory of Pattern Formation

Springer
Berlin
Heidelberg
New York
Barcelona
Budapest
Hong Kong
London
Milan
Paris
Santa Clara
Singapore
Tokyo

Springer Series in Synergetics Editor: Hermann Haken

An ever increasing number of scientific disciplines deal with complex systems. These are systems that are composed of many parts which interact with one another in a more or less complicated manner. One of the most striking features of many such systems is their ability to spontaneously form spatial or temporal structures. A great variety of these structures are found, in both the inanimate and the living world. In the inanimate world of physics and chemistry, examples include the growth of crystals, coherent oscillations of laser light, and the spiral structures formed in fluids and chemical reactions. In biology we encounter the growth of plants and animals (morphogenesis) and the evolution of species. In medicine we observe, for instance, the electromagnetic activity of the brain with its pronounced spatio-temporal structures. Psychology deals with characteristic features of human behavior ranging from simple pattern recognition tasks to complex patterns of social behavior. Examples from sociology include the formation of public opinion and cooperation or competition between social groups.

In recent decades, it has become increasingly evident that all these seemingly quite different kinds of structure formation have a number of important features in common. The task of studying analogies as well as differences between structure formation in these different fields has proved to be an ambitious but highly rewarding endeavor. The Springer Series in Synergetics provides a forum for interdisciplinary research and discussions on this fascinating new scientific challenge. It deals with both experimental and theoretical aspects. The scientific community and the interested layman are becoming ever more conscious of concepts such as self-organization, instabilities, deterministic chaos, nonlinearity, dynamical systems, stochastic processes, and complexity. All of these concepts are facets of a field that tackles complex systems, namely synergetics. Students, research workers, university teachers, and interested laymen can find the details and latest developments in the Springer Series in Synergetics, which publishes textbooks, monographs and, occasionally, proceedings. As witnessed by the previously published volumes, this series has always been at the forefront of modern research in the above mentioned fields. It includes textbooks on all aspects of this rapidly growing field, books which provide a sound basis for the study of complex systems.

A selection of volumes in the Springer Series in Synergetics:

Synergetics An Introduction
3rd Edition By H. Haken

The Fokker-Planck Equation
2nd Edition By H. Risken

Advanced Synergetics
2nd Edition By H. Haken

**Non-Equilibrium Dynamics
in Chemical Systems**
Editors: C. Vidal, A. Pacault

Self-Organization Autowaves and
Structures Far from Equilibrium
Editor: V. I. Krinsky

**Selforganization by Nonlinear
Irreversible Processes**
Editors: W. Ebeling, H. Ulbricht

**Temporal Disorder
in Human Oscillatory Systems**
Editors: L. Rensing, U. an der Heiden,
M. C. Mackey

The Physics of Structure Formation
Theory and Simulation
Editors: W. Guttinger, G. Dangelmayr

**From Chemical to Biological
Organization** Editors: M. Markus,
S. C. Müller, G. Nicolisov

**Propagation in Systems Far from
Equilibrium** Editors: J. E. Wesfreid,
H. R. Brand, P. Manneville, G. Albinet,
N. Boccara

**Optimal Structures in Heterogeneous
Reaction Systems** Editor: P. J. Plath

Foundations of Synergetics I
Distributed Active Systems
By A. S. Mikhailov

Foundations of Synergetics II
Complex Patterns
By A. S. Mikhailov, A.Yu. Loskutov

**Nonlinear Nonequilibrium
Thermodynamics I** Linear and Nonlinear
Fluctuation-Dissipation Theorems
By R. Stratonovich

**Nonlinear Nonequilibrium
Thermodynamics II** Advanced Theory
By R. Stratonovich

**Interdisciplinary Approaches
to Nonlinear Complex Systems**
Editors: H. Haken, A. Mikhailov

**Modelling the Dynamics
of Biological Systems**
Editors: E. Mosekilde, O. G. Mouritsen

Jian-Jun Xu

Interfacial Wave Theory of Pattern Formation

Selection of Dendritic Growth and Viscous Fingering in Hele–Shaw Flow

With 83 Figures and 11 Tables

 Springer

Professor Jian-Jun Xu

Department of Mathematics and Statistics
Burnside Hall, Room 1005, McGill University
805 Sherbrooks St. West, Montreal, Q.C., H3A 2K6
Canada

Series Editor:

Professor Dr. Dr. h.c.mult. Hermann Haken

Institut für Theoretische Physik und Synergetik der Universität Stuttgart
D-70550 Stuttgart, Germany
and
Center for Complex Systems, Florida Atlantic University
Boca Raton, FL 33431, USA

Library of Congress Cataloging-in-Publication Data
Xu, Jian-Jun, 1940-
Interfacial wave theory of pattern formation : selection of dendritic growth
and viscous fingering in Hele–Shaw flow / Jian-Jun Xu.
     p. cm. -- (Springer series in synergetics, ISSN 0172-7389)
Includes bibliographical references and index.
ISBN 3-540-63145-3 (alk. paper)
1. Crystalline interfaces. 2. Solid-liquid interfaces. 3. Liquid-liquid interfaces. 4. Dendritic crystals.
5. Crystal growth. 6. Fluid dynamics. 7. Nonlinear theories. I. Title. II. Series.
QC173.458.C78X82 1998   530.4'17--DC21   97-27289 CIP

ISSN 0172-7389

ISBN 3-540-63145-3 Springer-Verlag Berlin Heidelberg New York

Typesetting: Camera-ready copy from the author using a Springer T$_{\text{E}}$X macro package
Cover design: *design & production* GmbH, Heidelberg
SPIN 10552457   55/3144 - 5 4 3 2 1 0 - Printed on acid-free paper

Dedicated
to the young generation
of scientists

Preface

For the last several years, the study of interfacial instability and pattern formation phenomena has preoccupied many researchers in the broad area of nonlinear science. These phenomena occur in a variety of dynamical systems far from equilibrium. In many practically very important physical systems some fascinating patterns are always displayed at the interface between solid and liquid or between two liquids. Two prototypes of these phenomena are dendrite growth in solidification and viscous fingering in a Hele–Shaw cell. These two phenomena occur in completely different scientific fields, but both are described by similar nonlinear free boundary problems of partial-differential-equation systems; the boundary conditions on the interface for both cases contain a curvature operator involving the surface tension, which is nonlinear. Moreover, both cases raise the same challenging theoretical issues, interfacial instability mechanisms and pattern selection, and it is now found that these issues can be solved by the same analytical approach. Thus, these two phenomena are regarded as special examples of a class of nonlinear pattern formation phenomena in nature, and they are the prominent topics of the new interdisciplinary field of nonlinear science.

This research monograph is based on a series of lectures I have given at McGill University, Canada (1993–1994), Northwestern Polytechnical Institute, China (1994), Aachen University, Germany (1994), and the CRM summer school at Banff, Alberta, Canada (1995). I shall illustrate these phenomena, present the fundamental issues involved, and describe the recent theoretical developments. In particular, I shall discuss some long-standing problems in this field, such as the pattern selection principle and the interfacial instability mechanism, and systematically present a newly established, coherent, predictive theory of this subject: the interfacial wave theory.

This book will be useful for researchers, post-doctoral and graduate students in the fields of condensed matter physics, materials science, applied mathematics, mechanical engineering, and chemical engineering.

I am particularly indebted to Prof. J. D. Cole for teaching me the asymptotic analysis and various perturbation methods. I also thank him and Prof. M. E. Glicksman for introducing me to this challenging and exciting field and for their constant discussions and advise. During the past decade, I have also received invaluable help and advice from, and had discussions with, many other people in my investigation of this subject, as well as in the preparation of this book. This book could not have been written without their help; among them, Prof. C. C. Lin at MIT, Prof. S. H. Davis at Northwestern University, Prof. R. E. O'Malley at the University of Washington, and Dr. John R. Ockendon at Oxford University should be especially mentioned.

I thank my graduate student, Mr. Dong-Sheng Yu very much for carrying out the large amount of numerical computations and for preparing the figures. I also thank him and my graduate student, Mr. Mikhail Kharenko, for the careful verification of all the formulae. I also thank Miss Katina Michelis, a graduate student in my course, 'Topics in Applied Mathematics', for carefully proof-reading and checking the formulae in the manuscript. I would like to thank the production team at Springer-Verlag for their excellent editing work.

McGill University, Montreal   Jian-Jun Xu
May 1997

Table of Contents

1. Introduction

1.1 Interfacial Pattern Formations in Dendrite Growth and Hele–Shaw Flow

Pattern formation is a common phenomenon in nature. Generally speaking, when a system is driven away from its equilibrium state, a dynamic process will spontaneously take place. During this process, a certain type of pattern is often observed in the system. This pattern is characterized by the distribution of a certain physical quantity, such as streamlines, isothermal lines, isoconcentration lines for species, isopotential lines, etc. There are two major types of patterns that should be distinguished: (1) stationary patterns, and (2) fractal, or chaotic patterns. A stationary pattern may be either steady or time-periodic; while a fractal or chaotic pattern, like turbulence, is characterized by many different length scales. Stationary, or fractal patterns are often observed at the later stages of a process. The features of these patterns are determined by the nature of the corresponding dynamical system. These patterns are often irrelevant to the initial conditions of the system, and can persist for a long time. Before reaching the final stage, some transient patterns will be observed at the early stage of a process. These transient patterns are related to the initial conditions of the system, and are continually changing.

The essence and mechanism of fractal, chaotic pattern formation is a wide-open subject in all scientific areas. Our understanding of this complex phenomenon is obviously far from complete and in this book we will not deal with this topic. From a theoretical point of view, the formation of the stationary patterns is more fundamental, as well as simpler. The essence and mechanism of such formation is, of course, of great interest.

The most familiar examples of stationary patterns observed in fluid dynamics are the Rayleigh–Bénard convection cell and Taylor vortices in Taylor–Couette flow. But the most popular, prominent, and beautiful examples of pattern formation seen in mankind's experience are snowflakes and ice crystal growth. Figure 1.1 shows a photo of a typical snowflake in nature, and Fig. 1.2 shows an ice crystal growing in undercooled water. Both photos were taken by Yushinori Furukawa at Hokkaido University in Japan [1.1], [1.2].

Fig. 1.1. Photograph of a typical snowflake pattern taken by Furukawa

The patterns formed in these cases are somewhat different from the above-mentioned Rayleigh–Bénard convection cell or Taylor vortices. The patterns of snowflakes and ice crystals are interfacial, formed at the interface between the solid and vapor phase, or the solid and liquid phase, respectively. They are characterized by tree-like structures along the interface which are called dendritic patterns.

Dendritic patterns can appear in various growth systems for crystal materials. Figures 1.1 and 1.2 describe a typical case of dendrite growth. The system is originally setup in a metastable state, such as a liquid state with a given uniform undercooling temperature. The dynamical process is initiated by introducing a tiny seed or in other related ways. Once it starts, the crystal growth will proceed spontaneously according to the physical laws. There is no way to artificially control its tip growth velocity, or the formation of tree-like microstructure at its interface. It is for this reason that such a system is often called free dendrite growth in materials science, or spontaneous pattern formation in condensed matter physics. In a realistic free dendrite growth system, there will be a set of dendrites with different orientations growing simultaneously and interacting with each other. Eventually the system will form a so-called 'mushy zone'.

Fig. 1.2. The photo of an ice crystal taken by Furukuwa et al., which is growing at the undercooling temperature $\Delta T = 0.37\,\mathrm{K}$

Another typical example of dendrite growth occurs in unidirectional solidification. In Fig. 1.3 we show a sketch of a unidirectional solidification device, which consists of a thin sample in a Hele–Shaw cell and two uniform temperature zones at temperatures T_1 and T_2, which are, respectively, larger and smaller than the melting temperature of the material [1.3].

When the sample which is made of a binary mixture is at rest, the interface between solid and liquid is planar. The sample is pulled at a constant imposed velocity V. When the pulling velocity V increases, the interface will gradually develop the deep cellular structure shown in Fig. 1.4. When the pulling velocity V increases further and exceeds a critical value, the deep cellular structure will transform into arrays of dendrites, as shown in Fig. 1.5.

Dendrite growth also frequently appears in many other forms of material processing, such as alloy casting, metal ingot formation, and welding. Dendrite growth is one of the most profound subjects in the area of interfacial pattern formation. This is not only due to its underlying vital technical importance in the material processing industries, but also because dendrite growth represents a fascinating class of nonlinear phenomena occurring in inhomogeneous dynamical systems.

Theoretical investigation of dendrite growth at its current preliminary stage has mainly focused on single free dendrite growth, disregarding the interactions between neighboring dendrites in more complicated multiple dendrite growth, such as mushy zone and cellular dendrite formation.

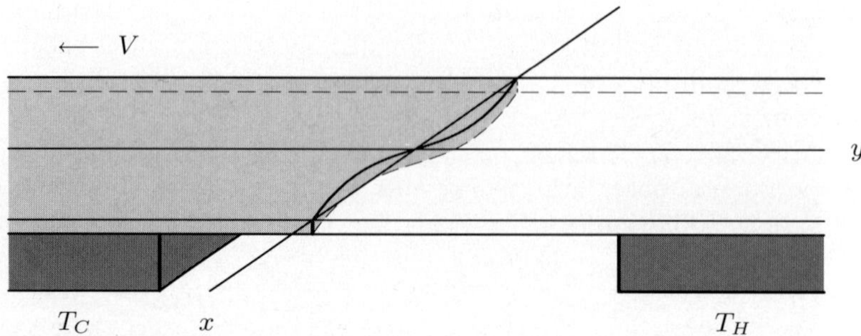

Fig. 1.3. Sketch of a unidirectional solidification device

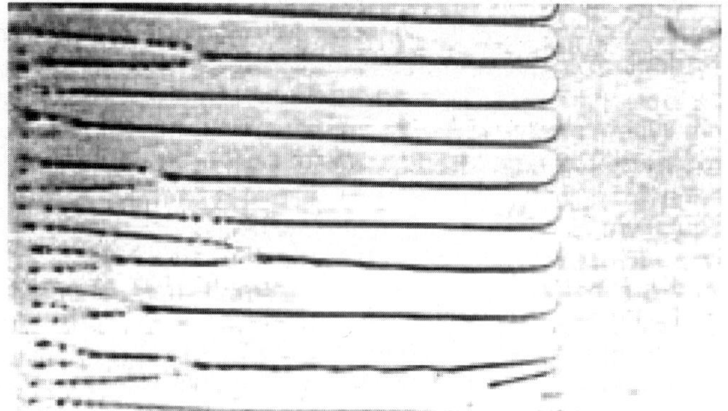

Fig. 1.4. Deep cellular structure formed during unidirectional solidification

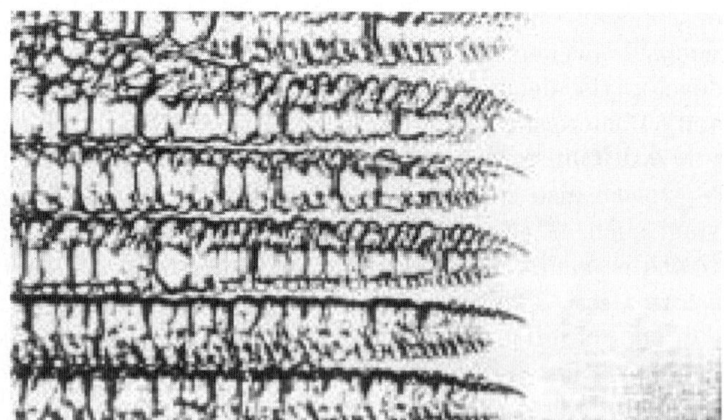

Fig. 1.5. Dendritic array structure formed during unidirectional solidification

A typical single free dendrite growth is shown in Fig. 1.6a, a photograph of single dendrite growth from a pure organic material melt succinonitrile (SCN) was made by Huang and Glicksman [1.4], while Fig. 1.6b shows the experimental recording curves for two-dimensional dendrite growth from a supersaturated NH_4Br solution, studied by Dougherty and Gollub [1.5].

Experimental observations show that at the later stage of growth a dendrite appears to have a smooth tip moving with a constant velocity. It emits a stationary wave-train, propagating along the interface towards the root. The essence and origin of this nonlinear interfacial phenomenon has been a fundamental subject in the field of condensed matter physics and material science for a long time [1.6]–[1.40]. It is believed that the understanding of this problem has a great significance for a much wider field including fluid dynamics, chemical engineering, biological science, etc., where similar pattern formation phenomena occur.

The first important contribution to this subject was the formulation of needle crystal growth and the Ivantsov solution (1947) [1.6]. Ivantsov considered the steady smooth needle growth from an undercooled pure melt, with zero surface tension at the interface between solid and liquid and found an exact similarity solution for this problem. This solution describes a continuous family of needle crystal growth, whose tip velocity is undetermined.

The second important contribution to this subject was the identification of the selection problem by Schaefer (1975) and Glicksman (1976) [1.11], [1.12]. Glicksman, Schaefer, and their co-workers first performed a series of careful experiments on dendrite growth, and accurately measured the tip velocities under various conditions. On the basis of this experimental data, it was confirmed that at the later stage of dendrite growth, the tip velocity is a uniquely determined function of the growth condition and the properties of the material.

The selection problem has motivated and inspired a broad range of theoretical and experimental research activities for over a generation. The basic questions of dendrite growth have been:

1. What mechanism determines the tip growth velocity?
2. What is the origin and essence of the microstructure?

Viscous fingering in the Hele–Shaw cell is another famous, classic, interfacial phenomenon which occurs in the field of fluid dynamics [1.41]–[1.45]. The Hele–Shaw cell is a thin, essentially 'two-dimensional', rectangular container in which the intrusion of one fluid into another leads to the formation of what are commonly called 'viscous fingers'. This simple device is an excellent environment for observing some fundamental features of interfacial dynamics, and the theory developed for viscous fingers provides a foundation for understanding many fluid–fluid systems. Observation of viscous fingers in Hele–Shaw cells has inspired over a generation of experimental and theoretical research since the first systematic study by Saffman and Taylor in

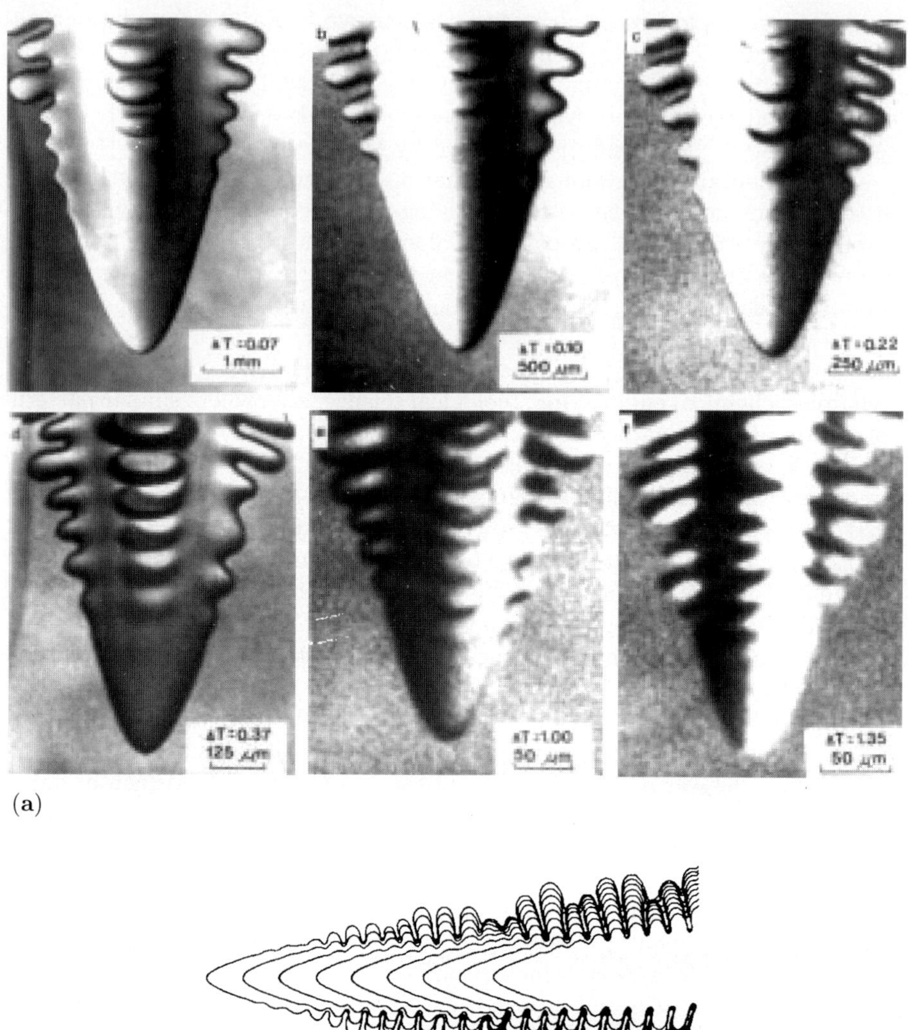

(a)

(b)

Fig. 1.6. (a) A photograph of a single free dendrite growth from pure organic material melt succinonitrile (SCN); (b) Experimental curve of two-dimensional dendrite growth from a supersaturated (NH_4Br) solution

1958 [1.42]. Experiments show that two types of stationary fingers can persist for long periods of time (see Fig. 1.7).

The first type, the subject of Saffman and Taylor's original 1958 research, are smooth, steady fingers, which occur if the surface tension parameter is large. When the surface tension parameter is small, another type of oscillatory and narrow, dendrite-like fingers may form. Oscillatory fingers were

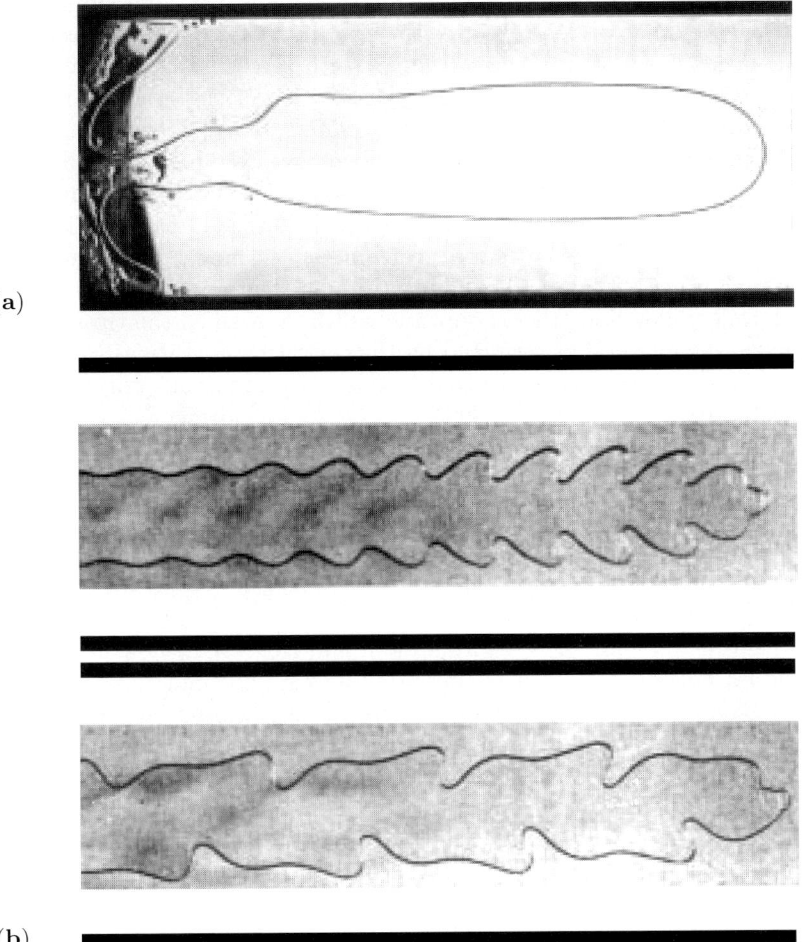

(a)

(b)

Fig. 1.7. Finger formation in a Hele–Shaw cell: **(a)** A smooth Saffman–Taylor finger; **(b)** An oscillatory finger

first discovered by Couder et al. (1986) [1.43], and later, by Kopf-Sill and Homsy [1.44]. The oscillatory fingers found by Couder et al. had a tiny bubble at the tip, while the oscillatory fingers found by Kopf-Sill and Homsy were formed naturally without a nose bubble. The presence of even a small bubble has a noticeable effect on the finger. However, once the surface tension parameter becomes sufficiently small, stationary fingers of any kind can no longer exist, causing splitting, shielding, and spreading into a chaotic pattern throughout the Hele–Shaw cell.

The study of finger formation in a Hele–Shaw cell has followed similar historical path to that of dendrite growth. The first solution to this problem was the solution for a steady smooth finger with zero surface tension. This solution is usually called the Saffman–Taylor solution (1958) [1.42]. However, it was actually found earlier by Zhuravlev in 1956 [1.41]. The Saffman–Taylor solution with zero surface tension, like the Ivantsov needle crystal solution, is a similarity solution. Hence, we cannot expect this solution to predict the tip velocity of the finger. As for dendrite growth, finger formation also has the selection problem. The selection problem for the finger formation was identified by Saffman–Taylor. They found through their experiments that, in a realistic Hele–Shaw flow, the asymptotic width, as well as the tip velocity of the finger was uniquely determined by the operating conditions.

So, the fundamental issues involved in the study of viscous fingering formation include:

1. Stability mechanisms,
2. Selection conditions for smooth fingers and both types of oscillatory fingers,
3. The transition point at which the oscillatory fingers with a nose bubble form chaotic or transient patterns.

The issues listed here for dendrite growth and Hele–Shaw flow have been long-standing, challenging problems in the fields of applied mathematics, fluid mechanics, condensed matters physics, chemical engineering, etc. for about half a century. These problems have preoccupied many investigators from various fields.

It has been recognized that these two distinct interfacial phenomena, dendrite growth in solidification and finger formation in a Hele–Shaw flow, are governed by some mathematically similar, entirely new instability mechanisms. The understanding of these new mechanisms evidently has a vital theoretical significance. It would also be of great importance for practical applications.

The present research monograph is devoted to the study of these interfacial phenomena.

It is seen that a tremendous amount of work has been done in this field. This book does not attempt to give an exhaustive summary of all significant works. Some well-written books and review articles with such purposes are available in the literature. Rather, this book intends to give a focused overview of those important theoretical results that have proved to be self-consistent, and have been supported by experimental data. We will address the major issues involved, summarize the main theories on the subject, and demonstrate the current understanding of the underlying mechanisms. In particular, the present book will describe the interfacial wave theory and compare the predictions of this theory with various experimental data.

1.2 A Brief Review of the Theories of Free Dendrite Growth

A first glance at dendrite growth phenomena reveals that the growth of dendrite-tip region is steady, having a monotonic interface shape, so that it can be well described by a steady state solution of the system; whereas the growth of the dendrite-stem region is oscillatory, forming a so-called side-branching structure, so that it is described by unsteady perturbed state solutions of the system. Therefore, in studying the phenomenon of dendritic growth, there are two major issues involved: (1) How does one specify the steady state of the system? (2) What are the stability properties of the system?

These issues have been investigated by many researchers for about a half of century. Many investigators have made positive contributions to this subject. A number of theories have been proposed, attempting, especially, to resolve the above-mentioned selection problem. In the following, I shall review some major theories, ignoring others, although they have all played a role at some stage of the research.

1.2.1 Maximum Velocity Principle (1976)

In 1974, Nash and Glicksman investigated the steady axially symmetric needle crystal growth with non-zero surface tension [1.10]. They attempted to find a selection principle for steady classic needle crystal growth by extending Ivantsov's problem to the case of non-zero isotropic surface tension ($\gamma \neq 0$). The classic needle crystal has a smooth tip and an infinitely long smooth root.

With the inclusion of surface tension, the needle crystal is nonisothermal due to the Gibbs–Thomson effect. The needle crystal growth problem becomes a highly nonlinear one, which is much more difficult to deal with than the isothermal case. Nash and Glicksman first attempted a complete mathematical formulation of the problem. They derived a nonlinear integro-differential equation and, as the boundary conditions for the interface shape, they proposed that the tip of the needle must have no cusp — the smooth tip condition, while the root of the needle must approach Ivantsov's paraboloidal needle — the *Nash–Glicksman far-field condition*. This mathematical problem is called the *Nash–Glicksman problem*.

The Nash–Glicksman problem has subsequently been extensively examined by many investigators. In particular, the Nash–Glicksman far-field condition has been applied by many researchers as a basic assumption, without proper justification. Unfortunately, it is now proven that this Nash–Glicksman far field condition is inadequate.

Nash and Glicksman did not justify the existence of a solution to their problem. They solved this problem near the tip by using an inaccurate numerical method. They found that under a given growth condition, the tip velocity has a unique maximum value. On the basis of this finding, they proposed that the selected tip velocity is just the maximum growth velocity. This selection mechanism was called the *maximum velocity principle* (MVP). The MVP was quickly ruled out by the authors themselves. In 1975–1976, Glicksman and his co-workers made precise measurements for dendrite growth of succinonitrile [1.11], [1.12]. They found that the calculated maximum velocity is larger than the real tip velocity by a factor of seven. They thought the discrepancy was too large to be acceptable. However, Glicksman and his co-workers did not realize that the Nash–Glicksman problem actually had no mathematical solution. Their numerical solution actually gave the wrong information.

1.2.2 Marginal Stability Hypothesis (1978)

Not long after Nash and Glicksman abandoned the maximum velocity principle as a solution of the selection problem, Müller-Krumbhaar and Langer proposed the *marginal stability hypothesis* (MSH) in 1978 [1.13]. The idea of Müller-Krumbhaar and Langer, in contrast to Nash and Glicksman, is that the selection must have something to do with the stability mechanism of the system.

Based on this idea, the authors introduced a new dimensionless stability parameter $\sigma = 2\ell_T \ell_c / \ell_t^2$, where ℓ_T is the thermal diffusion length determined by the thermal diffusivity, ℓ_c is the capillary length determined by the surface tension at the interface and ℓ_t is the tip radius of dendrite. They further deduced that the system of dendrite growth will be linearly unstable, as $\sigma < \sigma_*$ and concluded that the selected dendrite growth would be in the marginal stable state corresponding to this critical number σ_*.

Müller-Krumbhaar and Langer also proposed that the parameter σ_* can be written in the form:

$$\sigma_* = \left(\frac{\lambda_s}{2\pi\ell_t}\right)^2, \quad \lambda_s = 2\pi\sqrt{2\ell_T\ell_c}, \tag{1.1}$$

where λ_s is the shortest wavelength of a disturbance which would cause a plane interface, moving at velocity U, to suffer a Mullins–Sekerka instability. Then, they made an important assumption that in the marginal state, the tip radius of the dendrite is equal to the marginal stability wavelength λ_s of a plane interface. This assumption leads to the following stability criterion

$$\ell_t \equiv \lambda_s; \tag{1.2}$$

or

$$\sigma_* \approx \frac{1}{4\pi^2} \approx 0.0253. \tag{1.3}$$

Müller-Krumbhaar and Langer conducted numerical computations to solve the relevant linear eigenvalue problem. Their numerical results for the critical number σ_* are consistent with the above assumption.

The basic idea of the MSH is correct. It represents great progress compared with the maximum velocity principle and its results were in a good agreement with the experimental data in the small undercooling regime available at that time.

Nevertheless, from the theoretical point of view, the MSH theory is not satisfactory. At that time, Müller-Krumbhaar and Langer were unable to perform a global linear stability analysis for an inhomogeneous dynamical system to obtain the stability criterion for growth systems with a curved front. In the authors' words, "Unfortunately, we have not been able to obtain much information from the general equation of motion of F by analytic method, and have to resort to computational techniques. ..." [1.13]. It means that the MSH is not a self-consistent analytical theory. It is because of this theoretical weakness that the MSH was eventually abandoned by many researchers, including Langer himself, when a series of new challenges arose in the 1980s.

1.2.3 Microscopic Solvability Condition (MSC) Theory (1986–1990s)

In the 1980s, several groups of authors, such as Segur and Kruskal, Langer and co-workers, Kessler and Levine, Pelce and Pomeau, Ben Amer and Pomeau, Caroli and Caroli, and many others, reconsidered the steady needle crystal growth problem [1.15], [1.16] and [1.23]. Some simplified versions of the Nash–Glicksman problem, such as the so-called boundary layer model and geometry model, were proposed and carefully examined by using more powerful analytical approaches. It was very surprising when it emerged that, for the case of isotropic surface tension, the Nash–Glicksman problem has no solution! It was shown that a solution satisfying the Nash–Glicksman condition at the far field must have a cusp at the tip. When the surface tension is very small, the slope of this cusp is transcendentally small; so, a solution satisfying the smooth tip condition can never approach the Ivantsov paraboloid, satisfying the Nash–Glicksman far-field condition. In order for the system to have a steady needle growth solution, one may include a certain type of anisotropy in the surface tension. These unexpected results lead to the so-called *microscopic solvability condition* (MSC) theory. Hence, the MSC theory started with the study of the existence of steady needle solution, regarding side-branching structure formation as another subject, and was based on the acceptance of the Nash–Glicksman far-field condition. The MSC theory concluded that

(i) the anisotropy of surface tension is a necessary condition for dendritic growth. Without anisotropy, there will be no solution to the Nash–Glicksman problem;

(ii) with the inclusion of a small amount of surface tension anisotropy, the
system permits a discrete set of needle solutions to the Nash–Glicksman
problem.

Furthermore, MSC theory claimed that the solution with the largest tip velocity, among these steady classic needle solutions, is linearly stable, and so
is selected.

Although MSC theory raised some quite interesting mathematical issues
applying to the original phenomenon, it cannot be considered as a successful
physical theory. Disregarding the problem of not receiving much experimental
support (cf. [1.30]), the theory itself suffers some intrinsic difficulties. The key
is that the Nash–Glicksman far-field condition it adopts is over-prescribed
for the dendrite growth system. By looking for a steady solution with this
far-field condition, the MSC theory restricted itself to seeking a fixed point
limit solution for the dynamic system under investigation, while implicitly
excluding the possibility that the dynamic system may have other types of
limit solutions; hence, the pattern exhibited at the later stage of dendrite
growth may be time periodic, rather than steady.

H. Levine, one of the major contributors to the MSC theory, realized the
defect of this theory. He wrote in 1991: 'The resulting idea that one should
impose boundary conditions at infinity as a way of selecting the correct needle crystal, even though the steady state does not extend to infinity, is highly
non-trivial... We must realize that our approach may be incomplete. That is,
it may turn out that dendrite growth is an intrinsically time-dependent state.'
(see [1.51], p. 70). More seriously, it has been proven that when the anisotropy
of the surface tension is sufficiently small, the case of most interest, the steady
needle solution predicted by MSC theory, is actually linearly unstable. Therefore, it cannot be selected, except for a system with an anisotropy larger than
some critical value (cf. Chap.7).

1.2.4 Interfacial Wave (IFW) Theory (1990)

A more recent theory is the *interfacial wave* (IFW) theory. The IFW theory carries forward the stability idea contained in the previous MSH, and
is developed by using a unified matched asymptotic method and the multiple variable expansion approach. This theory modifies the Nash–Glicksman's
classic needle crystal problem for the steady state and redefines the steady
state for a dendrite growth system, by introducing the new idea of a generalized steady needle solution. Moreover, it explores the instability mechanisms
of the steady state, and on the basis of global stability analysis, establishes
the mechanisms of selection and pattern formation for dendritic growth, as
well as viscous fingering in a Hele–Shaw cell [1.27], [1.28], [1.40], [1.49].

The interfacial wave theory concludes that:

(i)   Dendritic growth is intrinsically a time-dependent wave phenomenon; the
dendrite growth system is subject to entirely new instability mechanisms

compared to a unidirectional solidification system. These instabilities are called *global instability mechanisms*; specifically, the global trapped wave instability and the low-frequency instability;

(ii) As a consequence, the anisotropy of surface tension is not a necessary condition for dendrite growth. The selection condition can be found even in the case of isotropic surface tension;

(iii) At the later stage of growth, the system permits a unique *global neutrally stable solution*. For systems with small anisotropy, the selected solution is not a stable steady state solution, but rather a time-periodic, oscillatory, neutrally stable solution. When the anisotropy is larger than a critical value, however, the selected solution is the steady needle solution.

The interfacial wave theory is a coherent and predictive analytical theory. It is in good agreement with all the available experimental data on dendrite growth, as well as on oscillatory fingering formation.

1.3 Macroscopic Continuum Model

In this book, we shall use the macroscopic continuum model. This implies that the liquid and solid bulk phases will be treated as continuous media, while the interface is considered as a geometric surface. This continuous medium model is well applicable to the pattern formation phenomena under investigation. The size of the microstructure that we study is on the micrometer scale, which is far larger than the mean-free path of the molecules. Furthermore, the time scale of the interface movement is far longer than the mean collision period of particles. Let us now imagine that, at first, we have a metastable undercooled liquid system. Then, due to nucleation or the presence of a seed, the solidification process starts. A front, which is described by the sharp interface between the solid phase and liquid phase, moves into the undercooled liquid phase and transforms more and more liquid into the solid phase. Eventually, the whole liquid phase is transformed into the more thermodynamically stable solid phase. During solidification, there may co-exist several interactive macroscopic transport processes in the system. First of all, due to the phase transition, the latent heat is released at the interface. Hence, a solidification system intrinsically has an inhomogeneous temperature field, and heat transfer through conduction is unavoidable. Furthermore, if the system involves two species, for instance, if it contains an impurity, then this binary mixture should be described not only by the temperature field, but also by the impurity concentration field. It is known that for a binary system with co-existing liquid and solid phase, the impurity concentration will have a jump at the interface because of the segregation effect. Therefore, the concentration field in a solidifying binary mixture system is always inhomogeneous. A mass transfer through mass diffusion in the concentration field will be present. Finally, the liquid phase in a solidification system may be in motion induced by various driving forces. For instance, the density change during phase transition will cause convection in the bulk liquid, because the interface acts as a mass sink when the liquid density is smaller than the solid density, or as a mass source when the liquid density is larger than the solid density. The buoyancy effect due to gravity may also cause convection, because the inhomogeneous temperature induces inhomogeneous density. Moreover, other forces on the body and external flow can all cause convection motion in the liquid phase. When convection exists in the liquid phase, there will be a momentum transfer in the system governed by fluid dynamics.

If the above three kinds of macroscopic transport processes co-exist, they will obviously interact with one another.

In this section, we shall give a general mathematical description of solidification on the basis of the macroscopic continuum model.

1.3.1 Macroscopic Transport Equations

Consider a front moving in an undercooled melt, with a characteristic velocity V in the reference frame in which the solid phase is at rest. For convenience, it may be called the rest frame. The melt may be a pure substance or a binary mixture. If it is a mixture, one of the species is dilute and may be regarded as an impurity. We assume that the liquid phase of the binary mixture has mass density ρ, thermal diffusivity $\kappa_{\rm T}$, specific heat $c_{\rm p}$, mass diffusivity of the impurity $\kappa_{\rm D}$; while the corresponding thermal characteristics in the solid phase are denoted by $\rho_{\rm S}$, $\kappa_{\rm TS}$ and $c_{\rm pS}$, $\kappa_{\rm DS}$, respectively. Very often, as a simplification, it is assumed that the mass diffusivity in the solid phase, $\kappa_{\rm DS}$, is zero.

Convective motion may occur in the liquid phase for a number of reasons as previously mentioned. We assume the liquid can be considered as an incompressible Newtonian fluid. The density inhomogeneity caused by the inhomogeneity of temperature is not important except in terms of a buoyancy effect. Thus, the Boussinesq approximation is applicable. The state of the system is described by the following macroscopic fields:

1. the temperature field in the liquid phase, $T(\mathbf{r}, t)$,
2. the temperature field in the solid phase, $T_{\rm S}(\mathbf{r}, t)$,
3. the concentration field of the impurity in the liquid phase, $C(\mathbf{r}, t)$,
4. the concentration field of the impurity in the solid phase, $C_{\rm S}(\mathbf{r}, t)$,
5. the absolute velocity field in the liquid phase, $\mathbf{U}(\mathbf{r}, t)$.

Here, we use $\mathbf{r}$ and t to denote the space vector and time, respectively. The governing equations consist of the heat conduction equation, mass diffusion equation and the Navier–Stokes equations in the Boussinesq model. These equations can be written as follows:

1. The heat conduction equation for the liquid phase. In the rest frame, the heat conduction equation can be written in the form:

$$c_{\mathrm{p}}\rho \left(\frac{\partial T}{\partial t} + \mathbf{U} \cdot \nabla T \right) = k_{\mathrm{T}} \nabla^2 T \tag{1.4}$$

or

$$\frac{\partial T}{\partial t} + \mathbf{U} \cdot \nabla T = \kappa_{\mathrm{T}} \nabla^2 T \tag{1.5}$$

where k_{T} is the heat conduction coefficient, while $\kappa_{\mathrm{T}} = \frac{k_{\mathrm{T}}}{c_{\mathrm{p}}\rho}$ is the thermal diffusivity in the liquid phase.

2. The heat conduction equation for the solid phase. In the rest frame, we similarly have

$$\frac{\partial T_{\mathrm{S}}}{\partial t} = \kappa_{\mathrm{TS}} \nabla^2 T_{\mathrm{S}} \tag{1.6}$$

where κ_{TS} is the thermal diffusivity in the solid phase.

3. The mass diffusion equation in the liquid phase:

$$\frac{\partial C}{\partial t} + \mathbf{U} \cdot \nabla C = \kappa_{\mathrm{D}} \nabla^2 C \tag{1.7}$$

where κ_{D} is mass diffusivity in the liquid phase.

4. The mass diffusion equation in the solid phase:

$$\frac{\partial C_{\mathrm{S}}}{\partial t} = \kappa_{\mathrm{DS}} \nabla^2 C \tag{1.8}$$

where κ_{DS} is mass diffusivity in the solid phase.

5. The continuity equation. For the incompressible liquid mixture, we have

$$\nabla \cdot \mathbf{U} = 0 . \tag{1.9}$$

6. The momentum equation:

$$\frac{\partial \mathbf{U}}{\partial t} + (\mathbf{U} \cdot \nabla)\mathbf{U} = -\frac{1}{\rho}\nabla P + \nu \nabla^2 \mathbf{U} + \beta(T - T_*)\mathbf{g} , \tag{1.10}$$

where P is the reduced pressure. If we introduce the vorticity Ω as

$$\Omega = \nabla \times \mathbf{U} , \tag{1.11}$$

the momentum equation can be replaced by the vorticity equation:

$$\frac{\partial \Omega}{\partial t} + \nabla \times (\Omega \times \mathbf{U}) = \nu \nabla^2 \Omega + \nabla \times \left\{ \beta(T - T_*)\mathbf{g} \right\} . \tag{1.12}$$

Here, we assume that there is no body force in the system, except for gravity; $\mathbf{g}$ is the acceleration of gravity; ν is the kinematic viscosity; β is the thermal expansion coefficient; and T_* is a reference temperature.

1.3.2 The Interface Conditions

The full set of boundary conditions must be specified for each specific problem. For instance, the far-field conditions will be different for different growth conditions. However, for the problems under investigation here the same type of interface conditions are, in general, always applied. We denote the interface shape in the rest frame by $S(\mathbf{r}, t) = 0$. Since the interface location during solidification is unknown, determining this interface shape function is a part of the solution. These problems, in mathematics, are all called free boundary problems or moving boundary problems.

To determine the interface condition, we assume that the system is in a local thermodynamic equilibrium at the interface. This implies the following. Let us take two control surfaces parallel and very close to the interface, one on the liquid side and the another on the solid side. The volume contained within these two control surfaces can be considered as the 'realistic interface domain', which we uniformly divide into many small volume elements by planes normal to the interface (cf. Fig. 1.8). At the same time, the interface is also divided into many small surface elements. When the division becomes finer and finer, the two control surfaces move closer and closer to one another, and the volume elements become smaller and smaller. As an approximation, one may assume that each sub-system, $(I) + (II) + (II)'$, consisting of the interface element (I), the volume element in liquid (II), and the volume element in solid $(II)'$, is in thermodynamic equilibrium. Certainly, different sub-systems are in different thermodynamic equilibrium states. This *local thermodynamic equilibrium state* assumption is a basic assumption that we shall employ throughout this monograph.

Based on this assumption, we can derive the following interface conditions in the rest frame used:

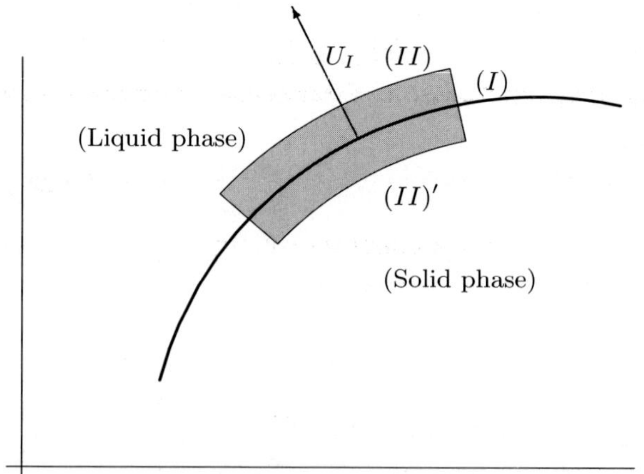

Fig. 1.8. Sketch of a moving front in an undercooled melt

1. *The thermodynamic equilibrium condition for temperature:*

$$T = T_{\mathrm{S}}\,.$$
(1.13)

2. Another thermodynamic equilibrium condition at the interface is the phase equilibrium condition. This implies that the chemical potential of the solid phase must equal the chemical potential of the liquid phase. This condition will lead to the so-called *Gibbs–Thomson condition*, which determines the temperature of the interface. The Gibbs–Thomson condition states that the solidification temperature at a general curved interface is different from the solidification temperature of the pure melt at a flat interface, say, T_{M0}. The correction comes from two sources. The first source is from the effect of the curvature of the interface on the surface energy. To demonstrate this effect, first consider a pure substance with a curved interface between the liquid and solid phases. Assume that this system is in thermodynamic equilibrium. From the minimum free energy principle at the local equilibrium state one finds that

$$T_{\mathrm{S}} = T_{\mathrm{M0}} \left(1 - \frac{\gamma}{\Delta H}\mathcal{K} \right),$$
(1.14)

where γ is the surface tension constant, ΔH is the latent heat per unit volume of the solid, T_{M0} is the melting temperature of a flat interface, and $\mathcal{K}$ is twice the local mean curvature of the interface.
The second source of the correction is due to the impurity at the interface. To study this effect, we consider a binary mixture with a flat interface between the liquid and solid phases, which is in thermodynamic equilibrium. From the thermodynamics, the equilibrium state of the system can be described by its phase diagram. A typical diagram is shown in Fig. 1.9. One can see that the melting temperature is determined by the liquidus curve and the concentration of the impurity in the liquid phase. When the impurity is dilute, the liquidus curve can be approximated by its tangent line at the impurity concentration $C = 0$. Suppose that the slope of the curve at $C = 0$ is $m < 0$. Then, as the impurity concentration on the liquid side of the interface is $C = C_{\mathrm{I}}$, the melting temperature will be different from the melting temperature of pure melt by an amount mC_{I}. Putting these two corrections together, we derive the interface temperature of a binary mixture as:

$$T_{\mathrm{S}} = T_{\mathrm{M0}} \left(1 - \frac{\gamma}{\Delta H}\mathcal{K} \right) + mC_{\mathrm{I}}\,.$$
(1.15)

This is the Gibbs–Thomson condition for a binary mixture, which relates the local melting temperature to the surface tension, the local mean curvature of the interface, and the local impurity concentration. This physical condition is the most important cause of the rich variety of pattern formation phenomena.

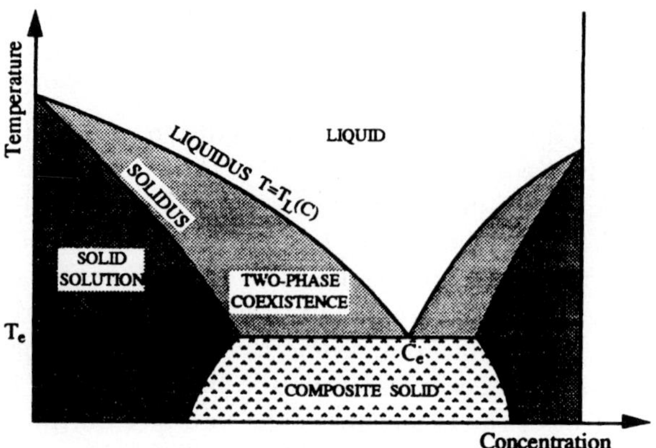

Fig. 1.9. Typical phase diagram of a binary mixture system. The shaded regions indicate what phases are in thermodynamic equilibrium in a sample of given bulk composition and uniform temperature. When the temperature T and bulk composition are between the liquidus and solidus curves, then liquid of the liquidus composition $C_L(T)$ is in equilibrium with solid of the solidus composition given by $\kappa C_L(T)$, which defines the local segregation coefficient $\kappa(C_L)$

3. *The energy conservation condition:* The enthalpy is conserved during the phase transition, so

$$\Delta H(U_{\mathrm{Sn}} - U_{\mathrm{I}}) = -\Delta H U_{\mathrm{I}} = \mathbf{n} \cdot \left[(k_{\mathrm{T}} \nabla T)_{\mathrm{liquid}} - (k_{\mathrm{T}} \nabla T)_{\mathrm{solid}} \right]$$

$$- \left[(c_{\mathrm{p}} T)_{\mathrm{liquid}} - (c_{\mathrm{p}} T)_{\mathrm{solid}} \right] \rho_{\mathrm{S}} U_{\mathrm{I}} , \qquad (1.16)$$

where

U_{Sn} is the normal component of the solid phase velocity at the interface, observed in the rest frame. We have assumed that it is zero.

U_{I} is the local growth velocity of the interface in the rest frame.

$\mathbf{n}$ is the normal vector of the interface.

The left-hand side of the formula (1.16) is the latent heat release per unit volume of solid per unit time. The first part of the right-hand side of (1.16) is the total enthalpy flux away from the interface to both the liquid and solid sides due to heat conduction; while the second part is the total enthalpy flux due to the convective motion.

4. *The segregation condition for the impurity:* According to the thermodynamics, the concentration of impurity will have a jump at the interface. The relationship of concentrations of the impurity on the two sides of the interface is derived from the phase diagram for the given binary mixture, namely,

$$C_{\mathrm{S}} = \kappa C , \qquad (1.17)$$

where κ is a material constant, the so-called segregation coefficient. Normally, $0 < \kappa < 1$. Hence, the impurity is normally rejected from the solid phase. As a result, it must build up in front of the advancing interface.

5. *The mass conservation condition for the impurity:* As we have seen, the interface acts as a mass source. The quantity of the impurity rejected from the interface into the liquid must be balanced by the mass flux in the liquid and solid due to the concentration gradient near the interface. Thus, one has

$$\Delta C U_{\mathrm{I}} = C U_{\mathrm{Ln}} - \mathbf{n} \cdot \left[(\kappa_{\mathrm{D}} \nabla \mathbf{C})_{\text{liquid}} - (\kappa_{\mathrm{D}} \nabla \mathbf{C})_{\text{solid}} \right] , \qquad (1.18)$$

where $\Delta C = C - C_{\mathrm{S}} = (1 - \kappa)C$ is the jump in the concentration of impurity at the interface and U_{Ln} is the normal component of the absolute velocity of a fluid element at the interface. In general, the mass diffusivity κ_{DS} in the solid is negligible.

6. *The total mass conservation condition:*

$$\rho(U_{\mathrm{Ln}} - U_{\mathrm{I}}) = \rho_{\mathrm{S}}(U_{\mathrm{Sn}} - U_{\mathrm{I}}) = -\rho_{\mathrm{S}} U_{\mathrm{I}} . \qquad (1.19)$$

7. *The continuity condition for the tangential component of velocity:*

$$U_{\mathrm{L}\tau} = U_{\mathrm{S}\tau} . \qquad (1.20)$$

As

$$U_{\mathrm{S}\tau} = 0 , \qquad (1.21)$$

one has

$$U_{\mathrm{L}\tau} = \mathbf{U} \cdot \mathbf{e}_{\tau} = 0 . \qquad (1.22)$$

In the above, the symbols have the meanings:
$\mathbf{e}_{\tau}$ the local unit tangent vector at the interface;
$U_{\mathrm{L}\tau}$ the tangential component of the absolute velocity of a liquid element at the interface;
$U_{\mathrm{S}\tau}$ the tangential component of the absolute velocity of a solid element at the interface.

Besides the above local thermodynamic equilibrium conditions and kinematic conditions, we also have some mechanical equilibrium conditions at the interface. These are the dynamic interface conditions derived from the momentum balance along the normal and tangential directions at the interface. Since the pattern formation problem is mostly not concerned with the stresses and strains in the solid phase, these dynamic conditions will not be needed.

One item in the above formulation that needs clarification is how to define and calculate the local growth velocity of the interface, U_{I}. Let the Cartesian coordinates (X_1, X_2, X_3) be those of the rest frame. We consider the motion of an arbitrary reference point $P:\{X_1(t), X_2(t), X_3(t)\}$ that always stays on the interface $S(X_1, X_2, X_3) = 0$. When the time t increases from $t \to (t + dt)$, the position of the point P will change from $\{X_1(t), X_2(t), X_3(t)\} \to \{X_1(t + dt), X_2(t + dt), X_3(t + dt)\}$. These increments must be subject to the relationship:

$$\mathrm{d}S = \frac{\partial S}{\partial t}\mathrm{d}t + \frac{\partial S}{\partial X_1}\mathrm{d}X_1 + \frac{\partial S}{\partial X_2}\mathrm{d}X_2 + \frac{\partial S}{\partial X_3}\mathrm{d}X_3 = 0\,. \tag{1.23}$$

The velocity of the motion of the point P is:

$$\mathbf{U}_\mathrm{I}(t) = \{X_1'(t), X_2'(t), X_3'(t)\} \tag{1.24}$$

and

$$\frac{\partial S}{\partial t} = -\nabla S \cdot \mathbf{U}_\mathrm{I}\,. \tag{1.25}$$

As a definition, we assume that the growth velocity of the interface remains along the direction normal to the interface, i.e., $\mathbf{U}_\mathrm{I} \parallel \mathbf{n}$. Thus, from (1.25), we have

$$\frac{\partial S}{\partial t} = -|\nabla S| \cdot |\mathbf{U}_\mathrm{I}| = -|\nabla S|U_\mathrm{I} \tag{1.26}$$

with

$$U_\mathrm{I} = -\frac{\frac{\partial S}{\partial t}}{|\nabla S|}\,. \tag{1.27}$$

For the sake of convenience, we often use a moving frame (x_1, x_2, x_3) with a velocity $\mathbf{V}$ as the reference frame, and set one of the coordinate axes, for instance, the x_3-axis whose unit vector is $\mathbf{e}_3$, parallel to $\mathbf{V}$. So, we have that

$$\mathbf{U} = \mathbf{u} + V\mathbf{e}_3\,. \tag{1.28}$$

In the moving frame, we denote the relative velocity field of a fluid by $\mathbf{u}$, and the relative growth velocity of the interface by $\mathbf{u}_\mathrm{I}$. From the moving frame to the rest frame, the following Galilean transformation holds:

$$\begin{cases} X_1 = x_1 \\ X_2 = x_2 \\ X_3 = x_3 + Vt \\ t = t\,. \end{cases} \tag{1.29}$$

The interface shape equation in the moving frame, $s(x_1, x_2, x_3, t) = 0$, can be obtained from that in the rest frame through the following relationships:

$$s(x_1, x_2, x_3, t) = S(x_1, x_2, x_3 + Vt, t) = 0 \tag{1.30}$$

and

$$S(X_1, X_2, X_3, t) = s(X_1, X_2, X_3 - Vt, t) = 0 \,. \tag{1.31}$$

Thus, we can calculate the growth velocity of the interface in terms of the interface shape equation in the moving frame. Namely,

$$U_{\mathrm{I}} = -\frac{\frac{\partial S}{\partial t}}{|\nabla S|} = -\frac{\frac{\partial s}{\partial t} - V\frac{\partial s}{\partial x_3}}{|\nabla s|} = \mathbf{V} \cdot \mathbf{n} - \frac{\frac{\partial s}{\partial t}}{|\nabla s|} = u_{\mathrm{I}} + V\mathbf{e}_3 \cdot \mathbf{n}\,, \tag{1.32}$$

where

$$u_{\mathrm{I}} = -\frac{\frac{\partial s}{\partial t}}{|\nabla s|} \tag{1.33}$$

is the local growth velocity of the interface observed in the moving frame and the vector normal to the interface

$$\mathbf{n} = \frac{\nabla s}{|\nabla s|} \,. \tag{1.34}$$

1.3.3 The Scaling and the Dimensionless System

As the first step of a mathematical treatment, one needs to make the system dimensionless by choosing a set of proper scales. The proper scales reflect the physical nature of the dynamical system. In many cases, using improper scales will cause difficulties in making either mathematical simplifications or physical interpretations.

For the problems formulated in the last section, we shall choose the characteristic growth velocity of the interface V as the scale for the velocity. The length scale can be chosen from the intrinsic length scales of the system. There are two intrinsic length scales in the system determined by the macroscopic transport processes: the thermal diffusion length $\ell_{\mathrm{T}} = \kappa_{\mathrm{T}}/V$ and the mass diffusion length $\ell_{\mathrm{D}} = \kappa_{\mathrm{D}}/V$. In general, the mass diffusion length is always much smaller than the thermal diffusion length, since

$$\lambda = \frac{\kappa_{\mathrm{D}}}{\kappa_{\mathrm{T}}} \ll 1 \quad \text{and} \quad \lambda_{\mathrm{S}} = \frac{\kappa_{\mathrm{D}}}{\kappa_{\mathrm{TS}}} \ll 1 \,. \tag{1.35}$$

Hence, when one deals with a binary mixture, the mass diffusion is the controlling factor for the macroscopic transport processes. Consequently, ℓ_{D} is chosen as the length scale. Of course, when one deals with a pure substance, where only thermal conduction is involved, the thermal diffusion length ℓ_{T} will be chosen as the length scale.

In the present system, the physical quantity $\Delta H/(c_{\mathrm{p}}\rho)$ has the scale of temperature, so it is naturally chosen as the unit of temperature. By using these scales, we define the following dimensionless quantities and variables:

$$(\bar{X}_1,\ \bar{X}_2,\ \bar{X}_3,\ \bar{t}) = (X_1/\ell_{\mathrm{D}},\ X_2/\ell_{\mathrm{D}},\ X_3/\ell_{\mathrm{D}},\ tV/\ell_{\mathrm{D}})$$

$$(\bar{x}_1,\ \bar{x}_2,\ \bar{x}_3) = (x_1/\ell_{\mathrm{D}},\ x_2/\ell_{\mathrm{D}},\ x_3/\ell_{\mathrm{D}})$$

$$\bar{T} = \frac{T - T_{\mathrm{M0}}}{\Delta H/(c_{\mathrm{p}}\rho)} \qquad\qquad (1.36)$$

$$\bar{\mathbf{u}} = \mathbf{u}/V$$

$$\bar{\omega} = \omega\kappa_{\mathrm{T}}/V^2 \quad (\omega = \nabla \times \mathbf{u}),$$

where T_{M0} is the melting temperature at a flat interface. From (1.28), we have

$$\mathbf{U}/V = \bar{\mathbf{U}} = \bar{\mathbf{u}} + \mathbf{e}_3\,. \qquad\qquad (1.37)$$

Then, the governing equations and related boundary conditions in the moving frame can be derived from those in the rest frame by applying the transformation:

$$\frac{\partial}{\partial t} \rightarrow \frac{\partial}{\partial t} - \mathbf{V}\cdot\nabla \qquad\qquad (1.38)$$

or

$$\frac{\partial}{\partial \bar{t}} \rightarrow \frac{\partial}{\partial \bar{t}} - \mathbf{e}_3\cdot\nabla\,. \qquad\qquad (1.39)$$

The dimensionless governing equations are obtained as follows.

1. Heat conduction equation in the liquid phase:

$$\lambda\left(\frac{\partial \bar{T}}{\partial \bar{t}} + \bar{\mathbf{u}}\cdot\nabla\bar{T}\right) = \nabla^2\bar{T}\,. \qquad\qquad (1.40)$$

2. Heat conduction equation in the solid phase:

$$\lambda\alpha_{\mathrm{T}}\left(\frac{\partial \bar{T}_{\mathrm{S}}}{\partial \bar{t}} + \mathbf{e}_3\cdot\nabla\bar{T}_{\mathrm{S}}\right) = \nabla^2\bar{T}_{\mathrm{S}}\,. \qquad\qquad (1.41)$$

3. Mass diffusion equation in the liquid phase:

$$\frac{\partial \bar{C}}{\partial \bar{t}} + \bar{\mathbf{u}}\cdot\nabla\bar{C} = \nabla^2\bar{C}\,. \qquad\qquad (1.42)$$

4. Continuity equation for the liquid state:

$$\nabla\cdot\bar{\mathbf{u}} = 0\,. \qquad\qquad (1.43)$$

5. Momentum equation for the liquid state:

$$\frac{\partial \bar{\omega}}{\partial t} + (\bar{\mathbf{u}} \cdot \nabla)\bar{\omega} - (\bar{\omega} \cdot \nabla)\bar{\mathbf{u}} = \mathrm{Pr}\nabla^2\bar{\omega} - \frac{\mathrm{Gr}}{\mathrm{T}_\infty}\nabla \times (T\mathbf{e_g}). \tag{1.44}$$

In the above, $\mathbf{e_g}$ is the unit vector along the direction of gravity and we introduce several parameters: $\mathrm{Pr} = \nu/\kappa_{\mathrm{T}}$, the Prandtl number; $T_\infty = -\left[T_{\mathrm{M}0} - (T_\infty)_{\mathrm{D}}\right]c_{\mathrm{p}}\rho/\Delta H = -\mathrm{St}$, the undercooling parameter, where $(T_\infty)_{\mathrm{D}}$ denotes the dimensional undercooling temperature; the parameter St is sometimes called the Stefan number; $\mathrm{Gr} = g\beta(T_{\mathrm{M}0} - (T_\infty)_{\mathrm{D}})\kappa_{\mathrm{T}}/V^3$ is the Grashof number; and $\alpha_{\mathrm{T}} = \kappa_{\mathrm{T}}/\kappa_{\mathrm{TS}}$ is the ratio of thermal diffusivities in the solid and liquid. Conventionally, the simplification with $\alpha_{\mathrm{T}} = 1$ is called the symmetric model; while the simplification with $\alpha_{\mathrm{T}} = 0$ is called the one-sided model.

The dimensionless conditions at the interface $\bar{S}(\mathbf{r}, t) = 0$ are:

(i) *Thermodynamic equilibrium for temperature:*

$$\bar{T} = \bar{T}_{\mathrm{S}}; \tag{1.45}$$

(ii) *Gibbs–Thomson condition:*

$$\bar{T}_{\mathrm{S}} = -\Gamma\{\bar{\mathcal{K}}\} - M\bar{C}; \tag{1.46}$$

where Γ is the surface tension parameter

$$\Gamma = \frac{\ell_{\mathrm{c}}}{\ell_{\mathrm{D}}} \tag{1.47}$$

and

$$\ell_{\mathrm{c}} = \frac{\gamma c_{\mathrm{p}} T_{\mathrm{M}0}\rho}{(\Delta H)^2} \tag{1.48}$$

is a length scale and usually called the capillary length; while M is the morphological parameter

$$M = -\frac{mC_\infty}{\Delta H/(c_{\mathrm{p}}\rho)}. \tag{1.49}$$

Since ℓ_{c} is determined by the interfacial energy γ, it is sometimes considered as a microscopic length scale. It will be seen later that the parameter Γ, the ratio of the macroscopic length ℓ_{D} and the microscopic length ℓ_{c}, is the most important parameter for the stability of the system.

(iii) *Enthalpy conservation:*

$$\left[1 - (1 - \beta)(1 + \alpha)(\bar{T} + \bar{T}_{\mathrm{MO}})\right]\bar{u}_{\mathrm{I}} + \mathbf{e}_3 \cdot \mathbf{n}$$

$$= \left[\frac{(1 + \alpha)}{\lambda_{\mathrm{S}}}\nabla\bar{T}_{\mathrm{S}} - \frac{1}{\lambda}\nabla\bar{T}\right] \cdot \mathbf{n}. \tag{1.50}$$

Hereby, we have defined

$$\begin{cases} \beta = \dfrac{c_{pS}}{c_p}, \quad \lambda_S = \dfrac{\kappa_D}{\kappa_{TS}}, \quad \lambda = \dfrac{\kappa_D}{\kappa_T}, \quad \alpha = \dfrac{\rho_S}{\rho} - 1, \\ \bar{T}_{MO} = \dfrac{T_{MO}}{\Delta H/(c_p\rho)}, \end{cases} \qquad (1.51)$$

which measure the changes of the thermodynamic characteristics in the phase transition.

(iv) *Mass conservation for the impurity:*

$$(1 - \kappa)\bar{C}(\bar{u}_I + \mathbf{e}_3 \cdot \mathbf{n}) = \bar{C}(\bar{\mathbf{u}} + \mathbf{e}_3) \cdot \mathbf{n} - \mathbf{n} \cdot \nabla\bar{C}; \qquad (1.52)$$

(v) *Conservation of total mass:*

$$\bar{\mathbf{u}} \cdot \mathbf{n} + \alpha\bar{u}_I + (1 + \alpha)\mathbf{e}_3 \cdot \mathbf{n} = 0; \qquad (1.53)$$

(vi) *Continuity of the tangential component of velocity:*

$$\bar{\mathbf{u}} \cdot \mathbf{e}_\tau = \mathbf{e}_3 \cdot \mathbf{e}_\tau. \qquad (1.54)$$

Later, for the sake of simplicity, we shall omit the bar over all dimensionless quantities without confusion.

References

1.1   T. Kobayashi and Y. Furukawa '*Snow crystals*', (Snow Crystal Museum Asahikawa, Hokkaido 1991).

1.2   K. Takahashi, Y. Furukawa and Y. Takahashi, '*Story of snow crystals*', (Koudansya, Tokyo 1995).

1.3   J. W. Rutter and B. Chalmers, "A Prismatic Substructure Formed During Solidification of Metals", Can. J. Phys. **31**, pp. 15–39, (1953).

1.4   S. C. Huang and M. E. Glicksman, "Fundamentals of Dendritic Solidification — I. Steady–State Tip Growth; II. Development of Sidebranch Structure", Acta Metall. **29**, pp. 701–734, (1981).

1.5   A. Dougherty and J. P. Gollub, "Steady-State Dendritic Growth of NH$_4$Br from Solution", Phys. Rev. A **38**, pp. 3043–3053, (1988).

1.6   G. P. Ivantsov, "Temperature Field around a Spheroidal, Cylindrical and Acicular Crystal Growing in a Supercooled Melt", Dokl. Akad. Nauk, SSSR. **58**, No. 4, pp. 567–569, (1947).

1.7   G. Horvay and J. W. Cahn, "Dendritic and Spheroidal Growth", Acta Metall. **9**, pp. 695–705, (1961).

1.8   W. W. Mullins and R. F. Sekerka, "Morphological Stability Of a Particle Growing by Diffusion or Heat Flow", J. Appl. Phys. **34**, pp. 323–329, (1963).

1.9   W. W. Mullins and R. F. Sekerka, "Stability of a Planar Interface During Solidification of a Dilute Binary Alloy", J. Appl. Phys. **35**, pp. 444–451, (1964).

1.10   G. E. Nash and M. E. Glicksman, "Capillarity-limited Steady-State Dendritic Growth I. Theoretical Development", Acta Metall. **22**, pp. 1283–1299, (1974).

1.11 M. E. Glicksman, R. J. Schaefer and J. D. Ayers, "High-Confidence Measurement of Solid/Liquid Surface Energy in a Pure Material", Philosophical Magazine **32**, pp. 725–743, (1975).

1.12 M. E. Glicksman, R. J. Schaefer and J. D. Ayers, "Dendrite Growth — A Test of Theory", Metall. Trans. **7A**, pp. 1747–1757, (1976).

1.13 J. S. Langer and H. Müller-Krumbhaar, "Theory of Dendritic Growth — I. Elements of a Stability Analysis; II. Instabilities in the Limit of Vanishing Surface Tension; III. Effects of Surface Tension", Acta Metall. **26**, pp. 1681–1708, (1978).

1.14 J. S. Langer, "Instability and Pattern Formation in Crystal Growth", Rev. Mod. Phys. **52**, pp. 1–28, January (1980).

1.15 J. S. Langer, *'Lectures in the Theory of Pattern Formation'*, USMG NATO AS Les Houches Session XLVI 1986 — Le hasard et la matiere/ chance and matter. Ed. by J. Souletie, J. Vannimenus and R. Stora, (Elsevier Science, Amsterdam 1986)

1.16 D. A. Kessler, J. Koplik and H. Levine, "Pattern Formation Far from Equilibrium: the free space dendritic crystal", in *'Proc. NATO A.R.W. on Patterns, Defects and Microstructures in Non-equilibrium Systems'*, (Austin, TX, March, 1986).

1.17 M. Kruskal and H. Segur, "Asymptotics Beyond All Orders in a Model of Crystal Growth", Stud. in Appl. Math. No. 85, pp. 129–181, (1991).

1.18 D. A. Kessler and H. Levine, "Stability of Dendritic Crystals", Phys. Rev. Lett. **57**, pp. 3069–3072, (1986).

1.19 P. Pelce and Y. Pomeau, "Dendrites in the small undercooling limit", Stud. Appl. Math. **74**, pp. 245–258, (1986).

1.20 J. S. Langer, "Dendritic Sidebranching in The Three-Dimensional Symmetric Model in The Presence Of Noise", Phys. Rev. A **36**, No. 7, pp. 3350–3358, (1987).

1.21 J. J. Xu, "Global Asymptotic Solution for Axi-symmetric Dendrite Growth with Small Undercooling" in *'Structure and Dynamics of Partially Solidified System'*, Ed. by D.E. Loper NATO ASI Series E. No. 125, pp. 97–109, (1987).

1.22 J. J. Xu, PhD. Thesis, Department of Mathematical Sciences, Rensselaer Polytechnic Institute, NY (1987).

1.23 P. Pelce, *'Dynamics of Curved Front'*, (Academic, New York 1988).

1.24 J. J. Xu, "Global Wave Mode Theory for Formation of Dendritic Structure on a Growing Needle Crystal" Physica Status Solidi (b) **157**, pp. 577–591, (1990).

1.25 J. J. Xu, "Global Neutral Stable State and Selection Condition of Tip Growth Velocity", J. Crystal Growth **100**, pp. 481–490, (1990).

1.26 J. J. Xu, "Asymptotic Theory of Steady Axisymmetric Needle-like Crystal Growth", Stud. Appl. Math. **82**, pp. 71–91, (1990).

1.27 J. J. Xu, "Interfacial Wave Theory of Solidification — Dendritic Pattern Formation and Selection of Tip Velocity", Phys. Rev. A15 **43**, No. 2, pp. 930–947, (1991).

1.28 J. J. Xu, "Two-Dimensional Dendritic Growth with Anisotropy of Surface Tension", Physica (D) **51**, pp. 579–595, (1991).

1.29 J. J. Xu, "Interfacial Wave Theory of Two-Dimensional Dendritic Growth with Anisotropy of Surface Tension", Canad. J. Phys. **69**, No. 7, pp. 789–800, (1991).

1.30 M. Muschol, D. Liu, and H. Z. Cummins, "Surface-Tension-Anisotropy Measurement of Succinonitrile and Pivalic Acid: Comparison with Microscopic Solvability Theory", Phys. Rev. A **46**, pp. 1038–1050, (1992).

1.31 J. S. Kirkaldy, 'Spontaneous Evolution of Spatiotemporal Patterns in Materials', Rep. Prog. Phys. **55**, pp. 723–795, (1992).

1.32 E. A. Brener and V. I. Melnikov, "Pattern Selection in Two Dimensional Dendritic Growth", Adv. Phys. Vol. **40**, pp. 53–97, (1991).

1.33 J. J. Xu, "The Effect of Convection Motion on Dendritic Growth", in '*Interactive Dynamics of Convection and Solidification*', pp. 101–103, Ed. by Davis et al. (Kluwer, Dordrecht 1992).

1.34 J. S. Langer, "Issues and Opportunities in Materials Research", Physics Today, October, 1992, pp. 24–31.

1.35 J. J. Xu, "Interfacial Wave Theory of Solidification — Dendritic Pattern Formation and Selection of Tip Velocity", Phys. Rev. A15 **43**, No. 2, pp. 930–947, (1991).

1.36 J. J. Xu, "Global Instability and Pattern Formation in Dendritic Solidification of Dilute Binary Alloy System", Canad. Appl. Math. Quar. **1**, No. 2, pp. 255–292, (1993).

1.37 J. J. Xu and Z. X. Pan, "Interfacial Wave Theory of Dendritic Growth from a Binary Mixture: A Comparison with Experiments", J. Crystal Growth, No. 129, pp. 666–682, (1993).

1.38 J. J. Xu, "Dendritic Growth From Melt with External Flow: Uniformly Valid Asymptotic Solution for the Steady State", J. Fluid Mechanics. **263**, pp. 227–243, (1994.)

1.39 J. J. Xu, "Effect of Convection Motion in Melt Induced by Density-Change on Dendritic Solidification", Canad. J. Phys. **72**, No. 3 & 4, pp. 120–125, (1994).

1.40 J. J. Xu, "Generalized Needle Solutions, Interfacial Instabilities and Pattern Formations", Phys. Rev. E **53**, No: 5, pp. 5051–5062, (1996).

1.41 P. A. Zhuravlev, "On the Motion of a Fluid in Channels", Zap. Leningr. Gorn. In-ta. **33**, No. 3, pp. 54–61, (1956).

1.42 P. G. Saffman and G.I. Taylor, "The Penetration of a Fluid into a Porous Medium or Hele–Shaw Cell Containing a More Viscous Liquid", Proc. R. Soc. London Ser. A. **245**, pp. 312–329, (1958).

1.43 Y. Couder, N. Gerard and M. Rabaud, "Narrow fingers in the Saffman–Taylor instability", Phys. Rev. A **34**, pp. 5175–5178, (1986).

1.44 A. R. Kopf-Sill and G. M. Homsy, "Narrow fingers in a Hele–Shaw cell", Phys. Fluids **30**, No. 9, pp. 2607–2609, (1987).

1.45 G. M. Homsy, "Viscous fingering in porous media", Ann. Rev. Fluid Mech. **19**, pp. 271–311, (1987).

1.46 D. Bensimon, P. Pelce and B. I. Shraiman, "Dynamics of curved fronts and pattern selection", J. Physique **48**, pp. 2081–2087, (1987).

1.47 S. Tanveer, "Analytic Theory For the Selection of a Symmetric Saffman–Taylor Finger in a Hele–Shaw Cell", Phys. Fluid **30**, No. 8, pp. 1589–1605, (1987).

1.48 J. J. Xu, "Global Instability of Viscous Fingering in Hele–Shaw Cell (I) — Formation of Oscillatory Fingers", Europ. J. Appl. Math. **2**, pp. 105–132, (1991).

1.49 J. J. Xu, "Interfacial Wave Theory for Oscillatory Finger's Formation in a Hele–Shaw Cell: a Comparison with Experiments" Europ. J. Appl. Math. **7** pp. 169–199, (1996).

1.50 J. J. Xu, "Interfacial Instabilities and Fingering Formation in Hele–Shaw Flow", IMA J. Appl. Math. **57**, pp. 101–135, (1996).

1.51 H. Segur, S. Tanveer and H. Levine (Eds.), '*Asymptotics Beyond All Orders*', NATO ASI Series, Series B: Physics, Vol. 284, (Plenum, New York 1991).

1.52 D. T. J. Hurle (Ed.), '*Handbook of Crystal Growth, Vol. 1: Fundamentals, Part B: Transport and Stability*', (Elservier Science, North–Holland, Amsterdam 1993).

2. Unidirectional Solidification and the Mullins–Sekerka Instability

Before we begin the study of dendritic growth, it is appropriate to examine a simple case first: the instability of a planar interface in unidirectional solidification. Mullins and Sekerka were the first, in 1963, to perform a systematic analysis of this system. Their linear stability analysis is now called the *Mullins–Sekerka instability* [2.1].

In what follows, we attempt to briefly review the results of Mullins and Sekerka on linear stability of unidirectional solidification, both from a pure melt and from a binary mixture. Our main purpose is to explain the physical essence of the phenomenon and demonstrate the mathematical methods of solving the problem. Thus only the simplest model problem and very basic results will be discussed. The approach used below is somehow different from that originally used by Mullins and Sekerka; but its spirit can be extended to solve more difficult problems, such as the stability of a curved front.

We assume that the system is two-dimensional, the gravity g and the ratio of densities α can be set to zero. Thus the system is free of convection. This is the case of solidification in a thin Hele–Shaw cell.

2.1 Solidification with Planar Interface from a Pure Melt

We assume that an originally flat, solidifying front advances into a pure undercooled melt with velocity V. We utilize moving two-dimensional Cartesian coordinates (x, y) with the x-axis fixed on the interface, as shown in Fig. 2.1. The y-axis is along the growth direction of the interface. Furthermore, we assume that the mass density ρ, the thermal diffusivity constant κ_T, and the specific heat per unit volume c_p are the same for both liquid and solid, namely, we adopt the symmetric model.

To make all physical properties dimensionless, we adopt the scale defined in Sect. 1.3.3 with the thermal length ℓ_T as the length scale. Thus, the dimensionless system can be derived from (1.40) in Sect. 1.3.1 by setting the relative velocity field $\mathbf{u} = -\mathbf{e_3} = -\mathbf{e_y}$, and $\lambda = \lambda_S = 1$ as follows:

$$\nabla^2 \bar{T} = \frac{\partial \bar{T}}{\partial \bar{t}} - \frac{\partial \bar{T}}{\partial \bar{y}}, \tag{2.1}$$

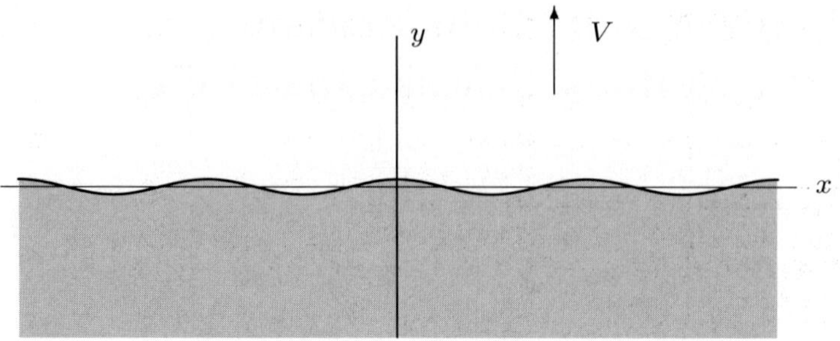

Fig. 2.1. A simple model for unidirectional solidification

where $\bar{T} = \bar{T}(x, y, t)$ is the temperature field in the liquid phase. For the solid phase we replace $\bar{T}$ by $\bar{T}_\mathrm{S}$.

The dimensionless boundary conditions are:

1. The up stream far-field condition: as $\bar{y} \to \infty$,

$$\bar{T} \to T_\infty = \frac{(T_\infty)_\mathrm{D} - T_{\mathrm{M0}}}{\Delta H / (c_\mathrm{p} \rho)} \, . \tag{2.2}$$

2. At the interface, $\bar{y} = \bar{h}(\bar{x}, \bar{t})$,

   (i) the thermodynamic equilibrium condition:

$$\bar{T} = \bar{T}_\mathrm{S} \, , \tag{2.3}$$

   (ii) the Gibbs–Thomson condition:

$$\bar{T}_\mathrm{S} = \Gamma \frac{\bar{h}_{\bar{x}\bar{x}}}{(1 + \bar{h}_{\bar{x}}^2)^{\frac{3}{2}}} \, , \tag{2.4}$$

   (iii) the enthalpy balance:

$$\frac{\partial}{\partial \bar{y}}(\bar{T} - \bar{T}_\mathrm{S}) - \bar{h}_{\bar{x}} \frac{\partial}{\partial \bar{x}}(\bar{T} - \bar{T}_\mathrm{S}) + \bar{h}_{\bar{t}} + 1 = 0 \, . \tag{2.5}$$

For the sake of convenience, we shall hereafter omit the bar over the dimensionless quantities.

2.1.1 Basic Steady State Solution

The above system allows a one-dimensional steady solution with a flat interface. The interface equation is $y = h_\mathrm{B} = 0$. The exact, one-dimensional steady state solution can be obtained by setting $\frac{\partial}{\partial x} = 0$ and $\frac{\partial}{\partial t} = 0$ in the above system. One thus finds

$$T_\mathrm{B} = \begin{cases} e^{-y} - 1, & y \geq 0 \ \ (\text{for the liquid}) \, ; \\ 0, & y \leq 0 \ \ (\text{for the solid}) \, , \end{cases} \tag{2.6}$$

where T_{B} is the basic state solution for temperature. From (2.6), it is seen that as $y \to \infty$, $T_{\mathrm{B}}(\infty) = T_\infty = -1$. Hence, this solution can only be applied to the special case: $T_\infty = -1$. For $T_\infty \neq -1$, the system cannot have a steady solution with a flat interface. More precisely, for the case $T_\infty > -1$, the interface of the steady state solution will be curved, e.g., a parabola; while for the case $T_\infty < -1$, the system will have no solution, as the interface can no longer be considered as being in local thermodynamic equilibrium.

2.1.2 Unsteady Perturbed Solutions and Mullins–Sekerka Instability

The unsteady solutions can be expressed in the form:

$$
\begin{aligned}
T(x, y, t) &= T_{\mathrm{B}}(y) + \tilde{T}(x, y, t)\,, \\
T_{\mathrm{S}}(x, y, t) &= T_{\mathrm{SB}}(y) + \tilde{T}_{\mathrm{S}}(x, y, t)\,, \\
h(x, t) &= h_{\mathrm{B}} + \tilde{h}(x, t)\,,
\end{aligned}
\tag{2.7}
$$

where $\tilde{T}, \tilde{T}_{\mathrm{S}},$ and $\tilde{h}$ are small perturbations around the steady state.

The governing equation for the perturbation is

$$
\nabla^2 \tilde{T} = \frac{\partial \tilde{T}}{\partial t} - \frac{\partial \tilde{T}}{\partial y}\,.
\tag{2.8}
$$

The boundary conditions are:

1. The up-steam far field condition: as $y \to \infty$,

$$
\tilde{T} \to 0\,.
\tag{2.9}
$$

2. The interface conditions: assume that the deformation $\tilde{h}(x, t)$ of the interface, measured using the characteristic amplitude $\delta \ll 1$, is very small. One can linearize the original boundary conditions on the interface $y = \tilde{h}(x, t)$ by expanding them in a Taylor series in δ around $y = h_{\mathrm{B}} = 0$. As the leading order approximation, we obtain the following linear interface conditions at $y = 0$

$$
\tilde{T} = \tilde{T}_{\mathrm{S}} - (\Delta G_1)\tilde{h}\,,
\tag{2.10}
$$

$$
\tilde{T}_{\mathrm{S}} = \Gamma \tilde{h}_{xx} - G_{1\mathrm{S}}\tilde{h}\,,
\tag{2.11}
$$

$$
\frac{\partial \left(\tilde{T} - \tilde{T}_{\mathrm{S}}\right)}{\partial y} + (\Delta G_2)\,\tilde{h} + \frac{\partial \tilde{h}}{\partial t} = 0\,,
\tag{2.12}
$$

where

$$G_{1L} = \frac{\partial T_B}{\partial y}(0) = -1 \; ; \quad G_{1S} = \frac{\partial T_{BS}}{\partial y}(0) = 0 \,, \tag{2.13}$$

$$\Delta G_1 = (G_{1L} - G_{1S}) = -1 \,, \tag{2.14}$$

$$\Delta G_2 = \left(\frac{\partial^2 T_B}{\partial y^2}\right)_{y=0} - \left(\frac{\partial^2 T_{SB}}{\partial y^2}\right)_{y=0} = 1 \,. \tag{2.15}$$

This linear system contains one parameter Γ. In practice, Γ is a very small parameter and it appears in front of the second derivatives $\tilde{h}_{xx}$ in (2.11). Thus, the above system gives rise to a singular perturbation problem. One can let $\Gamma \to 0$ and look for an asymptotic expansion solution. In the limiting process $\Gamma \to 0$, all perturbed quantities $\{\tilde{T}, \tilde{T}_S, \tilde{h}\}$ should have the same order of magnitude. From the interface condition (2.11), it is seen that in order for $\tilde{T}_S$ to have the same order magnitude as $\tilde{h}$, one must have $\tilde{h}_{xx} = O(\frac{\tilde{h}}{\Gamma})$. This is possible only when $\tilde{h}$ is also a function of the variable $\frac{x}{\varepsilon}$, where

$$\varepsilon = \sqrt{\Gamma} \,. \tag{2.16}$$

It will be seen that the parameter ε plays a vital role for interfacial stability. We call this parameter *the interfacial stability parameter*. The above argument provides us with an important hint that the solution must have the structure of multiple length scales; so one may derive an asymptotic expansion form of the solution by using the so-called *multiple variables expansion (MVE) method* (refer to [2.2]). The idea of the MVE method is that one may define a set of fast variables such that

$$x_+ = \frac{k(\varepsilon)x}{\varepsilon}$$

$$y_+ = \frac{g(\varepsilon)y}{\varepsilon} \tag{2.17}$$

$$t_+ = \frac{\sigma(\varepsilon)t}{\varepsilon} \,,$$

and consider the exact solution as a function of $(x_+, y_+, t_+, x, y, \varepsilon)$, where the faster variables (x_+, y_+, t_+) and the slow variables (x, y) are formally treated as the independent variables. Thus, as $\varepsilon \to 0$, the solution $\tilde{q} \equiv \{\tilde{T}, \tilde{T}_S, \tilde{h}\}$ is expanded in the following MVE form:

$$\tilde{q}(x, y, x_+, y_+, t_+, \varepsilon) \sim e^{t_+}\big\{\tilde{q}_0(x, y, x_+, y_+) + \varepsilon\tilde{q}_1(x, y, x_+, y_+) + \cdots\big\}$$

$$k(\varepsilon) \sim k_0 + \varepsilon k_1 + \cdots$$

$$g(\varepsilon) \sim k_0 + \varepsilon g_1 + \cdots \tag{2.18}$$

$$g_s(\varepsilon) \sim k_0 + \varepsilon g_{s1} + \cdots$$

$$\sigma(\varepsilon) \sim \sigma_0 + \varepsilon\sigma_1 + \cdots \,.$$

Note that in the above, we have used g_s for the solutions in the solid state, which may be different from g for the solutions in the liquid state. It implies that for the solutions in the solid state, we shall use the fast variable $y_+ = g_\mathrm{s} y/\varepsilon$. Moreover, to obtain the asymptotic solution, one needs to make expansions in ε for the parameter σ and the wave numbers, k and g. One can verify that the leading terms of the expansions for $k(\varepsilon)$, $g(\varepsilon)$, and $g_\mathrm{s}(\varepsilon)$ can be set the same, but their higher-order terms may be different. Now we need to transform the perturbed system (2.8)–(2.12) into the form with the above multiple variables. This can be done by replacing all the derivatives in (2.8)–(2.12) by using the following transformation:

$$
\begin{aligned}
\frac{\partial}{\partial x} &\Rightarrow \frac{\partial}{\partial x} + \frac{k}{\varepsilon}\frac{\partial}{\partial x_+}, \\
\frac{\partial}{\partial y} &\Rightarrow \frac{\partial}{\partial y} + \frac{g}{\varepsilon}\frac{\partial}{\partial y_+}, \\
\frac{\partial}{\partial t} &\Rightarrow \frac{\sigma}{\varepsilon}\frac{\partial}{\partial t_+}, \\
\frac{\partial^2}{\partial x^2} &\Rightarrow \frac{\partial^2}{\partial x^2} + \frac{2k}{\varepsilon}\frac{\partial^2}{\partial x\partial x_+} + \frac{k^2}{\varepsilon^2}\frac{\partial^2}{\partial x_+^2}, \\
\frac{\partial^2}{\partial y^2} &\Rightarrow \frac{\partial^2}{\partial y^2} + \frac{2g}{\varepsilon}\frac{\partial^2}{\partial y\partial y_+} + \frac{g^2}{\varepsilon^2}\frac{\partial^2}{\partial y_+^2}.
\end{aligned}
\tag{2.19}
$$

The multiple-variables form of the system is then obtained as follows:

$$
k^2\frac{\partial^2 \tilde{T}}{\partial x_+^2} + g^2\frac{\partial^2 \tilde{T}}{\partial y_+^2} = \varepsilon\left(\sigma\tilde{T} - g\frac{\partial \tilde{T}}{\partial y_+} - 2k\frac{\partial^2 \tilde{T}}{\partial x\partial x_+} - 2g\frac{\partial^2 \tilde{T}}{\partial y\partial y_+}\right)
$$
$$
- \varepsilon^2\left(\frac{\partial^2}{\partial x^2} + \frac{\partial^2}{\partial y^2} + \frac{\partial}{\partial y}\right)\tilde{T}\,;
\tag{2.20}
$$

$$
k^2\frac{\partial^2 \tilde{T}_\mathrm{S}}{\partial x_+^2} + g_\mathrm{s}^2\frac{\partial^2 \tilde{T}_\mathrm{S}}{\partial y_+^2} = \varepsilon\left(\sigma\tilde{T}_\mathrm{S} - g_\mathrm{s}\frac{\partial \tilde{T}_\mathrm{S}}{\partial y_+} - 2k\frac{\partial^2 \tilde{T}_\mathrm{S}}{\partial x\partial x_+} - 2g_\mathrm{s}\frac{\partial^2 \tilde{T}_\mathrm{S}}{\partial y\partial y_+}\right)
$$
$$
- \varepsilon^2\left(\frac{\partial^2}{\partial x^2} + \frac{\partial^2}{\partial y^2} + \frac{\partial}{\partial y}\right)\tilde{T}_\mathrm{S}\,,
\tag{2.21}
$$

with the boundary conditions:

1. As $y_+ \to \infty$,

$$
\tilde{T} \to 0.
\tag{2.22}
$$

2. As $y_+ \to -\infty$,

$$
\tilde{T}_\mathrm{S} \to 0.
\tag{2.23}
$$

3. At the interface $y = y_+ = 0$,

(i)

$$\tilde{T} = \tilde{T}_{\mathrm{S}} - (\Delta G_1)\tilde{h} \,, \tag{2.24}$$

(ii)

$$\tilde{T}_{\mathrm{S}} = k^2 \frac{\partial^2 \tilde{h}}{\partial x_+^2} + 2\varepsilon k \frac{\partial^2 \tilde{h}}{\partial x \partial x_+} + \varepsilon^2 \frac{\partial^2 \tilde{h}}{\partial x^2} - G_{1\mathrm{S}}\tilde{h} \,, \tag{2.25}$$

(iii)

$$g \frac{\partial \tilde{T}}{\partial y_+} - g_{\mathrm{s}} \frac{\partial \tilde{T}_{\mathrm{S}}}{\partial y_+} + \varepsilon \frac{\partial}{\partial y}\left(\tilde{T} - \tilde{T}_{\mathrm{S}}\right) + \sigma \tilde{h} + \varepsilon(\Delta G_2)\tilde{h} = 0\,. \tag{2.26}$$

Hereby, it is seen that as $\varepsilon \to 0$, in the leading order approximation, both the convective and unsteady heat transfer terms in the bulk are negligible, whereas the Gibbs–Thomson effect at the interface is important.

By substituting (2.18) into the above system (2.20)–(2.26), one can successively derive each order of approximation.

Zeroth-order approximation solutions. Assuming $k_0 \gg \varepsilon$, in the leading order approximation, we obtain the governing equation

$$k_0^2 \left(\frac{\partial^2}{\partial x_+^2} + \frac{\partial^2}{\partial y_+^2} \right) \tilde{T}_0 = 0\,, \tag{2.27}$$

and the boundary conditions:

1. As $y_+ \to \infty$,

$$\tilde{T}_0 \to 0\,. \tag{2.28}$$

2. As $y_+ \to -\infty$,

$$\tilde{T}_{\mathrm{S}0} \to 0\,. \tag{2.29}$$

3. At the interface, $y = y_+ = 0$,

$$\tilde{T}_0 = \tilde{T}_{\mathrm{S}0} - (\Delta G_1)\tilde{h}_0 \,, \tag{2.30}$$

$$\tilde{T}_{\mathrm{S}0} = k_0^2 \frac{\partial^2 \tilde{h}_0}{\partial x_+^2} - G_{1\mathrm{S}}\tilde{h}_0 \,, \tag{2.31}$$

$$k_0 \frac{\partial}{\partial y_+}\left(\tilde{T}_0 - \tilde{T}_{\mathrm{S}0}\right) + \sigma_0 \tilde{h}_0 = 0\,. \tag{2.32}$$

The system (2.27)–(2.32) allows the mode solutions:

$$\tilde{T}_0 = A_0(x,y)\mathrm{e}^{\mathrm{i}x_+ - y_+}$$

$$\tilde{T}_{\mathrm{S}0} = A_{\mathrm{S}0}(x,y)\mathrm{e}^{\mathrm{i}x_+ + y_+} \tag{2.33}$$

$$\tilde{h}_0 = \hat{D}_0 \mathrm{e}^{\mathrm{i}x_+} .$$

With the notation

$$\hat{A}_0 = A_0(x,0), \qquad \hat{A}_{\mathrm{S}0} = A_{\mathrm{S}0}(x,0) \tag{2.34}$$

and applying the boundary conditions (2.30)–(2.32) we derive

$$\hat{A}_0 = \hat{A}_{\mathrm{S}0} - \Delta G_1 D_0 ,$$
$$\hat{A}_{\mathrm{S}0} = -k_0^2 \hat{D}_0 - G_{1\mathrm{S}} \hat{D}_0 , \tag{2.35}$$
$$-k_0(\hat{A}_0 + \hat{A}_{\mathrm{S}0}) + \sigma_0 \hat{D}_0 = 0 .$$

From this system of homogeneous equations, one deduces that the amplitude functions $A_0, A_{\mathrm{S}0}$ must be independent of x and $\hat{A}_0, \hat{A}_{\mathrm{S}0}, \hat{D}_0$ must be constants. Moreover, for a nontrivial solution, one must have:

$$\Delta = \det \begin{pmatrix} 1 & -1 & \Delta G_1 \\ 0 & 1 & k_0^2 + G_{\mathrm{S}1} \\ -k_0 & -k_0 & \sigma_0 \end{pmatrix} = 0 . \tag{2.36}$$

Thus the wave number k_0 and the eigenvalue σ_0 in the mode solutions cannot be arbitrarily chosen. These two quantities must be subject to the following dispersion relation:

$$\sigma_0 = k_0 \left(-(G_{1\mathrm{L}} + G_{1\mathrm{S}}) - 2k_0^2 \right) . \tag{2.37}$$

For the system under consideration, we have

$$\sigma_0 = k_0 \left(1 - 2k_0^2 \right) . \tag{2.38}$$

This formula was first obtained by Mullins–Sekerka in 1963. It is now called the Mullins–Sekerka dispersion relation. From this dispersion relation, with a given eigenvalue σ_0, the corresponding wave numbers k_0 are determined (see Fig. 2.2). Note that if the wave number k_0 has an imaginary part, the perturbation will tend to infinity either as $x \to \infty$, or as $x \to -\infty$. Furthermore, if the wave number k_0 is a negative real number, the perturbed temperature, $\tilde{T}_0$, will tend to infinity as $y \to \infty$. Evidently, these results are physically unacceptable.

Thus, the eigenvalue σ_0 of the system must be real. There are three cases:

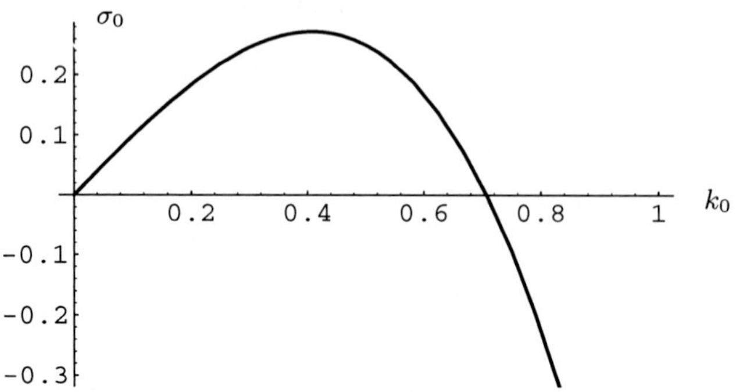

Fig. 2.2. Mullins–Sekerka dispersion relation for a pure melt system

(i) $0 < \sigma_0 < \frac{1}{3}\sqrt{\frac{2}{3}}$. From the local dispersion relation (2.38), one solves for the three real roots: $k_0^{(1)} > 0, k_0^{(2)} < 0, k_0^{(3)} > 0$. Only the positive roots $k_0^{(1)}, k_0^{(3)}$ are physically meaningful. The temperature field corresponding to $k_0^{(2)}$ will increase exponentially as $y \to \infty$, violating the boundary condition (2.28).
The general solution of the perturbed states is then

$$\tilde{h} \approx \Re\left\{ \left(D_0^{(1)} e^{\frac{i}{\varepsilon} k_0^{(1)} x} + D_0^{(3)} e^{\frac{i}{\varepsilon} k_0^{(3)} x} \right) e^{\frac{\sigma_0}{\varepsilon} t} \right\}. \tag{2.39}$$

These solutions are growing unstable modes. A special case is $\sigma_0 = 0$, for which we have $k_0^{(3)} = 0$ and $k_0^{(1)} > 0$. The corresponding solutions are neutrally stable modes.

(ii) If $\sigma_0 < 0$, the local dispersion relation (2.38) only allows one real root $k_0^{(1)}$ and the system has only one corresponding decaying mode.

(iii) If $\sigma > \frac{1}{3}\sqrt{\frac{2}{3}}$, the local dispersion relation (2.38) has no real root, so the system has no corresponding mode.

Returning to the original slow variables (x, y, t), the normal mode solution is written in the form:

$$\tilde{T}_0 e^{\sigma_0 t+} = A_0 e^{\tilde{k}_0 (ix - y) + \tilde{\sigma}_0 t}$$

$$\tilde{T}_{S0} e^{\sigma_0 t+} = A_{S0} e^{\tilde{k}_0 (ix + y) + \tilde{\sigma}_0 t} \tag{2.40}$$

$$\tilde{h}_0 e^{\sigma_0 t+} = \hat{D}_0 e^{i\tilde{k}_0 x + \tilde{\sigma}_0 t},$$

where

$$\tilde{k}_0 = \frac{k_0}{\varepsilon}; \quad \tilde{\sigma}_0 = \frac{\sigma_0}{\varepsilon}. \tag{2.41}$$

Thus, the dispersion relation (2.37) is written in the form:

$$\tilde{\sigma}_0 = \tilde{k}_0 \Big(-(G_{1\mathrm{L}} + G_{1\mathrm{S}}) - 2\varepsilon^2 \tilde{k}_0^2 \Big) . \tag{2.42}$$

It is clear from this formula that, when the mean temperature gradient $\bar{G}_1 = \frac{G_{1\mathrm{L}} + G_{1\mathrm{S}}}{2} < 0$, the heat conduction represented by the term $-2\bar{G}_1 \tilde{k}_0$ is an unstable factor; while the surface tension, represented by the term $\varepsilon^3 \tilde{k}_0^3$, is a stable factor. If the surface tension parameter is zero, the dispersion relation reduces to

$$\tilde{\sigma}_0 = -2\bar{G}_1 \tilde{k}_0 . \tag{2.43}$$

Hence, for any $\tilde{k}_0 > 0$, one always has $\tilde{\sigma}_0 > 0$. This implies that the system will always be unstable. When the surface tension parameter is nonzero, the surface tension suppresses perturbations with a short wavelength. But perturbations with long wavelengths all grow with time. More precisely, for $\tilde{k}_0 > \sqrt{-\bar{G}_1}/\varepsilon$, $\tilde{\sigma}_0 < 0$, and mode solutions are decaying and stable. For $0 < \tilde{k}_0 < \sqrt{-\bar{G}_1}/\varepsilon$, the mode solutions are growing and unstable. The critical number $\tilde{k}_{\mathrm{c}} = \sqrt{-\bar{G}_1}/\varepsilon$ corresponds to the so-called neutral modes.

From the above results, one can conclude that in unidirectional solidification from a pure melt, due to the Mullins–Sekerka instability, any perturbed states are either purely growing or purely decaying. The system allows a unique neutrally stable state, which represents a steady state. It does not allow an oscillatory state. When $(G_{1\mathrm{L}} + G_{1\mathrm{S}}) > 0$, the system will be stable for all $k_0 > 0$. This implies that when the solidification interface advances to a high temperature liquid region, the interface will always be smooth. It should be noted that so far these conclusions are drawn from the leading order approximation of the solution in the limit $\varepsilon \to 0$. The dispersion relation (2.37) is not exact. For more accurate information, one needs to examine the higher-order approximation solutions.

First-order approximation solutions. In the first-order approximation, we obtain the solution

$$k_0^2 \left\{ \frac{\partial^2}{\partial x_+^2} + \frac{\partial^2}{\partial y_+^2} \right\} \tilde{T}_1 = \sigma_0 \tilde{T}_0 - k_0 \frac{\partial \tilde{T}_0}{\partial y_+} - 2k_0 \frac{\partial^2 \tilde{T}_0}{\partial x \partial x_+} - 2k_0 \frac{\partial^2 \tilde{T}_0}{\partial y \partial y_+}$$

$$- 2k_0 k_1 \frac{\partial^2 \tilde{T}_0}{\partial x_+^2} - 2k_0 g_1 \frac{\partial^2 \tilde{T}_0}{\partial y_+^2} , \tag{2.44}$$

$$k_0^2 \left\{ \frac{\partial^2}{\partial x_+^2} + \frac{\partial^2}{\partial y_+^2} \right\} \tilde{T}_{\mathrm{S}1} = \sigma_0 \tilde{T}_0 - k_0 \frac{\partial \tilde{T}_{\mathrm{S}0}}{\partial y_+} - 2k_0 \frac{\partial^2 \tilde{T}_{\mathrm{S}0}}{\partial x \partial x_+} - 2k_0 \frac{\partial^2 \tilde{T}_{\mathrm{S}0}}{\partial y \partial y_+}$$

$$- 2k_0 k_1 \frac{\partial^2 \tilde{T}_{\mathrm{S}0}}{\partial x_+^2} - 2k_0 g_{\mathrm{s}1} \frac{\partial^2 \tilde{T}_{\mathrm{S}0}}{\partial y_+^2} . \tag{2.45}$$

Consequently,

$$\begin{cases} k_0^2 \left\{ \frac{\partial^2}{\partial x_+^2} + \frac{\partial^2}{\partial y_+^2} \right\} \tilde{T}_1 = a_0 \mathrm{e}^{(\mathrm{i}x_+ - y_+)} \\[2mm] k_0^2 \left\{ \frac{\partial^2}{\partial x_+^2} + \frac{\partial^2}{\partial y_+^2} \right\} \tilde{T}_{S1} = b_0 \mathrm{e}^{(\mathrm{i}x_+ + y_+)} , \end{cases} \tag{2.46}$$

where

$$\begin{cases} a_0 = 2k_0 \left\{ A_0'(y) + \left(\frac{1}{2} + \frac{\sigma_0}{2k_0} + k_1 - g_1 \right) A_0(y) \right\} \\[2mm] b_0 = -2k_0 \left\{ A_{S0}'(y) + \left(\frac{1}{2} - \frac{\sigma_0}{2k_0} - k_1 + g_{s1} \right) A_{S0}(y) \right\}. \end{cases} \tag{2.47}$$

In order to obtain a uniformly valid asymptotic solution, one must eliminate the secular terms by setting $a_0 = b_0 = 0$. Moreover, without loss of generality, we assume $A_0'(y) = A_{S0}'(y) = 0$, so that

$$A_0(y) = \hat{A}_0, \quad A_{S0}(y) = \hat{A}_{S0}. \tag{2.48}$$

This leads to

$$\begin{cases} g_1 - k_1 = \frac{1}{2}\left(1 + \frac{\sigma_0}{k_0}\right) \\[2mm] g_{s1} = g_1 - 1. \end{cases} \tag{2.49}$$

From $\nabla^2 \tilde{T}_1 = \nabla^2 \tilde{T}_{S1} = 0$, we obtain the solutions:

$$\tilde{T}_1 = A_1(y)\mathrm{e}^{\mathrm{i}x_+ - y_+}$$

$$\tilde{T}_{S1} = A_{S1}(y)\mathrm{e}^{\mathrm{i}x_+ + y_+} \tag{2.50}$$

$$\tilde{h}_1 = \hat{D}_1 \mathrm{e}^{\mathrm{i}x_+}.$$

Hereby, one can easily justify that the amplitude functions A_1 and A_{S1} are only dependent on y, as the coefficients of the system are constant. Let

$$\hat{A}_1 = A_1(0); \quad \hat{A}_{S1} = A_{S1}(0). \tag{2.51}$$

For the first-order approximation, we derive the following boundary conditions at the interface, $y = y_+ = 0$:

$$\tilde{T}_1 = \tilde{T}_{S1} - (\Delta G_1)\tilde{h}_1, \tag{2.52}$$

$$\tilde{T}_{S1} = k_0^2 \frac{\partial^2 \tilde{h}_1}{\partial x_+^2} + 2k_0 \frac{\partial^2 \tilde{h}_0}{\partial x \partial x_+} + 2k_0 k_1 \frac{\partial^2 \tilde{h}_0}{\partial x_+^2} - G_{1S}\tilde{h}_1, \tag{2.53}$$

$$k_0 \frac{\partial}{\partial y_+}(\tilde{T}_1 - \tilde{T}_{S1}) + g_1 \frac{\partial \tilde{T}_0}{\partial y_+} - g_{s1} \frac{\partial \tilde{T}_{S0}}{\partial y_+} + \frac{\partial}{\partial y}(\tilde{T}_0 - \tilde{T}_{S0})$$

$$+ \sigma_0 \tilde{h}_1 + \sigma_1 \tilde{h}_0 + (\Delta G_2)\tilde{h}_0 = 0. \tag{2.54}$$

In order for the mode solution (2.51) to satisfy the boundary conditions (2.52)–(2.54), we must have

$$\hat{A}_1 = \hat{A}_{S1} - (\Delta G_1)\hat{D}_1 \,, \tag{2.55}$$

$$\hat{A}_{S1} = -\left(k_0^2 + G_{S1}\right)\hat{D}_1 + I_2\hat{D}_0 \,, \tag{2.56}$$

$$-k_0\left(\hat{A}_1 + \hat{A}_{S1}\right) - g_1\left(\hat{A}_0 + \hat{A}_{S0}\right) + \hat{A}_{S0} +$$
$$+(\Delta G_2)\hat{D}_0 + \sigma_0\hat{D}_1 + \sigma_1\hat{D}_0 = 0 \,, \tag{2.57}$$

or

$$-k_0\left(\hat{A}_1 + \hat{A}_{S1}\right) + \sigma_0\hat{D}_1 = I_3\hat{D}_0 \,. \tag{2.58}$$

In the above, we have used the notation

$$\begin{cases} I_2 = -2k_0 k_1 \\ I_3 = k_0^2 + G_{1S} + \frac{\sigma_0 g_1}{k_0} - \sigma_1 - \Delta G_2 \,. \end{cases} \tag{2.59}$$

The determinant of the above inhomogeneous system is $\Delta = 0$. Hence, for a nontrivial solution $\{\hat{A}_1, \hat{A}_{S1}, \hat{D}_1\}$, the following solvability condition must hold:

$$\det \begin{pmatrix} 1 & -1 & 0 \\ 0 & 1 & I_2 \\ -k_0 & -k_0 & I_3 \end{pmatrix} = 0 \,, \tag{2.60}$$

or

$$I_3 + 2k_0 I_2 = 0 \,. \tag{2.61}$$

From (2.61), one obtains

$$\sigma_1 = g_1(1 - 2k_0^2) + k_0^2 - 1 - 4k_0^2 k_1 \,. \tag{2.62}$$

We set $k_1 = 0$. Thus, with $\Delta G_1 = -1$ and $\Delta G_2 = 1$, we get

$$\sigma_1 = g_1(1 - 2k_0^2) + k_0^2 - 1 \,. \tag{2.63}$$

It follows from (2.49) that

$$g_1 = \frac{1}{2}\left(1 + \frac{\sigma_0}{k_0}\right) \,. \tag{2.64}$$

From the above, we obtain the modified dispersion relation:

$$\begin{aligned} \sigma &= \sigma_0 + \varepsilon\sigma_1 + O(\varepsilon^2) \\ &= k_0(1 - 2k_0^2) - \varepsilon + \varepsilon k_0^2 + \varepsilon g_1(1 - 2k_0^2) + O(\varepsilon^2) \\ &= -\varepsilon + \varepsilon k^2 + \left(k + \frac{\varepsilon}{2} + \frac{\varepsilon\sigma_0}{2k}\right)(1 - 2k^2) + O(\varepsilon^2) \,. \end{aligned} \tag{2.65}$$

Note that we have replaced k_0 by k in the final expression. The above procedure can be continued to even higher-order approximations if necessary.

However, as indicated before, the above MVE solution (2.18) is not applicable in the long-wavelength range $k_0 = O(\varepsilon)$. In fact, as $k_0 \ll 1$, the effect of higher-order terms on the right-hand side of (2.20), as well as in the boundary condition (2.26), becomes important. They must be taken into account in the leading order approximation. Moreover, the MVE solution (2.18) is not valid in the extremely short-wavelength range $k_0 = O(\frac{1}{\varepsilon})$ either. For any given small $\varepsilon > 0$, as $k_0 \gg 1$, we have $|\sigma_0| \ll |\varepsilon\sigma_1|$ and $|g_0| \ll |\varepsilon g_1|$.

Therefore, in order to study the behaviors of solution in the ranges $k \ll 1$ and $k \gg 1$, one needs to make different asymptotic expansions. In what follows, we shall look into these cases separately.

2.1.3 Asymptotic Solutions in the Long-Wavelength Regime, $k = O(\varepsilon)$

In the long-wavelength regime, $\{k = O(\varepsilon); g = O(\varepsilon); \sigma = O(\varepsilon)\}$, the solution will not have the structure of multiple scales, as both the fast variables x_+, y_+, t_+ and the slow variables x, y, t are of the same order. We denote

$$\hat{k} = \frac{k}{\varepsilon}; \quad \hat{g} = \frac{g}{\varepsilon}; \quad \hat{\sigma} = \frac{\sigma}{\varepsilon} \tag{2.66}$$

and introduce the new slow variables

$$\begin{cases} \hat{x} = \hat{k}x \\ \hat{y} = \hat{g}y \\ \hat{t} = \hat{\sigma}t. \end{cases} \tag{2.67}$$

We expand the solution $\tilde{q} \equiv \{\tilde{T}, \tilde{T}_{\mathrm{S}}, \tilde{h}\}$ in the following asymptotic form:

$$\begin{aligned}
\tilde{q}(\hat{x}, \hat{y}, \hat{t}, \varepsilon) &\sim e^{\hat{t}}\left\{\tilde{q}_0(\hat{x}, \hat{y}) + \varepsilon^2 \tilde{q}_1(\hat{x}, \hat{y}) + \cdots\right\} \\
\hat{k}(\varepsilon) &\sim \hat{k}_0 + \varepsilon^2 \hat{k}_1 + \cdots \\
\hat{g}(\varepsilon) &\sim \hat{g}_0 + \varepsilon^2 \hat{g}_1 + \cdots \\
\hat{g}_{\mathrm{s}}(\varepsilon) &\sim \hat{g}_{\mathrm{s}0} + \varepsilon^2 \hat{g}_{\mathrm{s}1} + \cdots \\
\hat{\sigma}(\varepsilon) &\sim \hat{\sigma}_0 + \varepsilon^2 \hat{\sigma}_1 + \cdots.
\end{aligned} \tag{2.68}$$

Note that for solutions of the temperature distribution in the solid phase, a different slow variable

$$\hat{y} = \hat{g}_{\mathrm{s}}y \tag{2.69}$$

is used. In terms of the above variables, the linear system (2.8)–(2.12) is transformed to

$$\left\{ \hat{k}^2 \frac{\partial^2}{\partial \hat{x}^2} + \hat{g}^2 \frac{\partial^2}{\partial \hat{y}^2} \right\} \tilde{T} = \hat{\sigma} \tilde{T} - \hat{g} \frac{\partial \tilde{T}}{\partial \hat{y}} , \qquad (2.70)$$

with the boundary conditions:

1. As $\hat{y} \to \infty$,

$$\tilde{T} \to 0. \qquad (2.71)$$

2. As $\hat{y} \to -\infty$,

$$\tilde{T}_{\mathrm{S}} \to 0. \qquad (2.72)$$

3. At the interface, $\hat{y} = 0$,
   (i)

$$\tilde{T} = \tilde{T}_{\mathrm{S}} - (\Delta G_1)\tilde{h} , \qquad (2.73)$$

   (ii)

$$\tilde{T}_{\mathrm{S}} = \varepsilon^2 \hat{k}^2 \frac{\partial^2 \tilde{h}}{\partial \hat{x}} - G_{1\mathrm{S}}\tilde{h} , \qquad (2.74)$$

   (iii)

$$\hat{g} \frac{\partial \tilde{T}}{\partial \hat{y}} - \hat{g}_{\mathrm{s}} \frac{\partial \tilde{T}_{\mathrm{S}}}{\partial \hat{y}} + \hat{\sigma}\tilde{h} + \Delta G_2 \tilde{h} = 0. \qquad (2.75)$$

From the above, it is seen that as $\varepsilon \to 0$, in the leading order approximation for the solution in the long-wavelength regime, both heat convective and unsteady transfer terms in the bulk are important, but the Gibbs–Thomson effect on the interface is negligible.

By substituting (2.68) into the above system (2.70)–(2.75), one can successively derive each order of approximation.

$O(\varepsilon^0)$. In the leading order approximation, we derive the governing equation

$$\hat{k}_0^2 \frac{\partial^2 \tilde{T}_0}{\partial \hat{x}^2} + \hat{g}_0^2 \frac{\partial^2 \tilde{T}_0}{\partial \hat{y}^2} = \hat{\sigma}_0 \tilde{T}_0 - \hat{g}_0 \frac{\partial \tilde{T}_0}{\partial \hat{y}} . \qquad (2.76)$$

The above equation yields the solutions

$$\tilde{T}_0 = \hat{A}_0 \mathrm{e}^{\mathrm{i}\hat{x} - \hat{y}}$$

$$\tilde{T}_{\mathrm{S}0} = \hat{A}_{\mathrm{S}0} \mathrm{e}^{\mathrm{i}\hat{x} + \hat{y}} \qquad (2.77)$$

$$\tilde{h}_0 = \hat{D}_0 \mathrm{e}^{\mathrm{i}\hat{x}}$$

and we derive

$$\hat{g}_0^2 - \hat{g}_0 - (\hat{k}_0^2 + \hat{\sigma}_0) = 0 \,. \tag{2.78}$$

For the liquid phase,

$$\hat{g}_0 = \frac{1}{2} + \sqrt{\hat{k}_0^2 + \hat{\sigma}_0 + \frac{1}{4}} \,, \tag{2.79}$$

while for the solid phase,

$$\hat{g}_{s0} = \hat{g}_0 - 1 \,. \tag{2.80}$$

With $\Delta G_1 = -1, \Delta G_2 = 1$, and $G_{1S} = 0$, we derive from the boundary conditions (2.71)–(2.75),

$$\begin{cases} \hat{A}_0 = \hat{A}_{S0} + \hat{D}_0 \\[2mm] \hat{A}_{S0} = 0 \\[2mm] -\hat{g}_0 \hat{A}_0 - \hat{g}_{s0} \hat{A}_{S0} + (\hat{\sigma}_0 + 1)\hat{D}_0 = 0 \,. \end{cases} \tag{2.81}$$

Thus, it is found that

$$\hat{\sigma}_0 = -1 + \hat{g}_0 = -\frac{1}{2} + \sqrt{\hat{k}_0^2 + \hat{\sigma}_0 + \frac{1}{4}} \tag{2.82}$$

or

$$\hat{\sigma}_0 = \hat{k}_0 \,. \tag{2.83}$$

$O(\varepsilon^2)$. In the first-order approximation, we derive the governing equations

$$\hat{k}_0^2 \frac{\partial^2 \tilde{T}_1}{\partial \hat{x}^2} + \hat{g}_0^2 \frac{\partial^2 \tilde{T}_1}{\partial \hat{y}^2} - \hat{\sigma}_0 \tilde{T}_1 + \hat{g}_0 \frac{\partial \tilde{T}_1}{\partial \hat{y}}$$

$$= \left(2\hat{k}_0 \hat{k}_1 - 2\hat{g}_0 \hat{g}_1 + \hat{\sigma}_1 + \hat{g}_1\right)\tilde{T}_0 \tag{2.84}$$

and

$$\hat{k}_0^2 \frac{\partial^2 \tilde{T}_{S1}}{\partial \hat{x}^2} + \hat{g}_{s0}^2 \frac{\partial^2 \tilde{T}_{S1}}{\partial \hat{y}^2} - \hat{\sigma}_0 \tilde{T}_{S1} + \hat{g}_{s0} \frac{\partial \tilde{T}_{S1}}{\partial \hat{y}}$$

$$= \left(2\hat{k}_0 \hat{k}_1 - 2\hat{g}_{s0} \hat{g}_{s1} + \hat{\sigma}_1 - \hat{g}_{s1}\right)\tilde{T}_{S0} \,. \tag{2.85}$$

To eliminate the secular terms, we must set

$$\begin{cases} 2\hat{k}_0 \hat{k}_1 - 2\hat{g}_0 \hat{g}_1 + \hat{\sigma}_1 + \hat{g}_1 = 0 \\[2mm] 2\hat{k}_0 \hat{k}_1 - 2\hat{g}_{s0} \hat{g}_{s1} + \hat{\sigma}_1 - \hat{g}_{s1} = 0 \,. \end{cases} \tag{2.86}$$

We put $k_1 = 0$. It follows that

$$\hat{g}_1 = \hat{g}_{s1} = \frac{\hat{\sigma}_1}{2\hat{g}_0 - 1} . \tag{2.87}$$

The first-order approximation yields the solutions

$$\tilde{T}_1 = \hat{A}_1 e^{i\hat{x} - \hat{y}}$$

$$\tilde{T}_{S1} = \hat{A}_{S1} e^{i\hat{x} + \hat{y}} \tag{2.88}$$

$$\tilde{h}_1 = \hat{D}_1 e^{i\hat{x}} .$$

From the boundary conditions (2.71)–(2.75), we derive

$$\begin{cases} \hat{A}_1 = \hat{A}_{S1} + \hat{D}_1 \\ \hat{A}_{S1} = -\hat{k}_0^2 \hat{D}_0 \\ -\hat{g}_0 \hat{A}_1 - \hat{g}_{s0} \hat{A}_{S1} + (\hat{\sigma}_0 + 1)\hat{D}_1 = \hat{A}_S \hat{g}_1 + \hat{A}_{S0} \hat{g}_{s1} - \hat{\sigma}_1 \hat{D}_0 . \end{cases} \tag{2.89}$$

From this system, we obtain

$$\hat{\sigma}_1 = \hat{g}_1 + (1 - 2\hat{g}_0)\hat{k}_0^2 \tag{2.90}$$

or

$$\hat{\sigma}_1 = -\frac{\hat{k}_0}{2}(1 + 2\hat{k}_0)^2. \tag{2.91}$$

Combining (2.87) with (2.91), we derive

$$\hat{g}_1 = -\hat{k}_0^2 \left[1 + \frac{1}{2(\hat{g}_0 - 1)} \right] \tag{2.92}$$

or

$$\hat{g}_1 = -\frac{\hat{k}_0}{2}(1 + 2\hat{k}_0). \tag{2.93}$$

Thus, we have

$$\begin{aligned} \hat{\sigma} &= \hat{\sigma}_0 + \varepsilon^2 \hat{\sigma}_1 \\ &= (-1 + \hat{g}_0) + \varepsilon^2 \hat{g}_1 + \varepsilon^2 (1 - 2\hat{g}_0)\hat{k}_0^2 \\ &= -1 + \varepsilon^2 \hat{k}^2 + (\hat{g}_0 + \varepsilon^2 \hat{g}_1)(1 - 2\varepsilon^2 \hat{k}^2) + 2\varepsilon^4 \hat{g}_1 \hat{k}^2 . \end{aligned} \tag{2.94}$$

Therefore, expressing the equations in terms of the original parameters (k, σ), in the long-wavelength regime ($k \to 0$), we have the dispersion relation:

$$
\begin{cases}
\sigma_0 = -\frac{\varepsilon}{2} + \sqrt{k^2 + \varepsilon\sigma_0 + \frac{\varepsilon^2}{4}} \\[2mm]
g_0 = \frac{\varepsilon}{2} + \sqrt{k^2 + \varepsilon\sigma_0 + \frac{\varepsilon^2}{4}} \\[2mm]
\varepsilon^2 g_1 = -\varepsilon k^2 \left\{ 1 + \frac{\varepsilon}{2(g_0 - \varepsilon)} \right\} \\[2mm]
\sigma = -\varepsilon + \varepsilon k^2 + (g_0 + \varepsilon^2 g_1)(1 - 2k^2) + 2\varepsilon^2 g_1 k^2 .
\end{cases}
\tag{2.95}
$$

Note that in this case one can write

$$
G_1 = g_0 + \varepsilon^2 g_1 = \frac{\varepsilon}{2} + \sqrt{k^2 + \varepsilon\sigma + \frac{\varepsilon^2}{4}} + O(\varepsilon^3).
\tag{2.96}
$$

Comparing (2.65 with (2.95), it is seen that if one applies the approximate formula $\left(k + \frac{\varepsilon\sigma_0}{2k} \right) \approx \sqrt{k^2 + \varepsilon\sigma_0}$ to the dispersion relation (2.65) then the resultant formula

$$
\sigma = -\varepsilon + \varepsilon k^2 + \left(\frac{\varepsilon}{2} + \sqrt{k^2 + \varepsilon\sigma_0} \right)(1 - 2k^2) + O(\varepsilon^2)
\tag{2.97}
$$

will be applicable to the whole range $0 \leq k \ll \frac{1}{\varepsilon}$.

2.1.4 Asymptotic Solutions
in the Extremely Short-Wavelength Regime, $k = O(\frac{1}{\varepsilon})$

In the extremely short-wavelength regime, $\{ k = O(\frac{1}{\varepsilon}); g = O(\frac{1}{\varepsilon}) \}$, one can still find a MVE solution with the fast variable x_+, y_+, t_+ and the slow variables x, y, t. We write

$$
\begin{cases}
\tilde{k} = \varepsilon k = O(1) \\[2mm]
\tilde{g} = \varepsilon g = O(1) \\[2mm]
\tilde{\sigma} = \varepsilon^\alpha \sigma = O(1),
\end{cases}
\tag{2.98}
$$

where the exponent α is to be determined. The MVE system (2.20)–(2.26) is still applicable to the present case if we change k, g and σ to $\tilde{k}, \tilde{g}$ and $\tilde{\sigma}$. Thus, we have

$$
\left(\tilde{k}^2 \frac{\partial^2}{\partial x_+^2} + \tilde{g}^2 \frac{\partial^2}{\partial y_+^2} \right) \tilde{T} = \varepsilon^{3-\alpha} \tilde{\sigma}\tilde{T} - \varepsilon^2 \tilde{g} \frac{\partial \tilde{T}}{\partial y_+} - 2\varepsilon^2 \tilde{k} \frac{\partial^2 \tilde{T}}{\partial x \partial x_+}
$$

$$
- 2\varepsilon^2 \tilde{g} \frac{\partial^2 \tilde{T}}{\partial y \partial y_+} - \varepsilon^4 \left(\frac{\partial^2}{\partial x^2} + \frac{\partial^2}{\partial y^2} - \frac{\partial}{\partial y} \right) \tilde{T},
\tag{2.99}
$$

with the boundary conditions:

1. As $y_+ \to \infty$,

$$\tilde{T} \to 0. \tag{2.100}$$

2. As $y_+ \to -\infty$,

$$\tilde{T}_S \to 0. \tag{2.101}$$

3. At the interface, $y = y_+ = 0$,

(i)

$$\tilde{T} = \tilde{T}_S - (\Delta G_1)\tilde{h}, \tag{2.102}$$

(ii)

$$\tilde{T}_S = \frac{\tilde{k}^2}{\varepsilon^2}\frac{\partial^2 \tilde{h}}{\partial x_+^2} + 2\tilde{k}\frac{\partial^2 \tilde{h}}{\partial x \partial x_+} + \varepsilon^2 \frac{\partial^2 \tilde{h}}{\partial x^2} - G_{1S}\tilde{h}, \tag{2.103}$$

(iii)

$$\tilde{g}\frac{\partial}{\partial y_+}\left(\tilde{T} - \tilde{T}_S\right) + \varepsilon^2 \frac{\partial}{\partial y}\left(\tilde{T} - \tilde{T}_S\right) + \varepsilon^{1-\alpha}\tilde{\sigma}\tilde{h}$$

$$+\varepsilon^2 (\Delta G_2)\tilde{h} = 0. \tag{2.104}$$

One can see from (2.103) that $\tilde{h} = O(\varepsilon^2 \tilde{T})$. Furthermore, from (2.104) one can prove that the exponent α must be chosen as $\alpha = 3$. Hence the system allows the following asymptotic expansions:

$$\tilde{T} \sim \mathrm{e}^{t+}\left\{\tilde{T}_0(x, y, x_+, y_+) + \varepsilon^2 \tilde{T}_1(x, y, x_+, y_+) + \cdots\right\}$$

$$\tilde{T}_S \sim \mathrm{e}^{t+}\left\{\tilde{T}_{S0}(x, y, x_+, y_+) + \varepsilon^2 \tilde{T}_{S1}(x, y, x_+, y_+) + \cdots\right\}$$

$$\tilde{h} \sim \varepsilon^2 \mathrm{e}^{t+}\left\{\tilde{h}_0(x, y, x_+, y_+) + \varepsilon^2 \tilde{h}_1(x, y, x_+, y_+) + \cdots\right\}$$

$$\tilde{k}(\varepsilon) \sim \tilde{k}_0 + \varepsilon^2 \tilde{k}_1 + \cdots$$

$$\tilde{g}(\varepsilon) \sim \tilde{g}_0 + \varepsilon^2 \tilde{g}_1 + \cdots$$

$$\tilde{g}_s(\varepsilon) \sim \tilde{g}_{s0} + \varepsilon^2 \tilde{g}_{s1} + \cdots$$

$$\tilde{\sigma}(\varepsilon) \sim \tilde{\sigma}_0 + \varepsilon^2 \tilde{\sigma}_1 + \cdots . \tag{2.105}$$

By substituting (2.105) into the system (2.99)–(2.104), one can successively derive each order of approximation.

$O(\varepsilon^0)$. In the zeroth-order approximation, we have

$$\tilde{k}_0^2 \frac{\partial^2 \tilde{T}_0}{\partial x_+^2} + \tilde{g}_0^2 \frac{\partial^2 \tilde{T}_0}{\partial y_+^2} = \tilde{\sigma}_0 \tilde{T}_0 \,, \tag{2.106}$$

which allows the mode solutions:

$$\tilde{T}_0 = A_0(y)\mathrm{e}^{\mathrm{i}x_+ - y_+}$$

$$\tilde{T}_{\mathrm{S}0} = A_{\mathrm{S}0}(y)\mathrm{e}^{\mathrm{i}x_+ + y_+} \tag{2.107}$$

$$\tilde{h}_0 = \hat{D}_0 \mathrm{e}^{\mathrm{i}x_+}$$

with

$$\tilde{g}_0 = \sqrt{\tilde{\sigma}_0 + \tilde{k}_0^2} \,. \tag{2.108}$$

With the notation

$$\hat{A}_0 = A_0(0), \qquad \hat{A}_{\mathrm{S}0} = A_{\mathrm{S}0}(0), \tag{2.109}$$

from the boundary conditions (2.102)–(2.104), we derive

$$\begin{cases} \hat{A}_0 = \hat{A}_{\mathrm{S}0} \\[2mm] \hat{A}_{\mathrm{S}0} = -\tilde{k}_0^2 \hat{D}_0 \\[2mm] -\tilde{g}_0(\hat{A}_0 + \hat{A}_{\mathrm{S}0}) + \tilde{\sigma}_0 \hat{D}_0 = 0. \end{cases} \tag{2.110}$$

It is seen that in the leading order approximation for the solution in the extremely short-wavelength regime, the unsteady heat transfer term in the bulk is important, but the effect of interface displacement is negligible. We derive the following dispersion relation:

$$\tilde{\sigma}_0 = -2\tilde{g}_0 \tilde{k}_0^2 \,. \tag{2.111}$$

This may be rewritten as

$$\sigma_0 = -2k_0^2 \sqrt{k_0^2 + \varepsilon\sigma_0} \,. \tag{2.112}$$

$O(\varepsilon^2)$. In the first-order approximation, we have

$$\tilde{k}_0^2 \frac{\partial^2 \tilde{T}_1}{\partial x_+^2} + \tilde{g}_0^2 \frac{\partial^2 \tilde{T}_1}{\partial y_+^2} - \tilde{\sigma}_0 \tilde{T}_1 = \tilde{a}_0 \tilde{T}_0$$

$$\tilde{k}_0^2 \frac{\partial^2 \tilde{T}_{\mathrm{S}1}}{\partial x_+^2} + \tilde{g}_0^2 \frac{\partial^2 \tilde{T}_{\mathrm{S}1}}{\partial y_+^2} - \tilde{\sigma}_0 \tilde{T}_{\mathrm{S}1} = \tilde{b}_0 \tilde{T}_{\mathrm{S}0} \,, \tag{2.113}$$

where

$$\begin{cases} \tilde{a}_0 = 2\tilde{g}_0 A_0'(y) + \left(\tilde{\sigma}_1 + \tilde{g}_0 + 2\tilde{k}_0\tilde{k}_1 - 2\tilde{g}_0\tilde{g}_1\right) A_0(y) \\[2mm] \tilde{b}_0 = -2\tilde{g}_0 A_{\mathrm{S}0}'(y) + \left(\tilde{\sigma}_1 - \tilde{g}_0 + 2\tilde{k}_0\tilde{k}_1 - 2\tilde{g}_0\tilde{g}_{\mathrm{s}1}\right) A_{\mathrm{S}0}(y) \,. \end{cases} \tag{2.114}$$

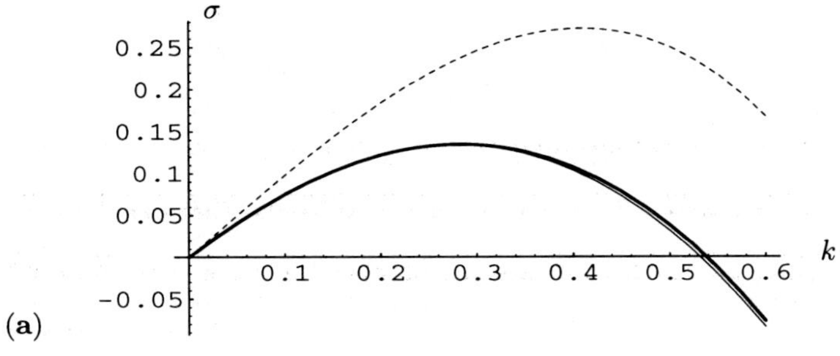

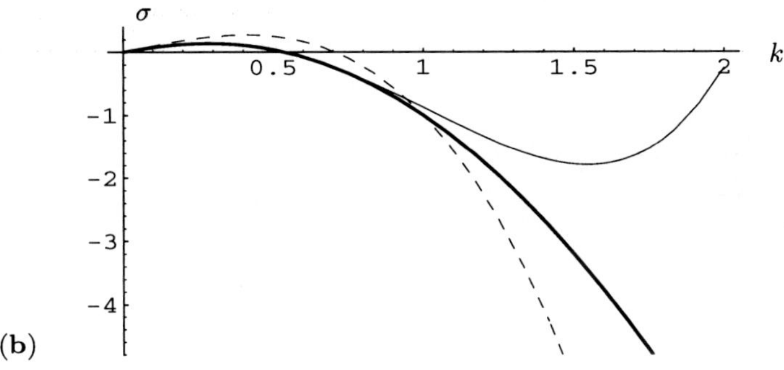

Fig. 2.3a,b. The dispersion curves for $\varepsilon = 0.5$. The dashed line is the zeroth-order approximation, the thin solid line is the modified first-order approximation, whereas the bold solid line is the exact solution: (**a**) for the range $0 \leq k \leq 0.6$; (**b**) for the range $0 \leq k \leq 2$

To eliminate the secular terms, we set $\tilde{a}_0 = \tilde{b}_0 = 0$. Moreover, we assume $A_0'(y) = A_{S0}'(y) = 0$, and consequently,

$$A_0 = \hat{A}_0, \quad A_{S0}(y) = \hat{A}_{S0}. \tag{2.115}$$

Thus, it follows that

$$\begin{cases} \tilde{g}_1 = \tfrac{1}{2} + \frac{\tilde{\sigma}_1}{2\tilde{g}_0} \\ \tilde{g}_{s1} = \tilde{g}_1 - 1 \end{cases} \tag{2.116}$$

and

$$\tilde{\sigma}_1 = -\tilde{g}_0 - 2\tilde{k}_0\tilde{k}_1 + 2\tilde{g}_0\tilde{g}_1. \tag{2.117}$$

As before, we can let $\tilde{k}_1 = 0$. Hence, we have

$$\tilde{\sigma}_1 = \tilde{g}_0(2\tilde{g}_1 - 1). \tag{2.118}$$

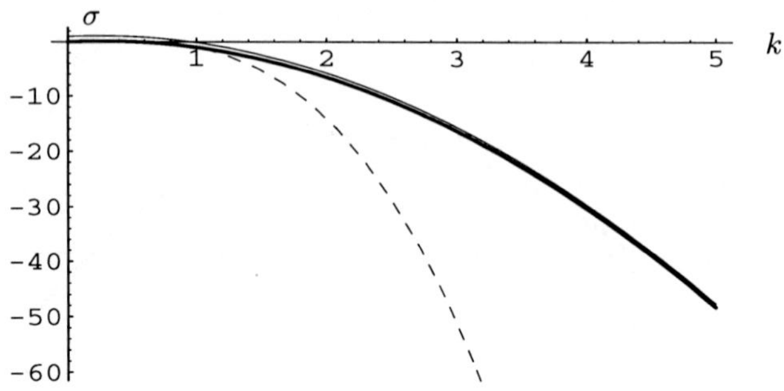

Fig. 2.4. Dispersion curves: The dashed line is for the zeroth-order approximation, the thin solid line is the extremely short-wavelength approximation, whereas the bold solid line is the exact solution

The first-order system allows the mode solutions:

$$\tilde{T}_1 = A_1(y)\mathrm{e}^{\mathrm{i}x_+ - y_+} \tag{2.119}$$

$$\tilde{T}_{S1} = A_{S1}(y)\mathrm{e}^{\mathrm{i}x_+ + y_+} \tag{2.120}$$

$$\tilde{h}_1 = \hat{D}_1\mathrm{e}^{\mathrm{i}x_+} . \tag{2.121}$$

We write

$$\hat{A}_1 = A_1(0), \qquad \hat{A}_{S1} = A_{S1}(0). \tag{2.122}$$

From the boundary conditions (2.102)–(2.104), we derive

$$\begin{cases} \hat{A}_1 = \hat{A}_{S1} + \hat{D}_0 \\ \hat{A}_{S1} = -\tilde{k}_0^2\hat{D}_1 + I_2\hat{D}_0 \\ -\tilde{g}_0(\hat{A}_1 + \hat{A}_{S1}) + \tilde{\sigma}_0\hat{D}_1 = I_3\hat{D}_0 , \end{cases} \tag{2.123}$$

where

$$\begin{cases} I_2 = -2\tilde{k}_0\tilde{k}_1 \\ I_3 = -2\tilde{k}_0^2\tilde{g}_1 + \tilde{k}_0^2 - \tilde{\sigma}_1. \end{cases} \tag{2.124}$$

For a nontrivial solution, the condition

$$I_3 + 2\tilde{g}_0 I_2 + \tilde{g}_0 = 0 \tag{2.125}$$

is necessary. Thus, it follows that

$$\tilde{k}_0^2 - \tilde{\sigma}_1 - 2\tilde{k}_0^2\tilde{g}_1 - 4\tilde{k}_0\tilde{g}_0\tilde{k}_1 + \tilde{g}_0 = 0 . \tag{2.126}$$

As $\tilde{k}_1 = 0$, from (2.117)–(2.126) we derive

$$\tilde{g}_1 = 1 - \frac{\tilde{k}_0^2}{2(\tilde{g}_0 + \tilde{k}_0^2)} \tag{2.127}$$

and

$$\begin{aligned}
\tilde{\sigma} &= \tilde{\sigma}_0 + \varepsilon^2 \tilde{\sigma}_1 = -2\tilde{g}_0 \tilde{k}_0^2 + \varepsilon^2 (\tilde{k}_0^2 + \tilde{g}_0 - 2\tilde{k}_0^2 \tilde{g}_1) \\
&= \varepsilon^2 \tilde{k}_0^2 + (\tilde{g}_0 + \varepsilon^2 \tilde{g}_1)(\varepsilon^2 - 2\tilde{k}_0^2) - \varepsilon^4 \tilde{g}_1 .
\end{aligned} \tag{2.128}$$

Returning to the original parameters $\{k, \sigma\}$, we derive the dispersion formula in the extremely short-wavelength regime as

$$\left\{ \begin{aligned}
\sigma_0 &= -2\left(\sqrt{k^2 + \varepsilon\sigma_0}\right) k^2 \\
g_0 &= \sqrt{k^2 + \varepsilon\sigma_0} \\
\varepsilon^2 g_1 &= \varepsilon - \frac{1}{2}\frac{\varepsilon^2 k^2}{g_0 + \varepsilon k^2} \\
\sigma &= \varepsilon k^2 + (g_0 + \varepsilon^2 g_1)(1 - 2k^2) - \varepsilon^2 g_1 .
\end{aligned} \right. \tag{2.129}$$

In the present case, one can show that

$$G_1 = g_0 + \varepsilon^2 g_1 = \frac{\varepsilon}{2} + \sqrt{k^2 + \varepsilon\sigma} + O(\varepsilon^3) . \tag{2.130}$$

It should be noted that the system (2.8)–(2.12) is a linear system with constant coefficients, so that it can be solved exactly. The exact solutions are:

$$\tilde{T} = \hat{A} e^{\left(\frac{\sigma t}{\varepsilon} + \frac{ikx}{\varepsilon} - \frac{qy}{\varepsilon}\right)} \tag{2.131}$$

$$\tilde{T}_{\mathrm{S}} = \hat{A}_{\mathrm{S}} e^{\left(\frac{\sigma t}{\varepsilon} + \frac{ikx}{\varepsilon} + \frac{gy}{\varepsilon} - y\right)} \tag{2.132}$$

$$\tilde{h} = \hat{D} e^{\left(\frac{\sigma t}{\varepsilon} + \frac{ikx}{\varepsilon}\right)}, \tag{2.133}$$

where

$$g = \frac{\varepsilon}{2} + \sqrt{k^2 + \varepsilon\sigma + \varepsilon^2/4}. \tag{2.134}$$

The exact dispersion relation is

$$\sigma = -\varepsilon + \varepsilon k^2 + \left(\frac{\varepsilon}{2} + \sqrt{k^2 + \varepsilon\sigma + \varepsilon^2/4}\right)(1 - 2k^2). \tag{2.135}$$

By comparing (2.65) with (2.135), one sees that the dispersion formula of the MVE solution obtained above, in the regime $(\varepsilon \ll k \ll \frac{1}{\varepsilon})$, is the regular perturbation expansion of the exact solution as $\varepsilon \to 0$. In Fig. 2.3, we have shown all these dispersion curves in the (σ, k) plane. It is seen that up to $\varepsilon = 0.5$, the agreement between the exact solution and the first-order MVE solution (2.97) is very good in the regime $0 \le k < 1$. However, as $k > 1.5$, the first-order MVE solution (2.65), as expected, is no longer close to the exact solution, even qualitatively. Nevertheless, the asymptotic solution in the extremely short-wavelength regime agrees with the exact solution (2.129) very well (see Fig. 2.4).

2.2 Unidirectional Solidification from a Binary Mixture

Although the simple model problem discussed in the last section demonstrated the onset of the Mullins–Sekerka instability mechanism very well, it is difficult to precisely implement in experiment. A more practical model of unidirectional solidification, which can be easily examined in experiment, is solidification of a binary mixture in a Hele–Shaw cell. The Mullins–Sekerka instability mechanism that we have explored for the system with a pure melt is still valid for this binary system and can be derived with the same mathematical approach. Consider a unidirectional solidification device with two uniform temperature zones and a thin sample material, as sketched in Fig. 1.3. The distance between the two zones is set as $(L)_{\mathrm{D}}$. The temperature of the hot zone is T_{H}; while the temperature of the cold zone is T_{C}. Furthermore, we have $T_{\mathrm{C}} < T_{\mathrm{M0}} < T_{\mathrm{H}}$.

The sample is being pulled at a constant imposed velocity V along the direction from the hot zone to the cold zone. The minor species in this binary mixture system, considered as an impurity, is dilute. Hence, the mathematical formulation given in Sect. 1.3 is applicable.

Again, for the sake of simplicity, we neglect the effect of convection. Thus the whole system is governed by thermodynamics.

2.2.1 Mathematical Formulation of the Problem

When the pulling velocity V is sufficiently small, the interface will be flat and located somewhere between the two zones. Assume the distances from the interface to the hot zone and the cold zone are L_1 and L_2, respectively, which are to be determined. Adopt the same coordinate system (x, y) as that defined in Sect. 2.1, whose origin is set at the interface. The rest frame is fixed to the solid phase, whereas the coordinate frame (x, y) is considered as a moving frame, moving together with the two zones with velocity $\mathbf{V}$ along the y-axis.

As specified in Sect. 1.3.3, we use the mass diffusion length ℓ_{D} as the length scale and the pulling velocity V as the velocity scale. The scales of the temperature T and concentration C are set as $\Delta H/(c_{\mathrm{p}}\rho)$ and C_{∞}, respectively.

Since the cell is very thin, the whole process can be treated as two-dimensional. Thus, the dimensionless governing equations for the present system can be written in the following form:

$$\nabla^2 \bar{T} = \lambda \left(\frac{\partial \bar{T}}{\partial \bar{t}} - \frac{\partial \bar{T}}{\partial \bar{y}} \right)$$
$$\nabla^2 \bar{C} = \left(\frac{\partial \bar{C}}{\partial \bar{t}} - \frac{\partial \bar{C}}{\partial \bar{y}} \right).$$
(2.136)

The dimensionless boundary conditions are:

1. At the edge of the hot zone, $\bar{y} = \bar{L}_1$,

$$\bar{T} = \bar{T}_{\mathrm{H}} = \frac{T_{\mathrm{H}} - T_{\mathrm{M0}}}{\Delta H / (c_{\mathrm{p}} \rho)} > 0.$$
(2.137)

2. At the edge of the cold zone, $\bar{y} = -\bar{L}_2$,

$$\bar{T} = \bar{T}_{\mathrm{C}} = \frac{T_{\mathrm{H}} - T_{\mathrm{M0}}}{\Delta H / (c_{\mathrm{p}} \rho)} < 0.$$
(2.138)

3. In the far field, as $\bar{y} \to \infty$,

$$\bar{C} \to 1.$$
(2.139)

4. At the interface, $\bar{y} = \bar{h}(\bar{x}, \bar{t})$,
   (i)

$$\bar{T} = \bar{T}_{\mathrm{S}},$$
(2.140)

   (ii)

$$\bar{T}_{\mathrm{S}} = \varepsilon^2 \frac{\bar{h}_{\bar{x}\bar{x}}}{(1 + \bar{h}_{\bar{x}}^2)^{\frac{3}{2}}} - M\bar{C},$$
(2.141)

   (iii)

$$\frac{\partial}{\partial \bar{y}}(\bar{T} - \bar{T}_{\mathrm{S}}) - \bar{h}_{\bar{x}} \frac{\partial}{\partial \bar{x}}(\bar{T} - \bar{T}_{\mathrm{S}}) + \lambda \left(\bar{h}_{\bar{t}} + 1 \right) = 0,$$
(2.142)

   (iv)

$$\frac{\partial \bar{C}}{\partial \bar{y}} - \bar{h}_{\bar{x}} \frac{\partial \bar{C}}{\partial \bar{x}} + \bar{C}(1 - \kappa) \left(\bar{h}_{\bar{t}} + 1 \right) = 0.$$
(2.143)

In the above, M is the morphological parameter defined by (1.49). In most cases, one has

$$\lambda = \frac{\ell_{\mathrm{D}}}{\ell_{\mathrm{T}}} \ll 1, \quad \text{and} \quad \bar{L}_1; \bar{L}_2 \gg 1.$$
(2.144)

For the sake of convenience, hereafter, we shall omit the bar '$^-$' over the dimensionless quantities.

2.2.2 Basic Steady State

The above system allows a one-dimensional steady-state solution with a flat interface for arbitrary temperature gradient and surface tension parameter $\varepsilon \geq 0$. The interface equation is taken to be $y = h_B = 0$. The exact solution of this one-dimensional steady state can be easily found as the following:

$$T_B = A_1 + A_2 e^{-\lambda y}$$

$$T_{BS} = A_{S1} + A_{S2} e^{\lambda y} \tag{2.145}$$

$$C_B = B_1 + B_2 e^{-y},$$

where T_B is the temperature distribution in the liquid phase, T_{BS} is the temperature distribution in the solid phase, and C_B is the concentration distribution in the liquid phase. From the boundary conditions (2.137), (2.138), we have

$$A_1 + A_2 e^{-\lambda L_1} = T_H \tag{2.146}$$

$$A_{S1} + A_{S2} e^{-\lambda L_2} = T_C. \tag{2.147}$$

On the other hand, from (2.139), we have

$$B_1 = 1. \tag{2.148}$$

Moreover, from the interface conditions (2.141)–(2.143), we have

$$\begin{cases} A_1 + A_2 = A_{S1} + A_{S2} \\ A_1 + A_2 = -M(B_1 + B_2) \\ A_2 + A_{S2} = 1 \\ B_2 - (1 - \kappa)(1 + B_2) = 0. \end{cases} \tag{2.149}$$

Finally, we have the relation

$$L_1 + L_2 = L. \tag{2.150}$$

From the above eight conditions one can determine the eight unknowns: $A_1, A_2, A_{S1}, A_{S2}, B_1, B_2, L_1, L_2$, and thus completely determine the basic steady-state solution. We obtain

$$C_B(y) = 1 + \frac{1 - \kappa}{\kappa} e^{-y}, \tag{2.151}$$

$$C_B(0) = \frac{1}{\kappa}, \tag{2.152}$$

$$T_B(0) = A_1 + A_2 = A_{S1} + A_{S2} = -\frac{M}{\kappa}, \tag{2.153}$$

$$\Delta G_1 = \frac{\partial T_B}{\partial y}(0) - \frac{\partial T_{BS}}{\partial y}(0) \quad = -\lambda, \tag{2.154}$$

$$\Delta G_2 = \frac{\partial^2 T_B}{\partial y^2}(0) - \frac{\partial^2 T_{BS}}{\partial y^2}(0) = -\lambda^2, \tag{2.155}$$

and

$$
\begin{aligned}
A_1 &= \frac{T_H + \frac{M}{\kappa}e^{-\lambda L_1}}{1 - e^{-\lambda L_1}} \\[2mm]
A_{S1} &= \frac{T_C + \frac{M}{\kappa}e^{\lambda L_2}}{1 - e^{-\lambda L_2}} \\[2mm]
A_2 &= -\frac{T_H + \frac{M}{\kappa}}{1 - e^{-\lambda L_1}} \\[2mm]
A_{S2} &= -\frac{T_C + \frac{M}{\kappa}}{1 - e^{-\lambda L_2}}.
\end{aligned}
\tag{2.156}
$$

Thus, it follows that

$$
\begin{aligned}
G_{1L} &= \frac{\partial T_B}{\partial y}(0) = \frac{\lambda}{1 - e^{-\lambda L_1}}\left(\frac{M}{\kappa} + T_H\right) \\[2mm]
G_{1S} &= \frac{\partial T_{BS}}{\partial y}(0) = \frac{-\lambda}{1 - e^{-\lambda L_2}}\left(\frac{M}{\kappa} + T_C\right).
\end{aligned}
\tag{2.157}
$$

Furthermore, from (2.154), one obtains

$$-\frac{T_H + \frac{M}{\kappa}}{1 - e^{-\lambda L_1}} - \frac{T_C + \frac{M}{\kappa}}{1 - e^{-\lambda L_2}} = 1. \tag{2.158}$$

From (2.158) and (2.150) one can find L_1 and L_2 for any given L, and hence determine the location of the interface.

As mentioned before, for most binary systems, the parameter $\lambda = 10^{-2} - 10^{-4}$, is very small. Letting $\lambda \to 0$ and keeping the $O(\lambda)$ terms, (2.158) becomes

$$\frac{T_H + \frac{M}{\kappa}}{L_1} + \frac{T_C + \frac{M}{\kappa}}{L_2} = 0. \tag{2.159}$$

Eliminating L_1 from (2.159) and (2.150), it follows that

$$\begin{cases} L_1 = \frac{L}{R} \\[2mm] L_2 = \left(1 - \frac{1}{R}\right)L, \end{cases} \tag{2.160}$$

where

$$R = \frac{T_H - T_C}{T_H + \frac{M}{\kappa}}. \tag{2.161}$$

Furthermore, as $\lambda \to 0$, from (2.157) one obtains

$$
\begin{aligned}
G_{1\mathrm{L}} &\approx \frac{1}{L_1}\left(\frac{M}{\kappa} + T_\mathrm{H}\right) = \frac{T_\mathrm{H} - T_\mathrm{C}}{L} \\
G_{1\mathrm{S}} &\approx -\frac{1}{L_2}\left(\frac{M}{\kappa} + T_\mathrm{C}\right) = \frac{T_\mathrm{H} - T_\mathrm{C}}{L}\,.
\end{aligned}
\tag{2.162}
$$

As a result, we derive that

$$
G = \frac{G_{1\mathrm{L}} + G_{1\mathrm{S}}}{2} \approx \frac{T_\mathrm{H} - T_\mathrm{C}}{L} \quad (\text{as } \lambda \to 0)\,.
\tag{2.163}
$$

2.2.3 Unsteady Perturbed Solutions

The unsteady solutions can be expressed in the forms:

$$
\begin{aligned}
T(x,y,t) &= T_\mathrm{B}(y) + \tilde{T}(x,y,t) \\
T_\mathrm{S}(x,y,t) &= T_\mathrm{SB}(y) + \tilde{T}_\mathrm{S}(x,y,t) \\
C(x,y,t) &= C_\mathrm{B}(y) + \tilde{C}(x,y,t) \\
h(x,t) &= h_\mathrm{B} + \tilde{h}(x,t)\,,
\end{aligned}
\tag{2.164}
$$

where $\tilde{T}, \tilde{T}_\mathrm{S}, \tilde{C}$, and $\tilde{h}$ are small perturbations around the basic steady state. The governing equation for the perturbation part is

$$
\begin{aligned}
\nabla^2 \tilde{T} &= \lambda\left(\frac{\partial \tilde{T}}{\partial t} - \frac{\partial \tilde{T}}{\partial y}\right) \\
\nabla^2 \tilde{C} &= \left(\frac{\partial \tilde{C}}{\partial t} - \frac{\partial \tilde{C}}{\partial y}\right)\,.
\end{aligned}
\tag{2.165}
$$

The boundary conditions are:

1. The external boundary condition:

$$
\begin{cases}
\tilde{C} \to 0, & \text{as } y \to \infty, \\
\tilde{T} = 0, & \text{as } y = L_1, \\
\tilde{T}_\mathrm{S} = 0, & \text{as } y = L_2\,.
\end{cases}
\tag{2.166}
$$

2. The interface conditions: since the deformation of the interface, $\tilde{h}(x,t)$, is very small, the original boundary conditions on the interface $y = \tilde{h}(x,t)$ can be expanded in a Taylor series in $\tilde{h}$ around $y = h_\mathrm{B} = 0$. Retaining

only the linear terms and omitting all higher-order small terms, we find that at $y = 0$

$$\tilde{T} = \tilde{T}_S - (\Delta G_1)\tilde{h}$$

$$\tilde{T}_S = \varepsilon^2 \tilde{h}_{xx} - M\tilde{C} + \left\{ \left(\frac{1-\kappa}{\kappa} \right) M - G_{1S} \right\} \tilde{h}$$

$$\frac{\partial \left(\tilde{T} - \tilde{T}_S \right)}{\partial y} + (\Delta G_2)\, \tilde{h} + \lambda \frac{\partial \tilde{h}}{\partial t} = 0 \tag{2.167}$$

$$\frac{\partial \tilde{C}}{\partial y} + \left(\frac{1-\kappa}{\kappa} \right) \frac{\partial \tilde{h}}{\partial t} + (1-\kappa)(\tilde{C} + \tilde{h}) = 0\,.$$

We solve this system by using the same MVE method described in the last section. The fast variables are still defined by (2.17), but the factor $g(\varepsilon)$ is used only for the solution of the concentration field. For the temperature field T in the liquid phase and the temperature field T_S in the solid phase, we use $q(\varepsilon)$ and $q_s(\varepsilon)$, respectively. The multiple variables form of this system is written as follows:

$$\left\{ k^2 \frac{\partial^2}{\partial x_+^2} + q^2 \frac{\partial^2}{\partial y_+^2} \right\} \tilde{T} = \lambda \varepsilon \left(\sigma \tilde{T} - q \frac{\partial \tilde{T}}{\partial y_+} \right) - 2\varepsilon \left(k \frac{\partial^2 \tilde{T}}{\partial x \partial x_+} + q \frac{\partial^2 \tilde{T}}{\partial y \partial y_+} \right)$$

$$-\varepsilon^2 \left(\frac{\partial^2 \tilde{T}}{\partial x^2} + \frac{\partial^2 \tilde{T}}{\partial y^2} + \lambda \frac{\partial \tilde{T}}{\partial y} \right), \tag{2.168}$$

$$\left\{ k^2 \frac{\partial^2}{\partial x_+^2} + q_s^2 \frac{\partial^2}{\partial y_+^2} \right\} \tilde{T}_S = \lambda \varepsilon \left(\sigma \tilde{T}_S - q_s \frac{\partial \tilde{T}_S}{\partial y_+} \right) - 2\varepsilon \left(k \frac{\partial^2 \tilde{T}_S}{\partial x \partial x_+} + q_s \frac{\partial^2 \tilde{T}_S}{\partial y \partial y_+} \right)$$

$$-\varepsilon^2 \left(\frac{\partial^2 \tilde{T}_S}{\partial x^2} + \frac{\partial^2 \tilde{T}_S}{\partial y^2} + \lambda \frac{\partial \tilde{T}_S}{\partial y} \right), \tag{2.169}$$

$$\left\{ k^2 \frac{\partial^2}{\partial x_+^2} + g^2 \frac{\partial^2}{\partial y_+^2} \right\} \tilde{C} = \varepsilon \left(\sigma \tilde{C} - g \frac{\partial \tilde{C}}{\partial y_+} - 2k \frac{\partial^2 \tilde{C}}{\partial x \partial x_+} - 2g \frac{\partial^2 \tilde{C}}{\partial y \partial y_+} \right)$$

$$-\varepsilon^2 \left(\frac{\partial^2 \tilde{C}}{\partial x^2} + \frac{\partial^2 \tilde{C}}{\partial y^2} + \frac{\partial \tilde{C}}{\partial y} \right), \tag{2.170}$$

with the boundary conditions:

1. As $y_+ = \frac{L_1}{\varepsilon}$,

$$\tilde{T} \rightarrow 0. \tag{2.171}$$

2. As $y_+ = -\frac{L_2}{\varepsilon}$,

$$\tilde{T}_S \rightarrow 0. \tag{2.172}$$

3. As $y_+ \to \infty$,

$$\tilde{C} \to 0. \tag{2.173}$$

4. At the interface $y = y_+ = 0$,

$$\tilde{T} = \tilde{T}_S - (\Delta G_1)\tilde{h} \,, \tag{2.174}$$

$$\tilde{T}_S = k^2 \frac{\partial^2 \tilde{h}}{\partial x_+^2} + 2\varepsilon k \frac{\partial^2 \tilde{h}}{\partial x \partial x_+} + \varepsilon^2 \frac{\partial^2 \tilde{h}}{\partial x^2} - M\tilde{C}$$

$$+ \left(M \tfrac{1-\kappa}{\kappa} - G_{1S} \right)\tilde{h} = 0 \,, \tag{2.175}$$

$$\left\{ g \frac{\partial}{\partial y_+} + \varepsilon \frac{\partial}{\partial y} \right\} (\tilde{T} - \tilde{T}_S) + \lambda \sigma \tilde{h} + \varepsilon(\Delta G_2)\tilde{h} = 0 \,, \tag{2.176}$$

$$g \frac{\partial \tilde{C}}{\partial y_+} + \varepsilon \frac{\partial \tilde{C}}{\partial y} + \tfrac{1-\kappa}{\kappa} \sigma \tilde{h} + \varepsilon(1-\kappa)(\tilde{C} + \tilde{h}) = 0 \,. \tag{2.177}$$

We assume that, as $\varepsilon \to 0$, the solution $\{\tilde{T}, \tilde{T}_S, \tilde{C}, \tilde{h}\}$ can be expanded in the following MVE form:

$$\tilde{T} \sim \left\{ \tilde{T}_0(x,y,x_+,y_+) + \varepsilon \tilde{T}_1(x,y,x_+,y_+) + \cdots \right\} e^{\sigma t_+}$$

$$\tilde{T}_S \sim \left\{ \tilde{T}_{S0}(x,y,x_+,y_+) + \varepsilon \tilde{T}_{S1}(x,y,x_+,y_+) + \cdots \right\} e^{\sigma t_+}$$

$$\tilde{C} \sim \left\{ \tilde{C}_0(x,y,x_+,y_+) + \varepsilon \tilde{C}_1(x,y,x_+,y_+) + \cdots \right\} e^{\sigma t_+}$$

$$\tilde{h} \sim \left\{ \tilde{h}_0(x,x_+) + \varepsilon \tilde{h}_1(x,x_+) + \cdots \right\} e^{\sigma t_+}$$

$$k(\varepsilon) \sim k_0 + \varepsilon k_1 + \cdots$$

$$g(\varepsilon) \sim k_0 + \varepsilon g_1 + \cdots \tag{2.178}$$

$$q(\varepsilon) \sim k_0 + \varepsilon q_1 + \cdots$$

$$q_s(\varepsilon) \sim k_0 + \varepsilon q_{s1} + \cdots$$

$$\sigma(\varepsilon) \sim \sigma_0 + \varepsilon \sigma_1 + \cdots \,.$$

By substituting the above expansion into the system (2.168)–(2.177), one can derive the solutions to different orders of ε.

Zeroth-order approximation solutions. For the zeroth-order approximation, we have the governing equations

$$\frac{\partial^2 \tilde{T}_0}{\partial x_+^2} + \frac{\partial^2 \tilde{T}_0}{\partial y_+^2} = 0$$

$$\tag{2.179}$$

$$\frac{\partial^2 \tilde{C}_0}{\partial x_+^2} + \frac{\partial^2 \tilde{C}_0}{\partial y_+^2} = 0$$

and the following boundary conditions:

1. As $y_+ \to \infty$,

$$\tilde{C}_0 \to 0 . \tag{2.180}$$

2. As $y_+ \to \infty$,

$$\tilde{T}_0 = 0 . \tag{2.181}$$

3. As $y_+ \to -\infty$,

$$\tilde{T}_{S0} = 0 . \tag{2.182}$$

Note that in deriving the above conditions (2.181) and (2.182), we have replaced the conditions (2.171) and (2.172) by

$$\text{as} \quad y_+ \to \infty, \quad \tilde{T} \to 0 \tag{2.183}$$

and

$$\text{as} \quad y_+ \to -\infty, \quad \tilde{T}_S \to 0 , \tag{2.184}$$

respectively. Such replacements will only cause some transcendentally small change in the asymptotic expansions.

4. At the interface, $y_+ = 0$,

$$\tilde{T}_0 = \tilde{T}_{S0} - (\Delta G_1)\tilde{h}_0 , \tag{2.185}$$

$$\tilde{T}_{S0} = k_0^2 \frac{\partial^2 \tilde{h}_0}{\partial x_+^2} - M\tilde{C}_0 + \left\{ M\frac{1-\kappa}{\kappa} - G_{1S} \right\}\tilde{h}_0 , \tag{2.186}$$

$$k_0 \frac{\partial}{\partial y_+}\left(\tilde{T}_0 - \tilde{T}_{S0} \right) + \lambda \sigma_0 \tilde{h}_0 = 0 , \tag{2.187}$$

$$k_0 \frac{\partial \tilde{C}_0}{\partial y_+} + \left(\frac{1-\kappa}{\kappa} \right)\sigma_0 \tilde{h}_0 = 0 . \tag{2.188}$$

The above system allows the following mode solutions:

$$\begin{aligned}
\tilde{T}_0 &= A_0(y)e^{ix_+ - y_+} \\
\tilde{T}_{S0} &= A_{S0}(y)e^{ix_+ + y_+} \\
\tilde{C}_0 &= B_0(y)e^{ix_+ - y_+} \\
\tilde{h}_0 &= \hat{D}_0 e^{ix_+} .
\end{aligned} \tag{2.189}$$

In order to satisfy the boundary conditions (2.185)–(2.188), one finds that the wave number k_0 and the eigenvalue σ_0 in the mode solutions must satisfy the following dispersion relation:

$$\sigma_0 = \frac{k_0}{\frac{\lambda}{2} + M\frac{1-\kappa}{\kappa}} \left\{ \frac{M(1-\kappa)}{\kappa} - \frac{G_{1L} + G_{1S}}{2} - k_0^2 \right\} . \tag{2.190}$$

It is seen that

$$\sigma_0 \le 0 \qquad \text{for all } k_0 \ge 0 \tag{2.191}$$

provided

$$G = \frac{G_{1L} + G_{1S}}{2} \ge M \frac{1-\kappa}{\kappa} . \tag{2.192}$$

Consequently, the system will be stable. If the above relation does not hold, the system will be unstable. This criterion was first found by Rutter and Chalmers in 1953 [2.3], and is referred to the *constitutional supercooling* criterion.

The above dispersion relation contains dimensionless parameters (G, M, λ, κ). Note that both the length scale, ℓ_D, and the time scale, ℓ_D/V, are related to the pulling velocity V, and that the parameter G depends on two operating conditions: the dimensional pulling velocity V and the design of the zones, (T_H, T_C, L). This is inconvenient, since one needs to consider how each of the operating conditions separately affects the stability of the system. In order to separately examine these effects, it is necessary to introduce some new dimensionless parameters, separately representing these two operating conditions, and re-express the above asymptotic solution and dispersion relation. Returning to the original dimensional quantities, it is easy to verify that

$$\frac{\varepsilon^2}{M\frac{1-\kappa}{\kappa}} = \frac{d_c}{\ell_D}$$

$$\frac{G}{M\frac{1-\kappa}{\kappa}} = \frac{\ell_D}{\ell_G} , \tag{2.193}$$

where d_c and ℓ_G both have length scales and are defined as

$$d_c = -\frac{\gamma \kappa T_{M0}}{m \Delta C \Delta H} , \tag{2.194}$$

$$\ell_G = \frac{(G)_D}{m \Delta C} . \tag{2.195}$$

Since d_c is proportional to the surface tension, it is also called the capillary length, like ℓ_c; whereas ℓ_G is also called the thermal length, as it is proportional to the dimensional temperature gradient $(G)_D$. Thus, it is very natural to define two new parameters

$$v = \frac{d_c}{\ell_D} , \tag{2.196}$$

$$\beta = \frac{d_c}{\ell_G} , \tag{2.197}$$

where v only depends on the pulling velocity V, while β only depends on the dimensional quantity $(G)_{\mathrm{D}}$, corresponding to the average temperature gradient $G = \frac{G_{1\mathrm{L}}+G_{1\mathrm{S}}}{2}$. Moreover, we replace the length scale ℓ_{D} and the time scale ℓ_{D}/V respectively by d_{c} and d_{c}/V. Accordingly, the normal mode solutions are expressed in the form:

$$\tilde{T}_0 e^{\tilde{\sigma}_0 \tilde{t}} = A_0 e^{\tilde{k}(\mathrm{i}\tilde{x}-\tilde{y})+\tilde{\sigma}_0 \tilde{t}}$$

$$\tilde{T}_{\mathrm{S}0} e^{\tilde{\sigma}_0 \tilde{t}} = A_{\mathrm{S}0} e^{\tilde{k}(\mathrm{i}\tilde{x}+\tilde{y})+\tilde{\sigma}_0 \tilde{t}}$$

$$\tilde{C}_0 e^{\tilde{\sigma}_0 \tilde{t}} = B_0 e^{\tilde{k}(\mathrm{i}\tilde{x}-\tilde{y})+\tilde{\sigma}_0 \tilde{t}} \tag{2.198}$$

$$\tilde{h}_0 = D_0 e^{\tilde{\sigma}_0 \tilde{t}} e^{\mathrm{i}\tilde{k}\tilde{x}+\tilde{\sigma}_0 \tilde{t}} \,,$$

where

$$k = \varepsilon \tilde{k}/v \,; \qquad \sigma_0 = \varepsilon \tilde{\sigma}_0/v \,. \tag{2.199}$$

Thus, the dispersion relation (2.190) is written in the form:

$$\tilde{\sigma}_0 = \frac{M\frac{1-\kappa}{\kappa}}{\frac{\lambda}{2}+M\frac{1-\kappa}{\kappa}} \tilde{k}\left(1 - \frac{\beta}{v} - \frac{\tilde{k}^2}{v}\right). \tag{2.200}$$

In the plane of operating parameters (v, β), one has the stability criterion:

$$\begin{cases} v > \beta : \text{unstable region;} \\[2mm] v < \beta : \text{globally, or, absolutely stable region.} \end{cases} \tag{2.201}$$

These two regions in the (β, v)-plane are separated by the straight line $v = \beta$.

First-order approximation solutions. In the first-order approximation, we derive the equations:

$$\begin{aligned}
k_0^2\left\{\frac{\partial^2}{\partial x_+^2} + \frac{\partial^2}{\partial y_+^2}\right\}\tilde{T}_1 &= a_0 e^{(\mathrm{i}x_+ - y_+)} \\[1mm]
k_0^2\left\{\frac{\partial^2}{\partial x_+^2} + \frac{\partial^2}{\partial y_+^2}\right\}\tilde{T}_{\mathrm{S}1} &= b_0 e^{(\mathrm{i}x_+ + y_+)} \\[1mm]
k_0^2\left\{\frac{\partial^2}{\partial x_+^2} + \frac{\partial^2}{\partial y_+^2}\right\}\tilde{C}_1 &= c_0 e^{(\mathrm{i}x_+ - y_+)},
\end{aligned} \tag{2.202}$$

where

$$\begin{cases} a_0 = 2k_0\left\{A_0'(y) + \left(\frac{\lambda}{2} + \frac{\lambda\sigma_0}{2k_0} + k_1 - q_1\right)A_0(y)\right\} \\[2mm] b_0 = -2k_0\left\{A_{\mathrm{S}0}'(y) + \left(\frac{\lambda}{2} - \frac{\lambda\sigma_0}{2k_0} - k_1 + q_{\mathrm{s}1}\right)A_{\mathrm{S}0}(y)\right\} \\[2mm] c_0 = 2k_0\left\{B_0'(y) + \left(\frac{1}{2} + \frac{\sigma_0}{2k_0} + k_1 - g_1\right)B_0(y)\right\}. \end{cases} \tag{2.203}$$

For the first-order approximation solution, we derive the following boundary conditions at the interface, $y = y_+ = 0$:

$$\tilde{T}_1 = \tilde{T}_{S1} - (\Delta G_1)\tilde{h}_1 \,, \tag{2.204}$$

$$\tilde{T}_{S1} = k_0^2 \frac{\partial^2 \tilde{h}_1}{\partial x_+^2} + 2k_0 \frac{\partial^2 \tilde{h}_0}{\partial x \partial x_+} + 2k_0 k_1 \frac{\partial^2 \tilde{h}_0}{\partial x_+^2} - M\tilde{C}_1$$

$$+ \left(M\frac{1-\kappa}{\kappa} - G_{1S}\right)\tilde{h}_1 \,, \tag{2.205}$$

$$k_0 \frac{\partial}{\partial y_+}(\tilde{T}_1 - \tilde{T}_{S1}) + q_1 \frac{\partial \tilde{T}_0}{\partial y_+} - q_{s1}\frac{\partial \tilde{T}_{S0}}{\partial y_+} + \frac{\partial}{\partial y}(\tilde{T}_0 - \tilde{T}_{S0})$$

$$+ \lambda\sigma_0\tilde{h}_1 + \lambda\sigma_1\tilde{h}_0 + (\Delta G_2)\tilde{h}_0 = 0 \,, \tag{2.206}$$

$$k_0\frac{\partial \tilde{C}_1}{\partial y_+} + g_1\frac{\partial \tilde{C}_0}{\partial y_+} + \frac{\partial \tilde{C}_0}{\partial y} + \frac{1-\kappa}{\kappa}(\sigma_0\tilde{h}_1 + \sigma_1\tilde{h}_0)$$

$$+ (1-\kappa)(\tilde{C}_0 + \tilde{h}_0) = 0\,. \tag{2.207}$$

To obtain a uniformly valid asymptotic solution, one must eliminate the secular terms by setting $a_0 = b_0 = c_0 = 0$. Moreover, we assume $k_1 = 0$ and $B_0'(y) = A_0'(y) = A_{S0}'(y) = 0$. We then let

$$B_0(y) = \hat{B}_0, \quad A_0(y) = \hat{A}_0, \quad A_{S0}(y) = \hat{A}_{S0}. \tag{2.208}$$

Thus, we derive

$$\begin{cases} g_1 = \tfrac{1}{2}\left(1 + \tfrac{\sigma_0}{k_0}\right) \\ q_1 = \tfrac{\lambda}{2}\left(1 + \tfrac{\sigma_0}{k_0}\right) \\ q_{s1} = g_1 - \lambda \end{cases} \tag{2.209}$$

and the solutions:

$$\tilde{T}_1 = A_1(y)e^{ix_+ - y_+}$$

$$\tilde{T}_{S1} = A_{S1}(y)e^{ix_+ + y_+}$$

$$\tilde{C}_1 = B_1(y)e^{ix_+ - y_+} \tag{2.210}$$

$$\tilde{h}_0 = \hat{D}_1 e^{ix_+}.$$

Setting

$$\hat{A}_1 = A_1(0), \quad \hat{A}_{S1} = A_{S1}(0), \quad \hat{B}_1 = B_1(0) \tag{2.211}$$

and substituting the mode solution (2.210) into the boundary conditions (2.204)–(2.207), one derives

$$\hat{A}_1 = \hat{A}_{S1} - (\Delta G_1)\hat{D}_1 \,, \tag{2.212}$$

$$\hat{A}_{S1} = -\left(k_0^2 - M\frac{1-\kappa}{\kappa} + G_{S1}\right)\hat{D}_1 - M\hat{B}_1 \,, \tag{2.213}$$

$$-k_0(\hat{A}_1 + \hat{A}_{S1}) - q_1(\hat{A}_0 + \hat{A}_{S0}) + \lambda\sigma_0\hat{D}_1 + \lambda\sigma_1\hat{D}_0$$
$$+ (\Delta G_2)\hat{D}_0 + \lambda\hat{A}_{S0} = 0 \,, \tag{2.214}$$

$$-k_0\hat{B}_1 - g_1\hat{B}_0 + \frac{1-\kappa}{\kappa}(\sigma_0\hat{D}_1 + \sigma_1\hat{D}_0)$$
$$+ (1-\kappa)(\hat{B}_0 + \hat{D}_0) = 0 \,. \tag{2.215}$$

Noting that

$$\hat{B}_0 = \frac{\sigma_0}{k_0}\frac{1-\kappa}{\kappa}\hat{D}_0$$
$$(\hat{A}_0 + \hat{A}_{S0}) = \frac{\lambda\sigma_0}{k_0}\hat{D}_0 \tag{2.216}$$
$$\lambda\hat{A}_{S0} = \frac{\lambda^2}{2}\left(\frac{\sigma_0}{k_0} - 1\right)\hat{D}_0 \,,$$

we obtain

$$\sigma_1\left(\lambda + 2M\frac{1-\kappa}{\kappa}\right) = -\Delta G_2 + \frac{1+\lambda}{4}\Delta G_1 - 2M(1-\kappa)$$
$$- 2M\frac{1-\kappa}{\kappa}(1-\kappa-g_1)\frac{\sigma_0}{k_0} + \frac{\lambda\sigma_0}{k_0}\left\{\frac{1+\lambda}{4} - \frac{\lambda}{2}\left(1+\frac{\sigma_0}{k_0}\right)\right\} . \tag{2.217}$$

The special case with $\lambda \to 0$ is particularly interesting. In this case, the results are greatly simplified, since $\Delta G_1 \to 0, \Delta G_2 \to 0$, and $(G)_{\mathrm{D}} = (T_{\mathrm{H}} - T_{\mathrm{C}})/L$. We obtain the following

$$\sigma_1 = -\kappa - (1-\kappa-g_1)\frac{\sigma_0}{k_0}. \tag{2.218}$$

Thus, the modified dispersion relation becomes

$$\sigma = \sigma_0 + \varepsilon\sigma_1 + O(\varepsilon^2)$$
$$= -\varepsilon\kappa + \left\{k + \varepsilon\left(\kappa + \frac{\sigma}{2k} - \frac{1}{2}\right)\right\}\left(1 - \frac{G}{M\frac{1-\kappa}{\kappa}} - \frac{k^2}{M\frac{1-\kappa}{\kappa}}\right)$$
$$+ O(\varepsilon^2). \tag{2.219}$$

As we have indicated in the case of a pure melt, the MVE solution obtained above is not valid in the long-wavelength regime $k = O(\varepsilon)$ and the extremely short-wavelength regime $k = O(\frac{1}{\varepsilon})$. Therefore, for more accurate information about the behavior of the solution in the full range $0 \le k < \infty$, one needs to seek different asymptotic expansions of solutions over different regimes, as we have done for the case of a pure melt. In what follows, we shall give the leading order asymptotic solutions in these two regimes.

2.2.4 Asymptotic Solutions in the Long-Wavelength Regime, $k = O(\varepsilon)$

As before, in the long-wavelength regime $\{k = O(\varepsilon); g = O(\varepsilon); \sigma = O(\varepsilon)\}$, we write

$$\hat{k} = \frac{k}{\varepsilon}; \quad \hat{g} = \frac{g}{\varepsilon}; \quad \hat{\sigma} = \frac{\sigma}{\varepsilon} \tag{2.220}$$

and introduce the new variables

$$\begin{cases} \hat{x} = \hat{k}x \\[2mm] \hat{y} = \hat{g}y \\[2mm] \hat{t} = \hat{\sigma}t, \end{cases} \tag{2.221}$$

where $\hat{g}$ is used for the solution of the concentration field $\tilde{C}$. For the solution $\tilde{T}$ in the liquid phase and $\tilde{T}_{\mathrm{S}}$ in the solid phase, we replace $\hat{g}$ by $\hat{q}$ and $\hat{q}_{\mathrm{s}}$, respectively.

We expand the solution $\tilde{q} \equiv \{\tilde{T}, \tilde{T}_{\mathrm{S}}, \tilde{C}, \tilde{h}\}$ into the following asymptotic form:

$$\tilde{q}(\hat{x}, \hat{y}, \hat{t}, \varepsilon) \sim e^{\hat{t}}\Big\{ \tilde{q}_0(\hat{x}, \hat{y}) + \varepsilon^2 \tilde{q}_1(\hat{x}, \hat{y}) + \cdots \Big\}$$

$$\hat{k}(\varepsilon) \sim \hat{k}_0 + \varepsilon^2 \hat{k}_1 + \cdots$$

$$\hat{g}(\varepsilon) \sim \hat{g}_0 + \varepsilon^2 \hat{g}_1 + \cdots$$

$$\hat{q}(\varepsilon) \sim \hat{q}_0 + \varepsilon^2 \hat{q}_1 + \cdots \tag{2.222}$$

$$\hat{q}_{\mathrm{s}}(\varepsilon) \sim \hat{q}_{\mathrm{s}0} + \varepsilon^2 \hat{q}_{\mathrm{s}1} + \cdots$$

$$\hat{\sigma}(\varepsilon) \sim \hat{\sigma}_0 + \varepsilon^2 \hat{\sigma}_1 + \cdots .$$

In terms of the above variables, the linear system (2.165)–(2.167) is transformed into

$$\left\{ \hat{k}^2 \frac{\partial^2}{\partial \hat{x}^2} + \hat{q}^2 \frac{\partial^2}{\partial \hat{y}^2} \right\} \tilde{T} = \lambda\left(\hat{\sigma}\tilde{T} - \hat{q}\frac{\partial \tilde{T}}{\partial \hat{y}} \right), \tag{2.223}$$

$$\left\{ \hat{k}^2 \frac{\partial^2}{\partial \hat{x}^2} + \hat{q}_{\mathrm{s}}^2 \frac{\partial^2}{\partial \hat{y}^2} \right\} \tilde{T}_{\mathrm{S}} = \lambda\left(\hat{\sigma}\tilde{T} - \hat{q}_{\mathrm{s}}\frac{\partial \tilde{T}_{\mathrm{S}}}{\partial \hat{y}} \right), \tag{2.224}$$

$$\left\{ \hat{k}^2 \frac{\partial^2}{\partial \hat{x}^2} + \hat{g}^2 \frac{\partial^2}{\partial \hat{y}^2} \right\} \tilde{C} = \left(\hat{\sigma}\tilde{C} - \hat{g}\frac{\partial \tilde{C}}{\partial \hat{y}} \right), \tag{2.225}$$

with the same form of external boundary conditions as (2.166), and the interface conditions, at $\hat{y} = 0$:

$$\tilde{T} = \tilde{T}_{\mathrm{S}} - (\Delta G_1)\tilde{h}$$

$$\tilde{T}_{\mathrm{S}} = \varepsilon^2 \hat{k}^2 \tilde{h}_{\hat{x}\hat{x}} - M\tilde{C} + \left\{ \left(\frac{1-\kappa}{\kappa}\right) M - G_{1S} \right\} \tilde{h}$$

$$\hat{q}\frac{\partial \tilde{T}}{\partial \hat{y}} - \hat{q}_{\mathrm{s}}\frac{\partial \tilde{T}_{\mathrm{S}}}{\partial \hat{y}} + \left(\Delta G_2 + \lambda\hat{\sigma}\tilde{h} \right) = 0 \tag{2.226}$$

$$\hat{g}\frac{\partial \tilde{C}}{\partial \hat{y}} + \left(\frac{1-\kappa}{\kappa}\right)\hat{\sigma}\tilde{h} + (1-\kappa)(\tilde{C} + \tilde{h}) = 0 \, .$$

Substituting (2.222) into (2.225) and (2.226), one can successively derive each order of approximation. In the leading order approximation, we obtain

$$\left\{ \hat{k}_0^2 \frac{\partial^2}{\partial \hat{x}^2} + \hat{q}_0^2 \frac{\partial^2}{\partial \hat{y}^2} \right\} \tilde{T}_0 = \lambda \left(\hat{\sigma}_0 \tilde{T}_0 - \hat{q}_0 \frac{\partial \tilde{T}_0}{\partial \hat{y}} \right), \tag{2.227}$$

$$\left\{ \hat{k}_0^2 \frac{\partial^2}{\partial \hat{x}^2} + \hat{g}_0^2 \frac{\partial^2}{\partial \hat{y}^2} \right\} \tilde{C}_0 = \left(\hat{\sigma}_0 \tilde{C}_0 - \hat{g}_0 \frac{\partial \tilde{C}_0}{\partial \hat{y}} \right), \tag{2.228}$$

and the interface conditions:

$$\tilde{T}_0 = \tilde{T}_{\mathrm{S}0} - (\Delta G_1)\tilde{h}_0$$

$$\tilde{T}_{\mathrm{S}0} = -M\tilde{C}_0 + \left\{ M\frac{1-\kappa}{\kappa} - G_{1S} \right\} \tilde{h}_0$$

$$\hat{q}_0\frac{\partial \tilde{T}_0}{\partial \hat{y}} - \hat{q}_{\mathrm{s}0}\frac{\partial \tilde{T}_{\mathrm{S}0}}{\partial \hat{y}} + (\Delta G_2)\tilde{h}_0 + \lambda\sigma_0\tilde{h}_0 = 0 \tag{2.229}$$

$$\hat{g}_0\frac{\partial \tilde{C}_0}{\partial \hat{y}} + \left(\frac{1-\kappa}{\kappa}\right)\hat{\sigma}_0\tilde{h}_0 + (1-\kappa)(\tilde{C}_0 + \tilde{h}_0) = 0.$$

The zeroth-order approximation solutions are

$$\tilde{T}_0 = \hat{A}_0 \mathrm{e}^{\mathrm{i}\hat{x}-\hat{y}}$$

$$\tilde{T}_{\mathrm{S}0} = \hat{A}_{\mathrm{S}0} \mathrm{e}^{\mathrm{i}\hat{x}+\hat{y}}$$

$$\tilde{C}_0 = \hat{B}_0 \mathrm{e}^{\mathrm{i}\hat{x}-\hat{y}} \tag{2.230}$$

$$\tilde{h}_0 = \hat{D}_0 \mathrm{e}^{\mathrm{i}\hat{x}}$$

with

$$\begin{cases} \hat{g}_0 = \tfrac{1}{2} + \sqrt{\hat{k}_0^2 + \hat{\sigma}_0 + \tfrac{1}{4}} \\[2mm] \hat{q}_0 = \tfrac{\lambda}{2} + \bar{q}_0 \\[2mm] \hat{q}_{\mathrm{s}0} = -\tfrac{\lambda}{2} + \bar{q}_0 \\[2mm] \bar{q}_0 = \sqrt{\hat{k}_0^2 + \lambda\hat{\sigma}_0 + \tfrac{\lambda^2}{4}} \, . \end{cases} \tag{2.231}$$

Substituting (2.230) into (2.229), we derive the dispersion relation in the long-wavelength regime:

$$\hat{\sigma}_0 = -\frac{2\bar{q}_0 M(1-\kappa)}{\lambda(-1+\kappa+\hat{g}_0) + 2\bar{q}_0 M\frac{1-\kappa}{\kappa}} + \frac{2\bar{q}_0(-1+\kappa+\hat{g}_0)}{\lambda(-1+\kappa+\hat{g}_0) + 2\bar{q}_0 M\frac{1-\kappa}{\kappa}}$$

$$\times \left\{ \frac{M(1-\kappa)}{\kappa} - \left(1+\frac{\lambda}{2\bar{q}_0}\right)\frac{G_{1L}+G_{1S}}{2} + \frac{\lambda G_{1S} - \Delta G_2}{2\bar{q}_0} \right\}. \qquad (2.232)$$

Returning to the parameters (σ_0, k_0) we obtain

$$\sigma_0 = -\frac{2\varepsilon q_0 M(1-\kappa)}{\lambda g_0 - \varepsilon\lambda(1-\kappa) + 2q_0 M\frac{1-\kappa}{\kappa}} + \frac{2q_0 g_0 - 2\varepsilon q_0(1-\kappa)}{\lambda g_0 - \varepsilon\lambda(1-\kappa) + 2q_0 M\frac{1-\kappa}{\kappa}}$$

$$\times \left\{ M\frac{1-\kappa}{\kappa} - \left(1+\frac{\varepsilon\lambda}{2q_0}\right)\frac{G_{1L}+G_{1S}}{2} + \varepsilon\frac{\lambda G_{1S} - \Delta G_2}{2q_0} \right\}, \qquad (2.233)$$

where

$$\begin{cases} g_0 = \varepsilon\hat{g}_0 = \frac{\varepsilon}{2} + \sqrt{k^2 + \varepsilon\sigma_0 + \frac{\varepsilon^2}{4}} \\[2mm] q_0 = \varepsilon\bar{q}_0 = \sqrt{k^2 + \varepsilon\lambda\sigma_0 + \frac{\lambda^2\varepsilon^2}{4}}\,. \end{cases} \qquad (2.234)$$

In particular, for the special case $\lambda \to 0$, we have

$$\sigma_0 = -\varepsilon\kappa + (-\varepsilon + \varepsilon\kappa + g_0)\left\{1 - \frac{G}{M(1-\kappa)/\kappa}\right\}. \qquad (2.235)$$

2.2.5 Asymptotic Solutions in the Extremely Short-Wavelength Regime, $\left\{ k = O(\frac{1}{\varepsilon}); \ g = O(\frac{1}{\varepsilon}) \right\}$

Similarly to the case of a pure melt, we can derive the MVE solution in the extremely short-wavelength regime. In doing so we define

$$\begin{cases} \tilde{k} = \varepsilon k \\[2mm] \tilde{g} = \varepsilon g \\[2mm] \tilde{q} = \varepsilon q \\[2mm] \tilde{\sigma} = \varepsilon^3 \sigma\,. \end{cases} \qquad (2.236)$$

Assume that as $\varepsilon \to 0$, the solution $\{\tilde{T}, \tilde{T}_{\mathrm{S}}, \tilde{C}, \tilde{h}\}$ is expanded in the MVE form:

$$\tilde{T} \sim \left\{ \tilde{T}_0(x, y, x_+, y_+) + \varepsilon \tilde{T}_1(x, y, x_+, y_+) + \cdots \right\} e^{\sigma t_+}$$

$$\tilde{T}_{\mathrm{S}} \sim \left\{ \tilde{T}_{\mathrm{S}0}(x, y, x_+, y_+) + \varepsilon \tilde{T}_{\mathrm{S}1}(x, y, x_+, y_+) + \cdots \right\} e^{\sigma t_+}$$

$$\tilde{C} \sim \left\{ \tilde{C}_0(x, y, x_+, y_+) + \varepsilon \tilde{C}_1(x, y, x_+, y_+) + \cdots \right\} e^{\sigma t_+} \qquad (2.237)$$

$$\tilde{h} \sim \varepsilon^2 \left\{ \tilde{h}_0(x, x_+) + \varepsilon \tilde{h}_1(x, x_+) + \cdots \right\} e^{\sigma t_+}$$

and

$$\tilde{k}(\varepsilon) = \tilde{k}_0 + \varepsilon^2 \tilde{k}_1 + \cdots$$

$$\tilde{g}(\varepsilon) = \tilde{g}_0 + \varepsilon^2 \tilde{g}_1 + \cdots$$

$$\tilde{q}(\varepsilon) = \tilde{q}_0 + \varepsilon^2 \tilde{q}_1 + \cdots \qquad (2.238)$$

$$\tilde{q}_{\mathrm{s}}(\varepsilon) = \tilde{q}_{\mathrm{s}0} + \varepsilon^2 \tilde{q}_{\mathrm{s}1} + \cdots$$

$$\tilde{\sigma}(\varepsilon) = \tilde{\sigma}_0 + \varepsilon^2 \tilde{\sigma}_1 + \cdots .$$

In leading order, we have the governing equations:

$$\tilde{k}_0^2 \frac{\partial^2 \tilde{T}_0}{\partial x_+^2} + \tilde{q}_0^2 \frac{\partial^2 \tilde{T}_0}{\partial y_+^2} = \lambda \tilde{\sigma}_0 \tilde{T}_0$$

$$\tilde{k}_0^2 \frac{\partial^2 \tilde{T}_0}{\partial x_+^2} + \tilde{q}_{\mathrm{s}0}^2 \frac{\partial^2 \tilde{T}_{\mathrm{S}0}}{\partial y_+^2} = \lambda \tilde{\sigma}_0 \tilde{T}_{\mathrm{S}0} \qquad (2.239)$$

$$\tilde{k}_0^2 \frac{\partial^2 \tilde{C}_0}{\partial x_+^2} + \tilde{g}_0^2 \frac{\partial^2 \tilde{C}_0}{\partial y_+^2} = \tilde{\sigma}_0 \tilde{C}_0 ,$$

and the mode solutions:

$$\tilde{T}_0 = A_0(y) e^{ix_+ - y_+}$$

$$\tilde{T}_{\mathrm{S}0} = A_{\mathrm{S}0}(y) e^{ix_+ + y_+}$$

$$\tilde{C}_0 = B_0(y) e^{ix_+ - y_+} \qquad (2.240)$$

$$\tilde{h}_0 = \hat{D}_0 e^{ix_+}$$

with $\tilde{q}_0 = \tilde{q}_{\mathrm{s}0} = \sqrt{\lambda \tilde{\sigma}_0 + \tilde{k}_0^2}$ and $\tilde{g}_0 = \sqrt{\tilde{\sigma}_0 + \tilde{k}_0^2}$. One can set

$$\hat{A}_0 = A_0(0) \quad \hat{A}_{\mathrm{S}0} = A_{\mathrm{S}0}(0) \quad \hat{B}_0 = B_0(0). \qquad (2.241)$$

Thus, the boundary conditions of the leading order approximation result in the following homogeneous system:

$$
\begin{cases}
\hat{A}_0 - \hat{A}_{S0} = 0 \\[2mm]
\hat{A}_{S0} + \tilde{k}_0^2 \hat{D}_0 + M\hat{B}_0 = 0 \\[2mm]
-\tilde{q}_0(\hat{A}_0 + \hat{A}_{S0}) + \lambda \tilde{\sigma}_0 \hat{D}_0 = 0 \\[2mm]
-\tilde{g}_0 \hat{B}_0 + \frac{1-\kappa}{\kappa}\tilde{\sigma}_0 \hat{D}_0 + (1-\kappa)\hat{B}_0 = 0,
\end{cases}
\tag{2.242}
$$

which gives rise to the dispersion relation:

$$
\tilde{\sigma}_0 = \frac{-2\tilde{q}_0 \tilde{k}_0^2 \left\{ \tilde{g}_0 - (1-\kappa) \right\}}{2\tilde{q}_0 M \frac{1-\kappa}{\kappa} + \lambda(\tilde{g}_0 - 1 + \kappa)}.
\tag{2.243}
$$

Returning to the original notation, this may be rewritten as

$$
\sigma_0 = \frac{-2k_0^2 \sqrt{k_0^2 + \lambda\varepsilon\sigma_0} \left\{ \varepsilon\sqrt{k_0^2 + \varepsilon\sigma_0} - (1-\kappa) \right\}}{-2\varepsilon M \frac{1-\kappa}{\kappa} \sqrt{k_0^2 + \lambda\varepsilon\sigma_0} + \lambda\left\{ \varepsilon\sqrt{k_0^2 + \varepsilon\sigma_0} - (1-\kappa) \right\}}.
\tag{2.244}
$$

For the limitting case, $\lambda \to 0$, we have

$$
\sigma_0 = \frac{-2k_0^2 \left\{ \varepsilon\sqrt{k_0^2 + \varepsilon\sigma_0} - (1-\kappa) \right\}}{2\varepsilon M \frac{1-\kappa}{\kappa}}.
\tag{2.245}
$$

The linear system with constant coefficients (2.165)–(2.167) can be solved exactly through various approaches. One can easily verify the following exact solutions for the special case $\lambda = 0$:

$$
\begin{cases}
\tilde{T} = \tilde{T}_S = 0 \\[2mm]
\tilde{C} = \hat{B}\mathrm{e}^{(ikx - qy + \sigma t)/\varepsilon} \\[2mm]
\tilde{h} = \hat{D}\mathrm{e}^{(ikx + \sigma t)/\varepsilon},
\end{cases}
\tag{2.246}
$$

where

$$
q = q_1 + \frac{1}{2}, \quad q_1 = \sqrt{k^2 + \varepsilon\sigma + \frac{\varepsilon^2}{2}}.
\tag{2.247}
$$

The exact dispersion relation is

$$
\sigma = -\varepsilon\kappa + \left(q_1 - \frac{\varepsilon}{2} + \varepsilon\kappa\right)\left(1 - \frac{G}{M\frac{1-\kappa}{\kappa}} - \frac{k^2}{M\frac{1-\kappa}{\kappa}}\right).
\tag{2.248}
$$

By comparing the dispersion formula of the exact solution (2.248) with the dispersion formula of the MVE solution (2.219), one sees that (2.219) is the

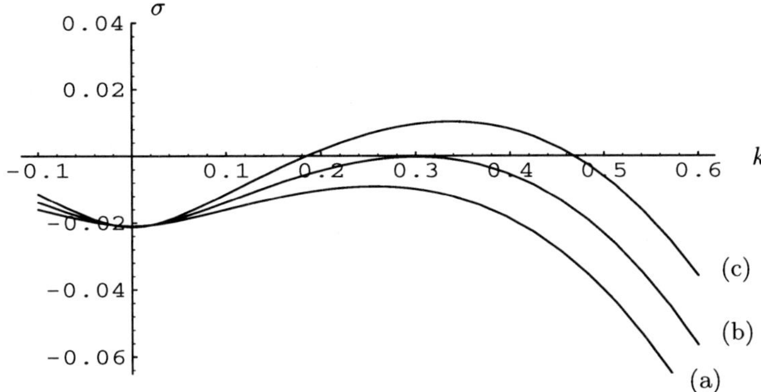

Fig. 2.5. Mullins–Sekerka instability for a binary alloy system. Three typical dispersion relation: (a) for the case $G = 1.572$, the system is absolutely stable; (b) for the case $G = 1.502$, the system is neutrally stable; (c) for the case $G = 1.432$, the system is unstable

regular perturbation expansion of (2.248) as $\varepsilon \to 0$ with any fixed k. Such an agreement confirms the effectiveness of our asymptotic approach.

In Fig. 2.5, we show three typical curves of the exact dispersion relation in the (σ, k)-plane, with fixed $\varepsilon = 0.2, \kappa = 0.103, M = 0.2$ and various G. Curve (a) corresponds to $G = 1.572$ and we see that the system is stable for the entire range of wave numbers $0 \le k < \infty$. It is called absolutely stable. Curve (b) corresponds to $G = 1.502$ and the system is neutrally stable. Curve (c) corresponds to $G = 1.432$ and the system is unstable.

In Fig. 2.6, we show the asymptotic solutions of the dispersion relation in the (σ, k)-plane for the above three cases. In these figures, the dashed lines represent the zeroth-order MVE solution, the thin, solid lines represent the first-order MVE solution, the dotted lines represent the zeroth-order long-wavelength approximate solution, and the bold, solid lines represent the exact solutions. Evidently in the long-wavelength region $(0 \le k < 0.1)$, as was expected, the MVE solution is not a good approximation to the exact solution. It approximates the exact solution very well in the region $(0.2 < k < 1)$. The figure also shows that the long-wavelength asymptotic solution (2.235) agrees with the exact solution in the region $(0 \le k < 0.1)$ very well.

For the case of unidirectional solidification, which allows an exact analytical solution, the asymptotic solution obtained does not appear superior and does not provide any information beyond the exact solution. Nevertheless, it should be pointed out that the asymptotic approach demonstrated in this chapter has a profound significance. It can be applied to more general inhomogeneous dynamical systems with nonconstant coefficients. Growth systems with a nonuniform curved interface, such as dendrite growth or the evolution of viscous fingering, belong to this category. Evidently, for these more complicated systems, there is no way to find the exact solutions. Our asymptotic

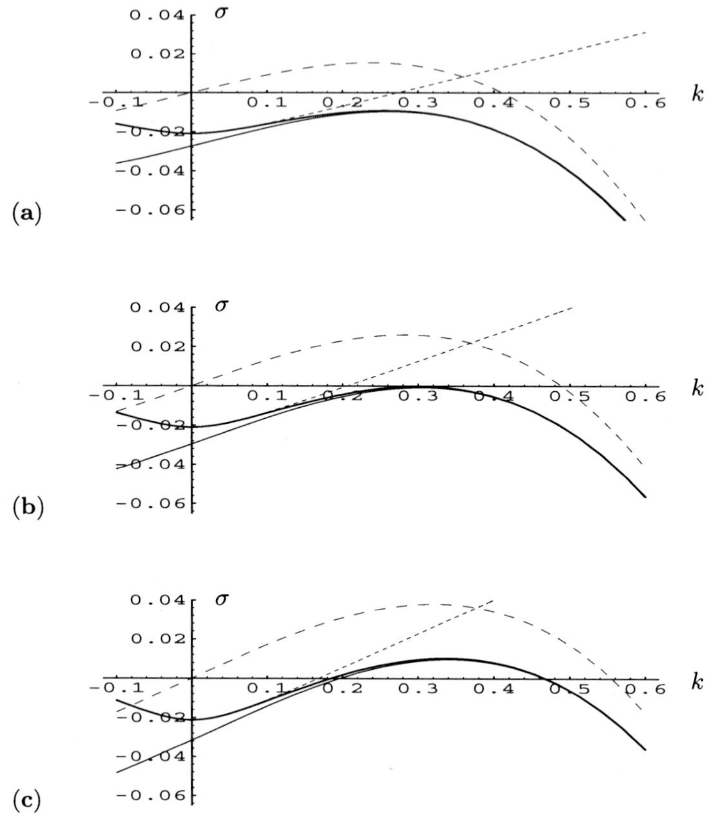

Fig. 2.6. Dispersion curves for unidirectional solidification from a binary mixture: (a) for the absolutely stable case; (b) for the neutrally stable case; (c) for the unstable case. Herein, the dashed lines represent the zeroth-order MVE solution, the thin solid lines represent the first-order MVE solution, the dotted lines represent the zeroth-order long-wavelength approximate solution, and the bold solid lines represent the exact solutions

approach turns out to be the only viable analytical tool for tackling these problems.

2.2.6 Some Remarks on Unidirectional Solidification

In this chapter, our intention was to demonstrate the asymptotic approach and we only discussed some very basic features of unidirectional solidification at the early stage of interface evolution. Many important aspects of this subject were left out. For instance, the effects of physical parameters, such as anisotropy and kinetic attachment on linear stability, the influence of convection flow in a liquid with solidification, the nonlinear bifurcation theory, and the study of pattern formation and selection at the later stage of interface evolution, such as formation of deep cell configurations, all are the issues of great interest which have been totally neglected. The scope of this book does not allow us to give a comprehensive review of this subject. Readers interested in studying these topics are referred to the excellent review articles [2.5]–[2.8]. Hereby, it should be pointed out that the results derived in this chapter have shown that, for unidirectional solidification, the neutrally stable, or say, marginally stable mode is a steady mode, corresponding to the zero eigenvalue $\sigma = 0$. Generally speaking, the system of unidirectional solidification will not allow an oscillatory neutrally stable state with a pure imaginary eigenvalue, unless some additional physical effect, for instance, kinetic attachment is included. In contrast, it will be seen later that a system with dendrite growth is quite different. The neutral mode of dendrite growth under most interesting circumstances is not a steady mode but an oscillatory mode with corresponding eigenvalue $\sigma = -i\omega \neq 0$.

References

2.1   W. W. Mullins, and R. F. Sekerka, "Morphological Stability of a Particle Growing by Diffusion or Heat Flow", J. Appl. Phys. **34**, pp. 323–329, (1963).

2.2   J. Kevorkian and J. D. Cole, '*Multiple Scale and Singular Perturbation Methods*', Applied Mathematical Sciences, Vol. 114, (Springer, Berlin, Heidelberg 1996).

2.3   J. W. Rutter, and B. Chalmers, "A Prismatic Substructure Formed During Solidification of Metals", Can. J. Phys. **31**, pp. 15–39, (1953).

2.4   W. W. Mullins, and R. F. Sekerka, "Stability of a Planar Interface during Solidification of a Dilute Binary Alloy", J. Appl. Phys. **35**, pp. 444–451, (1964).

2.5   S. H. Davis, "Hydrodynamic Interactions in Directional Solidification", J. Fluid Mech. **212**, pp. 241–262, (1990).

2.6   S. H. Davis, "Effect of Flow on Morphological Stability", in '*Handbook of Crystal Growth, Volume 1: Fundamentals, Part B: Transport and Stability*', Ed. by D.T.J. Hurle, (Elsevier Science, North–Holland, Amsterdam 1993).

2.7   B. Caroli, C. Caroli and B. Roulet, "Instability of Planar Solidification Fronts" in '*Solids Far from Equilibrium*', Ed. by C. Godreche, (Cambridge University Press, Cambridge, New York 1991).

2.8   B. Billia and R. Trivedi, "Pattern Formation in Crystal Growth", Chapter 14, in '*Handbook of Crystal Growth, Volume 1: Fundamentals, Part B: Transport and Stability*', Ed. by D.T.J. Hurle, (Elsevier Science, North–Holland, Amsterdam, 1993).

3. Mathematical Formulation
of Free Dendrite Growth from a Pure Melt

We first study the problem of free dendrite growth from a pure substance. Dendrite growth from a binary mixture will be studied in Chap. 8. Typical free dendrite growth was shown in Fig. 1.6. In general, the dendrite growth, starting from a tiny seed, proceeds via a very complex dynamic process until, at a later stage of evolution, a permanent pattern is displayed. The details of pattern formation during the entire evolution will obviously be affected by many factors, in particular by the details of the initial conditions. To describe and predict the whole history of growth is very difficult and not meaningful. From a theoretical point of view the most interesting thing, which is also the objective of the present monograph, is to study the behavior of the system during the later stages of evolution, when the dendrite is fully developed, with a sufficiently long stem, so that the effects of the initial growth conditions and the situation at the root will diminish to a minimum. Thus we assume that at the stage $t \gg 1$, a free dendrite grows with a characteristic velocity U into an undercooled pure melt with the undercooling temperature $T_\infty < T_{M0}$, where T_{M0} is the melting temperature of a flat interface. The characteristic velocity U may be chosen as the average velocity of the dendrite tip, which may depend on the time variable t and may finally become a constant at $t = \infty$.

For simplicity, we start by assuming that

1. the mass density ρ, the specific heat c_p, and other thermal characteristic constants of the solid state are the same as that of the liquid state (the so-called symmetric model);
2. gravity is negligible, so no convection is involved;
2. the surface tension at the interface is isotropic.

The thermal diffusion length $\ell_\mathrm{T} = \kappa_\mathrm{T}/U$ is used as the length scale and the quantity $\Delta H/(c_\mathrm{p}\rho)$ is used as the temperature scale. Here, κ_T is the thermal diffusivity, while ΔH is the latent heat released per unit volume of solid phase.

3.1 Three-Dimensional Axially Symmetric Free Dendrite Growth

Because the surface tension is assumed isotropic, one can consider three-dimensional axially symmetric dendrite growth. We adopt the paraboloidal coordinate system (ξ, η, θ), which can be defined through the cylindrical coordinate system (r, z, θ) by (see Fig. 3.1) :

$$\begin{cases} \dfrac{r}{\eta_0^2} = \xi\eta \\[2mm] \dfrac{z}{\eta_0^2} = \dfrac{1}{2}(\xi^2 - \eta^2) \end{cases} \tag{3.1}$$

or through the Cartesian coordinate system (x, y, z) by

$$\begin{cases} \dfrac{x}{\eta_0^2} = \xi\eta\cos\theta \\[2mm] \dfrac{y}{\eta_0^2} = \xi\eta\sin\theta \\[2mm] \dfrac{z}{\eta_0^2} = \dfrac{1}{2}(\xi^2 - \eta^2)\,, \end{cases} \tag{3.2}$$

where the constant η_0^2 is to be determined.

Note that the dynamical system under investigation is invariant under coordinate translation; we have the freedom to choose the origin of the

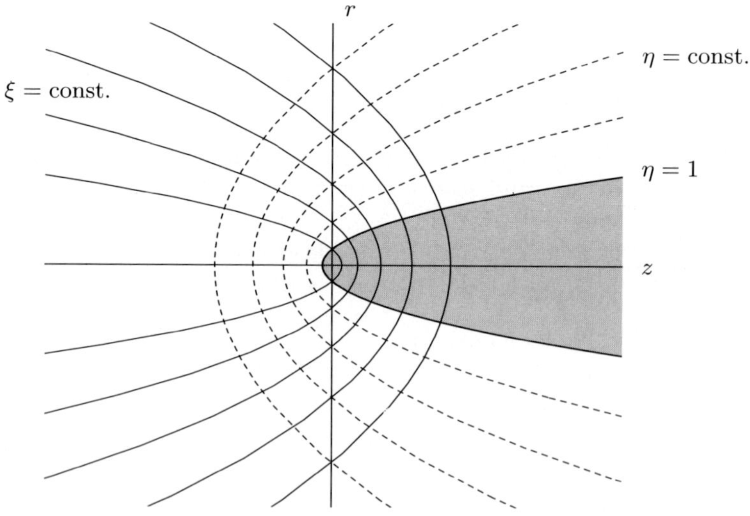

Fig. 3.1. The paraboloidal coordinate system (ξ, η, θ) for three-dimensional dendrite growth

paraboloidal system (3.1) at will. The constant η_0^2 in (3.1) can therefore be chosen so that the steady interface shape satisfies

$$\eta_{\mathrm{s}}(0) = 1 \,. \tag{3.3}$$

It will be seen later that this constant is just the Peclet number of the system with zero surface tension.

In order to describe the dendritic growth in the paraboloidal system, one needs to extensively apply some formulas of differential geometry. For the necessary background, readers are referred to [3.1].

The radius vector of any point with the coordinates (x, y, z) is expressed in the form:

$$\mathbf{r} = \eta_0^2 \left[\xi\eta \cos\theta \mathbf{i} + \xi\eta \sin\theta \mathbf{j} + \frac{1}{2}(\xi^2 - \eta^2)\mathbf{k} \right] , \tag{3.4}$$

where we have denoted the unit vectors along the x, y, z directions by $\mathbf{i}, \mathbf{j}, \mathbf{k}$, respectively. In the moving frame, the relative velocity of the liquid phase is obviously $\mathbf{u} = \mathbf{k}$. We assume that the vectors $\mathbf{e}_1, \mathbf{e}_2, \mathbf{e}_3$ are unit vectors along the ξ, η, θ directions, respectively. Namely,

$$\mathbf{e}_1 \left\| \frac{\partial \mathbf{r}}{\partial \xi} \right. , \quad \mathbf{e}_2 \left\| \frac{\partial \mathbf{r}}{\partial \eta} \right. , \quad \mathbf{e}_3 \left\| \frac{\partial \mathbf{r}}{\partial \theta} \right. . \tag{3.5}$$

Thus, we derive that

$$\mathbf{e}_1 = \frac{1}{\sqrt{\xi^2 + \eta^2}} (\eta \cos\theta \mathbf{i} + \eta \sin\theta \mathbf{j} + \xi \mathbf{k})$$

$$\mathbf{e}_2 = \frac{1}{\sqrt{\xi^2 + \eta^2}} (\xi \cos\theta \mathbf{i} + \xi \sin\theta \mathbf{j} - \eta \mathbf{k}) \tag{3.6}$$

$$\mathbf{e}_3 = (-\sin\theta \mathbf{i} + \cos\theta \mathbf{j}).$$

For an element of arc, $\mathrm{d}s$, we can write

$$(\mathrm{d}s)^2 = (\mathrm{d}x)^2 + (\mathrm{d}y)^2 + (\mathrm{d}z)^2 = (\alpha_1 \mathrm{d}\xi)^2 + (\alpha_2 \mathrm{d}\eta)^2 + (\alpha_3 \mathrm{d}\theta)^2, \tag{3.7}$$

where α_i are Lamé constants. From (3.2) one has

$$\begin{cases} \mathrm{d}x = \eta_0^2 (\xi \mathrm{d}\eta \cos\theta + \eta \mathrm{d}\xi \cos\theta - \xi\eta \sin\theta \mathrm{d}\theta) \\[2mm] \mathrm{d}y = \eta_0^2 (\xi \mathrm{d}\eta \sin\theta + \eta \mathrm{d}\xi \sin\theta + \xi\eta \cos\theta \mathrm{d}\theta) \\[2mm] \mathrm{d}z = \eta_0^2 (\xi \mathrm{d}\xi - \eta \mathrm{d}\eta). \end{cases} \tag{3.8}$$

Hence, it follows that the Lamé constants are

$$\alpha_1 = \alpha_2 = \eta_0^2 \sqrt{\xi^2 + \eta^2}, \quad \alpha_3 = \eta_0^2 \xi\eta \,. \tag{3.9}$$

The unknown functions for the present problem are the temperature fields $T(\xi, \eta, t)$, $T_S(\xi, \eta, t)$, and the interface shape $\eta_s(\xi, t)$. The dimensionless governing equation is simply the heat conduction equation. In order to write this governing equation in the paraboloidal coordinate system, one needs the formulas:

$$\nabla T = \frac{\mathbf{e}_1}{\alpha_1} \frac{\partial T}{\partial \xi} + \frac{\mathbf{e}_2}{\alpha_2} \frac{\partial T}{\partial \eta} + \frac{\mathbf{e}_3}{\alpha_3} \frac{\partial T}{\partial \theta}$$

$$\nabla \cdot \mathbf{U} = \frac{1}{\alpha_1 \alpha_2 \alpha_3} \left[\frac{\partial}{\partial \xi} \left(\alpha_2 \alpha_3 U_1 \right) + \frac{\partial}{\partial \eta} \left(\alpha_1 \alpha_3 U_2 \right) + \frac{\partial}{\partial \theta} \left(\alpha_1 \alpha_2 U_3 \right) \right]. \tag{3.10}$$

For the axially symmetric case we have

$$\nabla^2 T = \nabla \cdot (\nabla T)$$

$$= \frac{1}{\alpha_1 \alpha_2 \alpha_3} \left[\frac{\partial}{\partial \xi} \left(\frac{\alpha_2 \alpha_3}{\alpha_1} \frac{\partial}{\partial \xi} \right) + \frac{\partial}{\partial \eta} \left(\frac{\alpha_1 \alpha_3}{\alpha_2} \frac{\partial}{\partial \eta} \right) + \frac{\partial}{\partial \theta} \left(\frac{\alpha_1 \alpha_2}{\alpha_3} \frac{\partial}{\partial \theta} \right) \right] T$$

$$= \frac{1}{\eta_0^4 (\xi^2 + \eta^2)} \left[\frac{1}{\xi} \frac{\partial}{\partial \xi} \left(\xi \frac{\partial T}{\partial \xi} \right) + \frac{1}{\eta} \frac{\partial}{\partial \eta} \left(\eta \frac{\partial T}{\partial \eta} \right) \right] \tag{3.11}$$

and

$$\mathbf{k} \cdot \nabla T = \frac{1}{\eta_0^2 (\xi^2 + \eta^2)} \left[\frac{\partial T}{\partial \xi} (\mathbf{e}_1 \cdot \mathbf{k}) + \frac{\partial T}{\partial \eta} (\mathbf{e}_2 \cdot \mathbf{k}) \right]$$

$$= \frac{1}{\eta_0^2 (\xi^2 + \eta^2)} \left[\xi \frac{\partial T}{\partial \xi} - \eta \frac{\partial T}{\partial \eta} \right]. \tag{3.12}$$

In terms of (3.11) and (3.12), we write the governing equation

$$\frac{\partial^2 T}{\partial \xi^2} + \frac{\partial^2 T}{\partial \eta^2} + \frac{1}{\xi} \frac{\partial T}{\partial \xi} + \frac{1}{\eta} \frac{\partial T}{\partial \eta} = \eta_0^2 \left(\xi \frac{\partial T}{\partial \xi} - \eta \frac{\partial T}{\partial \eta} \right)$$

$$+ \eta_0^4 (\xi^2 + \eta^2) \frac{\partial T}{\partial t}. \tag{3.13}$$

We now write the boundary conditions in the paraboloidal coordinate system. In doing so, one needs the expressions for the vector normal to the interface and the curvature. Assume that the dimensionless, axially symmetric interface shape is denoted by $\eta_s(\xi, t)$, or, $s(\xi, \eta, t) = \eta - \eta_s(\xi, t) = 0$. Then, for any point on the interface, (ξ, η_s, θ), or say (x_s, y_s, z_s), the radius vector is

$$\mathbf{r} = x_s \mathbf{i} + y_s \mathbf{j} + z_s \mathbf{k} = \eta_0^2 \left[\xi \eta_s \cos \theta \mathbf{i} + \xi \eta_s \sin \theta \mathbf{j} + \frac{1}{2} (\xi^2 - \eta_s^2) \mathbf{k} \right]. \tag{3.14}$$

It follows that

$$
\mathbf{r}_1 = \frac{\partial \mathbf{r}}{\partial \xi} = \eta_0^2\Big[(\xi\eta_{\mathrm s})'\cos\theta\mathbf{i} + (\xi\eta_{\mathrm s})'\sin\theta\mathbf{j} + (\xi - \eta_{\mathrm s}\eta_{\mathrm s}')\mathbf{k}\Big]
$$

$$
\mathbf{r}_2 = \frac{\partial \mathbf{r}}{\partial \theta} = \eta_0^2\Big[-\xi\eta_{\mathrm s}\sin\theta\mathbf{i} + \xi\eta_{\mathrm s}\cos\theta\mathbf{j}\Big]
$$

$$
\mathbf{r}_{11} = \frac{\partial^2 \mathbf{r}}{\partial \xi^2} = \eta_0^2(\xi\eta_{\mathrm s})''\cos\theta\mathbf{i} + (\xi\eta_{\mathrm s})''\sin\theta\mathbf{j} + (1 - \eta_{\mathrm s}'^2 - \eta_{\mathrm s}\eta_{\mathrm s}'')\mathbf{k} \qquad (3.15)
$$

$$
\mathbf{r}_{12} = \frac{\partial^2 \mathbf{r}}{\partial \xi \partial \theta} = \eta_0^2\Big[-(\xi\eta_{\mathrm s})'\sin\theta\mathbf{i} + (\xi\eta_{\mathrm s})'\cos\theta\mathbf{j}\Big]
$$

$$
\mathbf{r}_{22} = \frac{\partial^2 \mathbf{r}}{\partial \theta^2} = \eta_0^2\Big[-\xi\eta_{\mathrm s}\cos\theta\mathbf{i} - \xi\eta_{\mathrm s}\sin\theta\mathbf{j}\Big]\,,
$$

where the prime "$'$" denotes the derivative $\frac{\partial}{\partial \xi}$. For an element of arc, ds, on the interface, we have

$$
ds^2 = \mathbf{r}_1^2 d\xi^2 + 2(\mathbf{r}_1\cdot\mathbf{r}_2)d\xi d\theta + \mathbf{r}_2^2 d\theta^2 = E d\xi^2 + 2F d\xi d\theta + G d\theta^2, \quad (3.16)
$$

where

$$
\begin{cases}
E = \mathbf{r}_1\cdot\mathbf{r}_1 = \eta_0^4(1 + \eta_{\mathrm s}'^2)(\xi^2 + \eta_{\mathrm s}^2) \\[2mm]
F = \mathbf{r}_1\cdot\mathbf{r}_2 = 0 \\[2mm]
G = \mathbf{r}_2\cdot\mathbf{r}_2 = \eta_0^4(\xi\eta_{\mathrm s})^2\,.
\end{cases}
\qquad (3.17)
$$

On the other hand, we can write the unit normal vector

$$
\mathbf{n} = \frac{\mathbf{r}_1\times\mathbf{r}_2}{|\mathbf{r}_1\times\mathbf{r}_2|} = \frac{1}{\sqrt{(1 + \eta_{\mathrm s}'^2)(\xi^2 + \eta_{\mathrm s}^2)}}
$$

$$
\times\Big[(\eta_{\mathrm s}\eta_{\mathrm s}' - \xi)\cos\theta\mathbf{i} + (\eta_{\mathrm s}\eta_{\mathrm s}' - \xi)\sin\theta\mathbf{j} + (\xi\eta_{\mathrm s})'\mathbf{k}\Big] \qquad (3.18)
$$

and calculate

$$
H = |\mathbf{r}_1\times\mathbf{r}_2| = \eta_0^4\xi\eta_{\mathrm s}\sqrt{(1 + \eta_{\mathrm s}'^2)(\xi^2 + \eta_{\mathrm s}^2)}
$$

$$
L = \mathbf{n}\cdot\mathbf{r}_{11} = \frac{\eta_0^2}{\sqrt{(\xi^2 + \eta_{\mathrm s}^2)(1 + \eta_{\mathrm s}'^2)}}
$$

$$
\times\Big[(\eta_{\mathrm s}\eta_{\mathrm s}' - \xi)(\xi\eta_{\mathrm s})'' + (1 - \eta_{\mathrm s}'^2 - \eta_{\mathrm s}\eta_{\mathrm s}'')(\xi\eta_{\mathrm s})'\Big] \qquad (3.19)
$$

$$
M = \mathbf{n}\cdot\mathbf{r}_{12} = 0
$$

$$
N = \mathbf{n}\cdot\mathbf{r}_{22} = \eta_0^2\frac{\xi\eta_{\mathrm s}(\xi - \eta_{\mathrm s}\eta_{\mathrm s}')}{\sqrt{(\xi^2 + \eta_{\mathrm s}^2)(1 + \eta_{\mathrm s}'^2)}}\,.
$$

Thus, according to differential geometry, one derives twice the mean curvature of the interface as follows:

$$\mathcal{K}_c = \frac{1}{H^2}(EN - 2FM + GL) = \frac{1}{\eta_0^2}\mathcal{K}\left\{\frac{d}{d\xi}, \frac{d^2}{d\xi^2}\right\}\eta_s$$

$$\mathcal{K} = -\frac{1}{\sqrt{\xi^2 + \eta_s^2}}\left(\frac{\eta_s''}{(1 + \eta_s'^2)^{\frac{3}{2}}} - \frac{1}{\eta_s(1 + \eta_s'^2)^{\frac{1}{2}}}\right. \tag{3.20}$$

$$\left. + \frac{\eta_s'(\eta_s^2 + 2\xi^2) - \xi\eta_s}{\xi(\xi^2 + \eta_s^2)(1 + \eta_s'^2)^{\frac{1}{2}}}\right).$$

In terms of the above formulas, we can write the boundary conditions as follows:

1. The up-stream far-field condition:

$$T \to T_\infty = \frac{(T_\infty)_D - T_{M0}}{\Delta H/(c_p\rho)} < 0 \quad \text{as} \quad \eta \to \infty. \tag{3.21}$$

2. The regularity condition:

$$T_S = O(1) \quad \text{as} \quad \eta \to 0. \tag{3.22}$$

3. The interface conditions, at $\eta = \eta_s(\xi, t)$:
   (i) the thermodynamic equilibrium condition:

$$T = T_S, \tag{3.23}$$

   (ii) the Gibbs–Thomson condition:

$$T_S = -\frac{\Gamma}{\eta_0^2}\mathcal{K}\left\{\frac{d}{d\xi}, \frac{d^2}{d\xi^2}\right\}\eta_s, \tag{3.24}$$

   (iii) the heat balance condition:

$$\left(\frac{\partial}{\partial\eta} - \eta_s'\frac{\partial}{\partial\xi}\right)(T - T_S) + \eta_0^4(\xi^2 + \eta^2)\frac{\partial\eta_s}{\partial t} + \eta_0^2(\xi\eta_s)' = 0, \tag{3.25}$$

where, as defined in Chap. 1,

$$\Gamma = \frac{\ell_c}{\ell_T} \tag{3.26}$$

is the surface tension parameter,

$$\ell_c = \frac{\gamma c_p T_{M0}}{(\Delta H)^2} \tag{3.27}$$

is the capillary length, and the constant γ is the coefficient of isotropic surface tension.

For reasons that will become clear later, in performing the stability analysis, the interfacial stability parameter of isotropic surface tension ε will often be used. The relationship between the two parameters ε and Γ is

$$\varepsilon = \frac{\sqrt{\Gamma}}{\eta_0^2}. \tag{3.28}$$

Until now, the mathematical formulation remains incomplete. A complete mathematical formulation should also include the boundary conditions which describe the behavior of the solution at the tip of the dendrite $\xi = 0$, as well as at its root, $\xi = L \gg 1$. In addition to the above boundary conditions, one may also need to impose the initial conditions.

A complete mathematical formulation of the pattern formation problem at an advanced stage of evolution will be given later.

3.2 Two-Dimensional Free Dendrite Growth

We also study two-dimensional free dendrite growth, as can be obtained in a Hele–Shaw cell. The mathematical treatment is simpler than that for the three-dimensional case and the results for both cases are very similar. We adopt the parabolic cylindrical coordinate system (ξ, η) moving with the characteristic velocity U defined as follows:

$$\begin{cases} \dfrac{x}{\eta_0^2} = \dfrac{1}{2}(\xi^2 - \eta^2) \\[2mm] \dfrac{y}{\eta_0^2} = \xi\eta \\[2mm] \dfrac{z}{\eta_0^2} = \zeta \end{cases} \tag{3.29}$$

and is illustrated in Fig. 3.2.

The radius vector of any point (x, y, z) can be expressed in the form:

$$\mathbf{r} = \eta_0^2 \left[\frac{1}{2}(\xi^2 - \eta^2)\mathbf{i} + \xi\eta\mathbf{j} + \zeta\mathbf{k} \right]. \tag{3.30}$$

In the moving frame, the relative velocity of the liquid phase for this case is now $\mathbf{u} = \mathbf{i}$.

Assume that the unit vectors are

$$\mathbf{e}_1 \Big\| \frac{\partial \mathbf{r}}{\partial \xi}, \quad \mathbf{e}_2 \Big\| \frac{\partial \mathbf{r}}{\partial \eta}, \quad \text{and} \quad \mathbf{e}_3 \Big\| \frac{\partial \mathbf{r}}{\partial \zeta}. \tag{3.31}$$

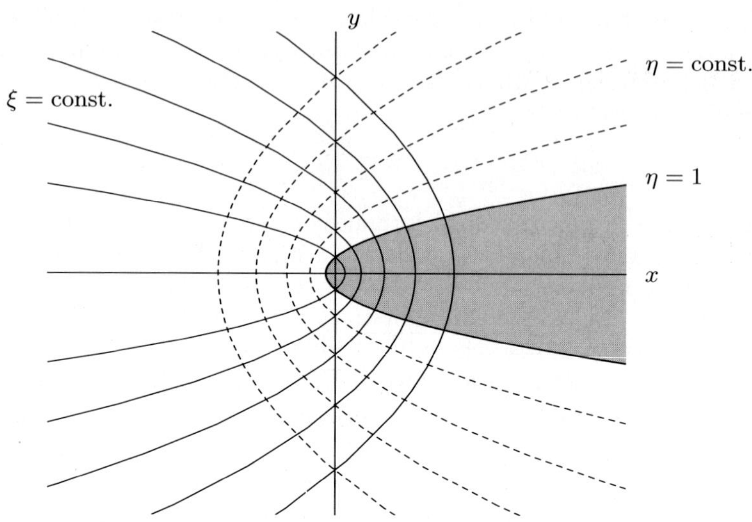

Fig. 3.2. The parabolic cylindrical coordinate system (ξ, η) for two-dimensional dendrite growth

Then it follows that

$$\mathbf{e}_1 = \frac{1}{\sqrt{\xi^2 + \eta^2}}(\xi\mathbf{i} + \eta\mathbf{j})$$

$$\mathbf{e}_2 = \frac{1}{\sqrt{\xi^2 + \eta^2}}(-\eta\mathbf{i} + \xi\mathbf{j}) \qquad (3.32)$$

$$\mathbf{e}_3 = \mathbf{k}.$$

For the element of arc, $\mathrm{d}s$, we can write

$$(\mathrm{d}s)^2 = (\mathrm{d}x)^2 + (\mathrm{d}y)^2 + (\mathrm{d}z)^2 = (\alpha_1 \mathrm{d}\xi)^2 + (\alpha_2 \mathrm{d}\eta)^2 + (\alpha_3 \mathrm{d}\zeta)^2, \quad (3.33)$$

where α_i are Lamé constants. From (3.29) one has

$$\begin{cases} \mathrm{d}x = \eta_0^2(\xi \mathrm{d}\xi - \eta \mathrm{d}\eta) \\ \mathrm{d}y = \eta_0^2(\eta \mathrm{d}\xi + \xi \mathrm{d}\eta) \\ \mathrm{d}z = \eta_0^2 \mathrm{d}\zeta. \end{cases} \qquad (3.34)$$

Subsequently, we derive that the Lamé constants are

$$\alpha_1 = \alpha_2 = \eta_0^2\sqrt{\xi^2 + \eta^2}, \quad \alpha_3 = \eta_0^2. \qquad (3.35)$$

Moreover, in the parabolic cylindrical system, we have

$$\nabla T = \frac{\mathbf{e}_1}{\alpha_1}\frac{\partial T}{\partial \xi} + \frac{\mathbf{e}_2}{\alpha_2}\frac{\partial T}{\partial \eta} + \frac{\mathbf{e}_3}{\alpha_3}\frac{\partial T}{\partial \zeta}$$

$$\nabla \cdot \mathbf{U} = \frac{1}{\alpha_1\alpha_2\alpha_3}\left[\frac{\partial}{\partial \xi}\left(\alpha_2\alpha_3 U_1\right) + \frac{\partial}{\partial \eta}\left(\alpha_1\alpha_3 U_2\right) + \frac{\partial}{\partial \zeta}\left(\alpha_1\alpha_2 U_3\right)\right], \qquad (3.36)$$

so that

$$\nabla^2 T = \nabla \cdot (\nabla T)$$

$$= \frac{1}{\alpha_1 \alpha_2 \alpha_3} \left[\frac{\partial}{\partial \xi} \left(\frac{\alpha_2 \alpha_3}{\alpha_1} \right) \frac{\partial}{\partial \xi} + \frac{\partial}{\partial \eta} \left(\frac{\alpha_1 \alpha_3}{\alpha_2} \right) \frac{\partial}{\partial \eta} + \frac{\partial}{\partial \zeta} \left(\frac{\alpha_1 \alpha_2}{\alpha_3} \right) \frac{\partial}{\partial \zeta} \right] T$$

$$= \frac{1}{\eta_0^4 (\xi^2 + \eta^2)} \left(\frac{\partial^2 T}{\partial \xi^2} + \frac{\partial^2 T}{\partial \eta^2} \right) + \frac{1}{\eta_0^2} \frac{\partial^2 T}{\partial \zeta^2} \tag{3.37}$$

and

$$\mathbf{i} \cdot \nabla T = \frac{1}{\eta_0^2 (\xi^2 + \eta^2)} \left(\xi \frac{\partial T}{\partial \xi} - \eta \frac{\partial T}{\partial \eta} \right) . \tag{3.38}$$

With (3.37) and (3.38), the governing equation is written in the form:

$$\left(\frac{\partial^2 T}{\partial \xi^2} + \frac{\partial^2 T}{\partial \eta^2} \right) = \eta_0^2 \left(\xi \frac{\partial T}{\partial \xi} - \eta \frac{\partial T}{\partial \eta} \right) + \eta_0^4 (\xi^2 + \eta^2) \frac{\partial T}{\partial t} . \tag{3.39}$$

To write the boundary conditions in the parabolic cylindrical coordinate system, one needs to find the expressions for the vector normal to the interface and the mean curvature operator. Assume that the two-dimensional interface shape is denoted by $\eta_\mathrm{s}(\xi, t)$, or, $s(\xi, \eta, t) = \eta - \eta_\mathrm{s}(\xi, \eta, t) = 0$. Then,

$$\mathbf{n} = \frac{\nabla s}{|\nabla s|} = \frac{1}{\sqrt{1 + \eta_\mathrm{s}'^2}} (-\eta_\mathrm{s}' \mathbf{e}_1 + \mathbf{e}_2) . \tag{3.40}$$

On the other hand, for any point on the interface, we can write

$$\mathbf{r} = \eta_0^2 \left[\frac{1}{2} (\xi^2 - \eta_\mathrm{s}^2) \mathbf{i} + \xi \eta_\mathrm{s} \mathbf{j} \right] + \eta_0^2 \zeta \mathbf{k} . \tag{3.41}$$

Thus, it follows that

$$\mathbf{r}_1 = \frac{\partial \mathbf{r}}{\partial \xi} = \eta_0^2 \left[(\xi - \eta_\mathrm{s} \eta_\mathrm{s}') \mathbf{i} + (\eta_\mathrm{s} + \xi \eta_\mathrm{s}') \mathbf{j} \right]$$

$$\mathbf{r}_2 = \frac{\partial \mathbf{r}}{\partial \zeta} = \eta_0^2 \mathbf{k}$$

$$\mathbf{r}_{11} = \frac{\partial^2 \mathbf{r}}{\partial \xi^2} = \eta_0^2 \left[(1 - \eta_\mathrm{s}'^2 - \eta_\mathrm{s} \eta_\mathrm{s}'') \mathbf{i} + (\xi \eta_\mathrm{s})'' \mathbf{j} \right] \tag{3.42}$$

$$\mathbf{r}_{12} = \frac{\partial^2 \mathbf{r}}{\partial \xi \partial \zeta} = 0$$

$$\mathbf{r}_{22} = \frac{\partial^2 \mathbf{r}}{\partial \zeta^2} = 0 .$$

Therefore, we obtain

$$\begin{cases} E = \mathbf{r}_1 \cdot \mathbf{r}_1 = \eta_0^4 (\xi^2 + \eta_\mathrm{s}^2)(1 + \eta_\mathrm{s}'^2) \\[2mm] F = \mathbf{r}_1 \cdot \mathbf{r}_2 = 0 \\[2mm] G = \mathbf{r}_2 \cdot \mathbf{r}_2 = \eta_0^4 \end{cases} \tag{3.43}$$

and

$$
L = -\mathbf{n} \cdot \mathbf{r}_{11} = \frac{\eta_0^2}{\sqrt{(\xi^2 + \eta_{\mathrm{s}}^2)(1 + \eta_{\mathrm{s}}'^2)}}
$$
$$
\times \left[-(\xi^2 + \eta_{\mathrm{s}}^2)\eta_{\mathrm{s}}'' - \xi\eta_{\mathrm{s}}' + \eta_{\mathrm{s}}\eta_{\mathrm{s}}'^2 - \xi\eta_{\mathrm{s}}'^3 + \eta_{\mathrm{s}} \right]
\tag{3.44}
$$

$$
M = \mathbf{n} \cdot \mathbf{r}_{12} = 0
$$

$$
N = \mathbf{n} \cdot \mathbf{r}_{22} = 0 \,.
$$

Applying some differential geometry, the mean curvature operator is derived as

$$
\mathcal{K}_c = \frac{L}{E} = \frac{1}{\eta_0^2} \mathcal{K}\left\{ \frac{d}{d\xi}, \frac{d^2}{d\xi^2} \right\} \eta_{\mathrm{s}}
$$
$$
\mathcal{K} = -\frac{1}{\sqrt{\xi^2 + \eta_{\mathrm{s}}^2}} \left(\frac{\eta_{\mathrm{s}}''}{(1 + \eta_{\mathrm{s}}'^2)^{\frac{3}{2}}} + \frac{\xi\eta_{\mathrm{s}}' - \eta_{\mathrm{s}}}{(\xi^2 + \eta_{\mathrm{s}}^2)(1 + \eta_{\mathrm{s}}'^2)^{\frac{1}{2}}} \right).
\tag{3.45}
$$

In terms of the above formulas, the boundary conditions for the problem are expressed as follows:

1. The up-stream far-field condition:

$$
T \to T_\infty = \frac{(T_\infty)_D - T_{\mathrm{M0}}}{\Delta H / (c_{\mathrm{p}}\rho)} < 0 \quad \text{as} \quad \eta \to \infty \,.
\tag{3.46}
$$

2. The regularity condition:

$$
T_{\mathrm{S}} = O(1) \quad \text{as } \eta \to 0 \,.
\tag{3.47}
$$

3. The interface conditions, at $\eta = \eta_{\mathrm{s}}(\xi, t)$:
   (i) the thermodynamic equilibrium condition:

$$
T = T_{\mathrm{S}} \,,
\tag{3.48}
$$

   (ii) the Gibbs–Thomson condition:

$$
T = -\frac{\Gamma}{\eta_0^2} \mathcal{K}\left\{ \frac{d}{d\xi}, \frac{d^2}{d\xi^2} \right\} \eta_{\mathrm{s}} \,,
\tag{3.49}
$$

   (iii) the heat balance condition:

$$
\left(\frac{\partial}{\partial \eta} - \eta_{\mathrm{s}}'\frac{\partial}{\partial \xi} \right)(T - T_{\mathrm{S}}) + \eta_0^4(\xi^2 + \eta^2)\frac{\partial \eta_{\mathrm{s}}}{\partial t} + \eta_0^2(\xi\eta_{\mathrm{s}})' = 0 \,.
\tag{3.50}
$$

The tip smoothness conditions at $\xi = 0$ and the root conditions at $\xi = L \gg 1$ will be specified later.

Reference

3.1   M. M. Lipshutz, '*Schaum's Outline Series: Theory and Problems of Differential Geometry*', (McGraw–Hill, New York 1969).

4. Steady State of Dendrite Growth with Zero Surface Tension and Its Regular Perturbation Expansion

4.1 The Ivantsov Solution and Unsolved Fundamental Problems

For zero surface tension ($\varepsilon = 0$) and arbitrary undercooling, the three-dimensional system (3.13)–(3.25) allows the following steady similarity solution

$$T = T_*(\eta) = T_\infty + \frac{\eta_0^2}{2} e^{\frac{\eta_0^2}{2}} E_1\left(\frac{\eta_0^2 \eta^2}{2}\right)$$

$$T_{\rm S} = T_{\rm S*} = T_*(1) = 0$$

$$\eta_* = 1 \tag{4.1}$$

$$T_\infty = -\frac{\eta_0^2}{2} e^{\frac{\eta_0^2}{2}} E_1\left(\frac{\eta_0^2}{2}\right)$$

$$(0 \leq \xi < \infty),$$

where $E_1(x)$ is the exponential function defined as

$$E_1(x) = \int_x^\infty \frac{e^{-t}}{t} dt \tag{4.2}$$

(see [4.1]). This solution was first found by Ivantsov in 1946 (cf. [4.2] and [4.3]) and is now called the Ivantsov solution.

For the two-dimensional case, the system (3.39)–(3.50) also allows the steady similarity solution,

$$T = T_*(\eta) = T_\infty + \eta_0^2 e^{\frac{\eta_0^2}{2}} \int_\eta^\infty e^{-\frac{\eta_0^2 \eta_1^2}{2}} d\eta_1$$

$$T_{\rm S} = T_{\rm S*} = T_*(1) = 0$$

$$\eta_* = 1 \tag{4.3}$$

$$T_\infty = -\eta_0^2 e^{\frac{\eta_0^2}{2}} \int_1^\infty e^{-\frac{\eta_0^2 \eta_1^2}{2}} d\eta_1$$

$$(0 \leq \xi < \infty).$$

This solution may also be called the Ivantsov solution in the 2D case.

In the above, the constant η_0^2 is uniquely determined as a function of the undercooling T_∞. The radius of curvature of the parabolic interface $\eta_* = 1$ at the tip $\xi = 0$ is calculated as

$$\ell_t = \eta_0^2 \ell_T. \tag{4.4}$$

Some investigators define the Peclet number as the ratio of the tip radius to the thermal diffusion length, i.e. $\mathrm{Pe} = \ell_t/\ell_T$. Obviously, the Peclet number is generally a function of ε as well as T_∞, i.e., $\mathrm{Pe} = \mathrm{Pe}(\varepsilon)$. Equation (4.4) shows that η_0^2 for the Ivantsov problem is actually the Peclet number with zero surface tension, i.e., $\eta_0^2 = \mathrm{Pe}_0 = \mathrm{Pe}(0)$. Returning to the dimensional tip radius ℓ_t, formula (4.4) shows:

$$\ell_t U = (\mathrm{Pe}_0)\kappa_T. \tag{4.5}$$

The Ivantsov solution has the following properties:

1. Being a smooth needle crystal solution, its interface has no microstructure.
2. Being a similarity solution, it represents a continuous family of dimensional, physical solutions for needle growth, with arbitrary tip velocity for given growth conditions and material properties.

Comparing the Ivantsov solution with the original dendrite growth phenomenon, it is immediately seen that the Ivantsov solution cannot be applied to the dendrite stem region, since it gives no information about microstructure formation at the interface. The Ivantsov solution cannot, as was hoped, be locally applied to the dendrite tip region for a sufficient information on the growth of the dendrite tip either, since it cannot fully determine the tip velocity itself.

However, when the tip velocity is determined correctly through experiments, the Ivantsov solution does describe the shape of the dendrite tip with a high accuracy. Therefore, the Ivantsov solution has been one of the most significant results on this subject. It has provided an important background to all further research work.

After Ivantsov, many researchers studied dendrite growth phenomena. The most important issue they faced was whether or not, in practice, the tip velocity of a dendrite at the later stages of growth was actually uniquely determined by the growth condition and material properties. If the answer is yes, the selection problem arises: How is the tip velocity actually selected? The answer to this question is not trivial. Many earlier researchers in this field presumed, without theoretical justification, nor accurate experimental evidence, that at the later stages of dendrite growth, the tip velocity would be uniquely determined. For instance, in the 1960s and 1970s, Temkin ([4.4]), Bolling and Tiller ([4.5]), Trivedi ([4.6]), Sekerka, and others all recognized that in realistic dendrite growth, the surface tension would give a sharp upper-bound to the dendrite tip velocity. They made great efforts to determine this upper-bound, by using some ad hoc assumption. Their conjecture cannot be considered as having identified the selection problem of the tip velocity — one of the most significant scientific issues in pattern formation.

The selection problem was logically proposed based on the two independent experimental results by Schaefer and co-workers in 1975 [4.7] and by Glicksman and co-workers in 1976 [4.8]. By using the transparent organic material SCN, these authors first performed a series of careful experiments on dendrite growth and accurately measured the tip velocities under various conditions. On the basis of their experimental data, it was confirmed that at the later stages of dendrite growth, the dendrite's tip velocity is a uniquely determined function of the growth condition and the properties of the material.

This selection problem has been at the center of broad theoretical and experimental research activities during the past decades. Since then, research on dendrite growth has focused on the following basic problems:

1. What is the mechanism which determines the tip growth velocity?
2. What is the origin and essence of the microstructure; specifically, the highly visible lateral foliation?

These problems have been long-standing, fundamental subjects in the field of condensed matter physics and material sciences. It was recognized early on that the key to resolving these problems is understanding the role of surface tension at the interface. The surface tension at the interface between the solid and liquid phases is usually a very small quantity. However, it is precisely this extremely small quantity that plays a vital role in interfacial pattern formation phenomenon. It is found that the above two problems are related to each other. To resolve them, one needs to study the steady-state of the system with nonzero surface tension and its linear stability.

4.2 Three-Dimensional Axially Symmetric Steady Needle Growth

This chapter will discuss the effect of isotropic surface tension on steady needle crystal growth. We shall first study the three-dimensional axially symmetric case, then consider the two-dimensional case.

4.2.1 Mathematical Formulation

For the steady, three-dimensional axially symmetric needle crystal growth, we have the equation

$$\frac{\partial^2 T}{\partial \xi^2} + \frac{\partial^2 T}{\partial \eta^2} + \frac{1}{\xi}\frac{\partial T}{\partial \xi} + \frac{1}{\eta}\frac{\partial T}{\partial \eta} = \eta_0^2\Big(\xi\frac{\partial T}{\partial \xi} - \eta\frac{\partial T}{\partial \eta}\Big) \tag{4.6}$$

with the following boundary conditions:

1. The up-stream condition:

$$T \to T_\infty \quad \text{(exponentially)}, \qquad \text{as } \eta \to \infty. \tag{4.7}$$

2. The regularity condition:

$$T_{\mathrm{S}} \quad \text{is regular}, \qquad \text{as } \eta \to 0. \tag{4.8}$$

3. On the interface, $\eta = \eta_{\mathrm{s}}(\xi)$, we must satisfy
    (i) the thermodynamic-equilibrium condition

$$T = T_{\mathrm{S}}, \tag{4.9}$$

(ii) the Gibbs–Thomson condition

$$T_{\mathrm{S}} = -\varepsilon^2 \eta_0^2 \mathcal{K}\left\{\eta_{\mathrm{s}}(\xi)\right\}, \tag{4.10}$$

(iii) and the heat-balance condition

$$\frac{\partial}{\partial \eta}\left(T - T_{\mathrm{S}}\right) - \eta_{\mathrm{s}}'\frac{\partial}{\partial \xi}\left(T - T_{\mathrm{S}}\right) + \eta_0^2\left(\xi\eta_{\mathrm{s}}\right)' = 0. \tag{4.11}$$

This problem is not yet well-posed. In order to specify a certain type of
solution, one also needs to introduce the additional conditions at the tip, as
well as in the root field. Different tip conditions and root conditions will give
rise to different types of solutions. Based on physical reasoning, we can first
specify the following tip condition:

4. The smooth tip condition: at $\xi = 0$,

$$\eta_{\mathrm{s}}'(0) = 0, \tag{4.12}$$

$$\eta_{\mathrm{s}}(0) = 1. \tag{4.13}$$

Condition (4.13) selects the location of the dendrite tip.

5. The far-field condition: how to specify the far-field condition is a subtle
    issue that must be handled carefully. A definite form of the root condition
    will be specified in the next chapter.

4.2.2 The Regular Perturbation Expansion Solutions (RPE) as $\varepsilon \to 0$

The interfacial stability parameter ε is, in practice, very small. Its numerical magnitude is about ≈ 0.1–0.2. Thus, in order to examine the steady
perturbation induced by the surface tension, it is very natural to consider a
regular perturbation expansion (RPE) around the Ivantsov solution in the
limit $\varepsilon \to 0$. Assume that

$$T = T(\xi, \eta) = T_0(\eta) + \varepsilon^2 \eta_0^2 T_1(\xi, \eta) + \dots$$

$$T_{\mathrm{S}} = T_{\mathrm{S}}(\xi, \eta) = T_{\mathrm{S}0} + \varepsilon^2 \eta_0^2 T_{\mathrm{S}1}(\xi, \eta) + \dots \tag{4.14}$$

$$\eta_{\mathrm{B}}(\xi) = 1 + \varepsilon^2 \eta_1(\xi) + \dots.$$

We substitute (4.14) into the system (4.6)–(4.13) and equate coefficients of like powers of ε to zero, then we can derive the approximations at each order of ε.

$\boldsymbol{O(\varepsilon^0)}$. The zeroth-order approximation solution is the Ivantsov solution.

$\boldsymbol{O(\varepsilon^2)}$. In the first-order approximation, we derive

$$L\{T_1\} = \left\{ \frac{\partial^2}{\partial \xi^2} + \frac{\partial^2}{\partial \eta^2} + \left(\frac{1}{\xi} - \eta_0^2 \xi \right) \frac{\partial}{\partial \xi} + \left(\frac{1}{\eta} + \eta_0^2 \eta \right) \frac{\partial}{\partial \eta} \right\} T_1 = 0, \quad (4.15)$$

with the boundary conditions

1. As $\eta \to \infty$,

$$T_1 \to 0 \quad \text{(exponentially)}. \qquad (4.16)$$

2. As $\xi \to \infty$,

$$T_1 \to 0 \quad \text{(algebraically)}. \qquad (4.17)$$

3. As $\eta \to 0$,

$$T_{\mathrm{S1}} \quad \text{regular}. \qquad (4.18)$$

4. At $\eta = 1$,

$$T_1 = T_{\mathrm{S1}} + \eta_1 , \qquad (4.19)$$

$$T_{\mathrm{S1}} = -\mathcal{K}_0(\xi) = -\frac{\xi^2 + 2}{\left(1 + \xi^2 \right)^{\frac{3}{2}}} , \qquad (4.20)$$

$$\frac{\partial}{\partial \eta}(T_1 - T_{\mathrm{S1}}) + \left(2 + \eta_0^2 \right) \eta_1 + \xi \frac{d\eta_1}{d\xi} = 0 . \qquad (4.21)$$

5. The tip-regularity condition, at $\xi = 0$,

$$\eta_1'(0) = 0 \qquad (4.22)$$

and

$$\eta_1(0) = 0. \qquad (4.23)$$

The solution for (4.15)–(4.23) can be derived by using separation of variables. Let

$$T_1(\xi, \eta) = X(\xi)Y(\eta). \qquad (4.24)$$

From (4.15) it follows that

$$X'' + \left(\frac{1}{\xi} - \eta_0^2\xi\right)X' + \eta_0^2\lambda_1{}^2 X = 0 \,, \tag{4.25}$$

$$Y'' + \left(\frac{1}{\eta} + \eta_0^2\eta\right)Y' - \eta_0^2\lambda_1{}^2 Y = 0 \,. \tag{4.26}$$

By letting

$$\sigma = \frac{\eta_0^2\xi^2}{2}, \quad X = X(\sigma), \tag{4.27}$$

equation (4.25) is transformed into the Kummer equation [4.1]

$$\sigma X''(\sigma) + (1 - \sigma)X'(\sigma) + \frac{\lambda_1^2}{2}X = 0 \,, \tag{4.28}$$

whose fundamental solutions are

$$X(\sigma) = \begin{cases} M\left(-\dfrac{\lambda_1^2}{2}, 1, \sigma\right) & \text{(regular at } \xi = \sigma = 0\text{)}\,; \\[2ex] U\left(-\dfrac{\lambda_1^2}{2}, 1, \sigma\right) & \text{(with a logarithmic singularity} \\[1ex] & \text{at } \xi = \sigma = 0\text{)}\,. \end{cases} \tag{4.29}$$

Here M and U are the confluent hypergeometric functions (see [4.1]). We choose

$$X(\sigma) = M\left(-\frac{\lambda_1^2}{2}, 1, \sigma\right). \tag{4.30}$$

On the other hand, as $\sigma \to \infty$, $M\left(-\frac{\lambda_1^2}{2}, 1, \sigma\right)$ grows algebraically only for

$$\frac{\lambda_1^2}{2} = n = 0, 1, 2, \ldots \quad . \tag{4.31}$$

For other values of $\frac{\lambda_1^2}{2}$, $M\left(-\frac{\lambda_1^2}{2}, 1, \sigma\right)$ grows exponentially as $\sigma \to \infty$. From the boundary condition (4.17) one has to set $\lambda_1^2/2 = n$, so that

$$X(\sigma) = M(-n, 1, \sigma) = L_n\left(\frac{\eta_0^2\xi^2}{2}\right), \tag{4.32}$$

where L_n is the Laguerre polynomial. We also need to solve (4.26). In the solid phase region, we let

$$\hat{\tau} = -\frac{\eta_0^2\eta^2}{2} \,, \tag{4.33}$$

so that (4.26) is transformed into

$$\hat{\tau}Y''(\hat{\tau}) + (1 - \hat{\tau})Y'(\hat{\tau}) + nY = 0. \tag{4.34}$$

The solution that is regular at $\hat{\tau} = \eta = 0$ is

$$Y(\hat{\tau}) = L_n\left(-\frac{\eta_0^2 \eta^2}{2}\right).$$
(4.35)

Therefore, the general solution for the temperature in the solid phase is

$$T_{S1}(\xi, \eta) = \sum_{n=0}^{\infty} \frac{\alpha_n}{L_n\left(-\frac{\eta_0^2}{2}\right)} L_n\left(\frac{\eta_0^2 \xi^2}{2}\right) L_n\left(-\frac{\eta_0^2 \eta^2}{2}\right).$$
(4.36)

In the liquid phase region, we introduce

$$\tau = \frac{\eta_0^2 \eta^2}{2}, \quad Y(\eta) = \frac{e^{-\frac{\tau}{2}}}{\tau^{\frac{1}{2}}} Z(\tau).$$
(4.37)

Equation (4.26) can be transformed to the Whittaker equation [4.1]

$$Z''(\tau) + \left(-\frac{1}{4} + \frac{\lambda}{\tau} + \frac{\frac{1}{4} - \mu^2}{\tau^2}\right) Z(\tau) = 0,$$
(4.38)

where

$$\lambda = -\left(n + \frac{1}{2}\right), \quad \mu = 0.$$
(4.39)

The fundamental solutions of (4.38) are known:

$$W_{\lambda,\mu}(\tau) = \begin{cases} e^{-\frac{\tau}{2}} \tau^{\mu+\frac{1}{2}} M\left(\mu - \lambda + \frac{1}{2}, \quad 2\mu + 1, \quad \tau\right) \\ e^{-\frac{\tau}{2}} \tau^{\mu+\frac{1}{2}} U\left(\mu - \lambda + \frac{1}{2}, \quad 2\mu + 1, \quad \tau\right). \end{cases}$$
(4.40)

Accordingly, we obtain as two basic solutions for $Y(\eta)$:

$$Y(\eta) = \begin{cases} e^{-\frac{\eta_0^2 \eta^2}{2}} M\left(n+1, \quad 1, \quad \frac{\eta_0^2 \eta^2}{2}\right) & \text{(vanishes algebraically} \\ & \qquad \text{as} \quad \eta \to \infty) \\ e^{-\frac{\eta_0^2 \eta^2}{2}} U\left(n+1, \quad 1, \quad \frac{\eta_0^2 \eta^2}{2}\right) & \text{(vanishes exponentially} \\ & \qquad \text{as} \quad \eta \to \infty). \end{cases}$$
(4.41)

The boundary condition (4.16) determines

$$Y(\eta) = e^{-\frac{\eta_0^2 \eta^2}{2}} U\left(n+1, \quad 1, \quad \frac{\eta_0^2 \eta^2}{2}\right).$$
(4.42)

Therefore, the general solution for the temperature in the liquid is

$$T_1(\xi, \eta) = \sum_{n=0}^{\infty} \beta_n L_n\left(\frac{\eta_0^2 \xi^2}{2}\right) \frac{e^{-\frac{\eta_0^2 \eta^2}{2}} U\left(n+1, 1, \frac{\eta_0^2 \eta^2}{2}\right)}{e^{-\frac{\eta_0^2}{2}} U\left(n+1, 1, \frac{\eta_0^2}{2}\right)}.$$
(4.43)

We can expand the function $\eta_1(\xi)$ as the following Laguerre series:

$$\eta_1(\xi) = \sum_{n=0}^{\infty} \gamma_n L_n\left(\frac{\eta_0^2 \xi^2}{2}\right).$$

(4.44)

Thus we must determine the coefficients $\{\alpha_n, \beta_n, \gamma_n \ (n = 0, 1, 2 \cdots)\}$ such that the boundary conditions (4.19)–(4.21) are satisfied. In what follows, we shall give the analytical forms for these coefficients.

(1) It follows from the boundary condition (4.19) that

$$\beta_n = \alpha_n + \gamma_n.$$

(4.45)

(2) The boundary condition (4.20) gives

$$\sum_{n=0}^{\infty} \alpha_n L_n\left(\frac{\eta_0^2 \xi^2}{2}\right) = -\mathcal{K}_0(\xi) = -\left[\frac{1}{(1+\xi^2)^{\frac{1}{2}}} + \frac{1}{(1+\xi^2)^{\frac{3}{2}}}\right].$$

(4.46)

Due to the orthogonality of the functions $L_n(x)$ $(n = 0, 1, 3, \cdots)$, the coefficients α_n in the above expansion can be determined from the following integral:

$$\alpha_n = -\int_0^{\infty} e^{-x} L_n(x) \mathcal{K}_0\left(\sqrt{2x}/\eta_0\right) dx.$$

(4.47)

By applying the formulas [4.1]

$$e^{-x} L_n(x) = \frac{2}{n!} \int_0^{\infty} e^{-z^2} z^{2n+1} J_0(2z\sqrt{x}) dz$$

$$\int_0^{\infty} \frac{J_0(2z\sqrt{x})}{(x + \frac{\eta_0^2}{2})^{\frac{1}{2}}} dx = 2 \int_0^{\infty} \frac{J_0(2zy)}{(y^2 + \frac{\eta_0^2}{2})^{\frac{1}{2}}} y \, dy = \frac{e^{-\sqrt{2}\eta_0 z}}{z}$$

$$\int_0^{\infty} \frac{J_0(2z\sqrt{x})}{(x + \frac{\eta_0^2}{2})^{\frac{3}{2}}} dx = 2 \int_0^{\infty} \frac{J_0(2zy)}{(y^2 + \frac{\eta_0^2}{2})^{\frac{3}{2}}} y \, dx = \frac{2\sqrt{2}}{\eta_0} e^{-\sqrt{2}\eta_0 z},$$

(4.48)

we find that

$$\alpha_n = -\frac{e^{\frac{\eta_0^2}{2}}}{n!}\left[\frac{\sqrt{2}}{\eta_0} \int_{\frac{\eta_0}{\sqrt{2}}}^{\infty} e^{-t^2}\left(t - \frac{\eta_0}{\sqrt{2}}\right)^{2n} dt + 2 \int_{\frac{\eta_0}{\sqrt{2}}}^{\infty} e^{-t^2}\left(t - \frac{\eta_0}{\sqrt{2}}\right)^{2n+1} dt\right]$$

$$= -\frac{\sqrt{\pi}}{2} \frac{(2n)!}{n!} \eta_0^2 e^{\frac{\eta_0^2}{2}}$$

$$\times \left[\frac{\sqrt{2}}{\eta_0} i^{2n} \operatorname{erfc}\left(\frac{\eta_0}{\sqrt{2}}\right) + 2(2n+1) i^{2n+1} \operatorname{erfc}\left(\frac{\eta_0}{\sqrt{2}}\right)\right],$$

(4.49)

where

$$i^n \operatorname{erfc}(a) = \frac{2}{\sqrt{\pi}} \int_a^\infty \frac{(t-a)^n}{n!} e^{-t^2}\, dt \,. \tag{4.50}$$

Therefore, given η_0^2, the numerical values of α_n can then be calculated, in terms of the recurrence formulas [4.1]

$$
\begin{aligned}
i^{-1} \operatorname{erfc}(a) &= \frac{2}{\sqrt{2}} e^{-a^2} \\
i^0 \operatorname{erfc}(a) &= \operatorname{erfc}(a) \\
i^n \operatorname{erfc}(a) &= -\frac{a}{n} i^{n-1} \operatorname{erfc}(a) + \frac{1}{2n} i^{n-2} \operatorname{erfc}(a) \\
&\quad (n = 1, 2, 3, \ldots) \,.
\end{aligned}
\tag{4.51}
$$

(3) We are now going to apply the boundary condition (4.21). From (4.36) and (4.43), one can write

$$\frac{\partial T_{\mathrm{S1}}}{\partial \eta}(\xi, 1) = \sum_{n=0}^\infty b_n L_n\!\left(\frac{\eta_0^2 \xi^2}{2}\right) \tag{4.52}$$

and

$$\frac{\partial T_1}{\partial \eta}(\xi, 1) = \sum_{n=0}^\infty a_n L_n\!\left(\frac{\eta_0^2 \xi^2}{2}\right) , \tag{4.53}$$

where

$$a_n = 2\beta_n A_n \,, \qquad b_n = 2n\alpha_n B_n \,, \tag{4.54}$$

$$
\begin{cases}
A_n = (n+1)^2 \dfrac{U\!\left(n+2, 1, \frac{\eta_0^2}{2}\right)}{U\!\left(n+1, 1, \frac{\eta_0^2}{2}\right)} - (n+1) - \dfrac{\eta_0^2}{2} \\[2.5ex]
B_n = 1 - \dfrac{L_{n-1}\!\left(\frac{-\eta_0^2}{2}\right)}{L_n\!\left(\frac{-\eta_0^2}{2}\right)} \,.
\end{cases}
\tag{4.55}
$$

In the above, we have used the following formulas:

$$
\begin{aligned}
xU'(n+1, 1, x) &= (n+1)^2 U(n+2, 1, x) - (n+1)U(n+1, 1, x) \\
xL_n'(x) &= n\{ L_n(x) - L_{n-1}(x) \} \,.
\end{aligned}
\tag{4.56}
$$

Thus, from the boundary condition (4.21), we have

$$\sum_{n=0}^\infty (a_n - b_n) L_n\!\left(\frac{\eta_0^2 \xi^2}{2}\right) + (2 + \eta_0^2) \sum_{n=0}^\infty \gamma_n L_n\!\left(\frac{\eta_0^2 \xi^2}{2}\right)$$

$$+ 2\sum_{n=0}^\infty n\gamma_n \left\{ L_n\!\left(\frac{\eta_0^2 \xi^2}{2}\right) - L_{n-1}\!\left(\frac{\eta_0^2 \xi^2}{2}\right) \right\} = 0 \tag{4.57}$$

and

$$(a_n - b_n) + (2 + \eta_0^2 + 2n)\gamma_n - 2(n+1)\gamma_{n+1} = 0, \qquad (4.58)$$

$$(n = 0, 1, 2, 3, \cdots).$$

From (4.54) and (4.59), we find that

$$\gamma_{n+1} = g_n \gamma_n + f_n \alpha_n, \quad (n = 0, 1, 2 \ldots), \qquad (4.59)$$

where

$$g_n = (n+1)\frac{U\left(n+2, 1, \frac{\eta_0^2}{2}\right)}{U\left(n+1, 1, \frac{\eta_0^2}{2}\right)}, \qquad (4.60)$$

$$f_n = \left\{ (n+1)\frac{U\left(n+2, 1, \frac{\eta_0^2}{2}\right)}{U\left(n+1, 1, \frac{\eta_0^2}{2}\right)} - 1 - \frac{\eta_0^2}{2(n+1)} \right\} - \frac{nB_n}{(n+1)\eta_0^2}$$

$$= \left\{ \frac{1 - \frac{\eta_0^2}{2}}{n+1} - 2 + \frac{n}{n+1}\frac{L_{n-1}\left(-\frac{\eta_0^2}{2}\right)}{L_n\left(-\frac{\eta_0^2}{2}\right)} + g_n \right\}. \qquad (4.61)$$

For any given γ_0, the formula (4.59) allows us to generate the series

$$\{\gamma_0, \gamma_1, \gamma_2, \cdots \gamma_n \cdots\}.$$

Thus, the function $\eta_1(\gamma_0, \xi)$ can be evaluated as

$$\eta_1(\gamma_0, \xi) = \sum_{n=0}^{\infty} \gamma_n L_n\left(\frac{\xi^2}{2}\right). \qquad (4.62)$$

The tip-regularity condition (4.22) is automatically satisfied. The value of γ_0 is then uniquely determined by the tip condition (4.23)

$$\eta_1(\gamma_0, 0) = \sum_{n=0}^{\infty} \gamma_n = 0. \qquad (4.63)$$

In the above, through (4.45), (4.46), and (4.59)–(4.63), we have obtained an analytical solution for the problem.

The solutions depend on the number η_0^2, which is related to T_∞. For $\frac{\eta_0^2}{2} = 0.005$, we obtain $\gamma_0 = -0.1681$. The solution $\eta_1(\xi)$ is displayed in Fig. 4.1. Thus, with $\varepsilon > 0$, we have

$$\eta_s(\xi, \varepsilon) \approx 1 + \varepsilon^2 \eta_1(\xi, \eta_0^2). \qquad (4.64)$$

The dendrite shapes with $\varepsilon = 0.1$ and 0.2 are shown in Fig. 4.2(a) and Fig. 4.2(b), respectively, in which the dashed lines represent the Ivantsov paraboloid with $\varepsilon = 0$.

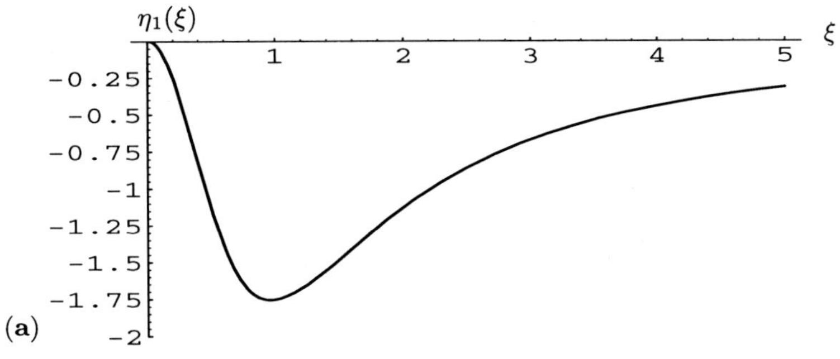

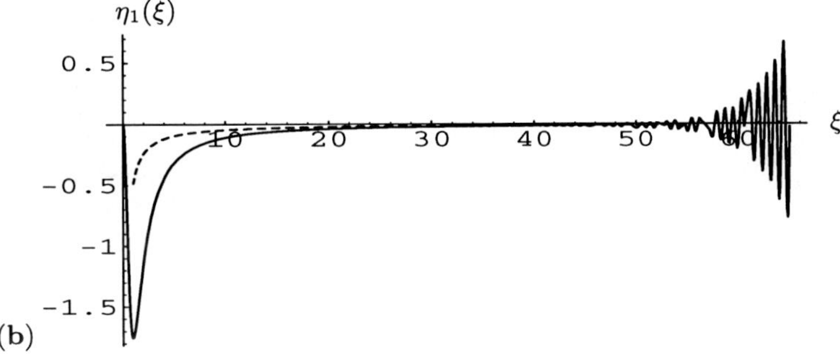

Fig. 4.1. The solution $\eta_1(\xi)$ for $\frac{\eta_0^2}{2} = 0.005$: (a) in the region of $0 \le \xi < 5$; (b) in the region of $0 \le \xi < 65$. The dashed line represents the asymptotic solution in the far field, as $\xi \to \infty$

The mean curvature of the interface at the tip is

$$\mathcal{K}(0) \approx 2 - \varepsilon^2 \eta_1''(0). \tag{4.65}$$

On the other hand, the dimensionless tip radius, $\mathrm{Pe} = \ell_{\mathrm{t}}/\ell_{\mathrm{T}}$ can be calculated with the formula of curvature, namely

$$\frac{\mathcal{K}(0)}{\eta_0^2} = \frac{2}{\mathrm{Pe}}. \tag{4.66}$$

Therefore, we get

$$\mathrm{Pe} = \frac{\ell_{\mathrm{t}}}{\ell_{\mathrm{T}}} = \frac{\eta_0^2}{1 - \varepsilon^2 \eta_1''(0)/2}. \tag{4.67}$$

The derivatives $\eta_1''(0, \eta_0^2)$ can be numerically computed as

$$\eta_1''(0) = \lim_{\xi \to 0} \frac{2\eta_1(\xi)}{\xi^2}. \tag{4.68}$$

This solution have been evaluated over a large range of undercooling temperature T_∞, or say, $\frac{\eta_0^2}{2}$. It should be pointed out that, when $\frac{\eta_0^2}{2}$ becomes

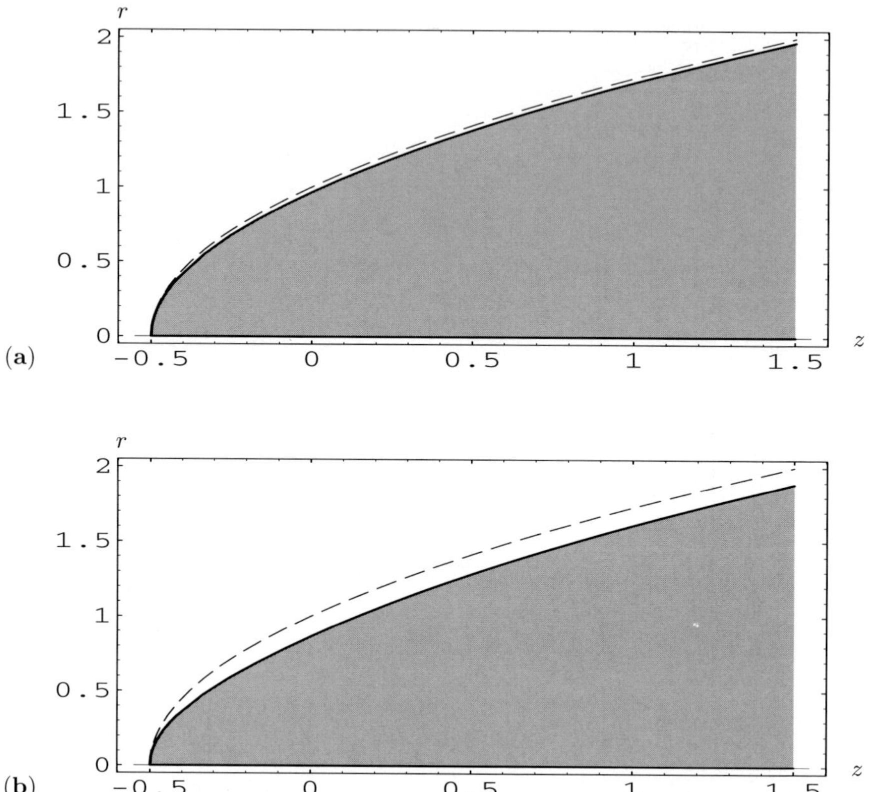

Fig. 4.2. The shape of dendrite with $\frac{\eta_0^2}{2} = 0.005$: **(a)** for $\varepsilon = 0.1$; **(b)** for $\varepsilon = 0.2$

smaller and smaller, the numerical calculations become increasingly difficult, as the convergence of the Laguerre series expansions, such as in (4.46)
and (4.62), becomes increasingly slow. For instance, using a Silicon Graphics
workstation for the case $\frac{\eta_0^2}{2} = 0.001$, to obtain three significant digits for
$\eta_1''(0)$, we have to take more than $N = 20\,000$ terms.

For the case $\frac{\eta_0^2}{2} = 0.005$, we obtain $\eta_1''(0) = -15.838$. The variations of
$\eta_1''(0)$ with the parameters T_∞ are shown in Table 4.1, as well as in Fig. 4.3.

The variations of the Peclet number $\mathrm{Pe} = \ell_t/\ell_T$ with the parameters T_∞
and ε are shown in Fig. 4.4. This steady correction for the Ivantsov needle
solution due to the isotropic surface tension is of practical interest. Such a
correction has been observed experimentally.

One can continue the above procedure further to determine higher-order
approximations. For the higher-order approximate system, the governing
equations are the same as (4.15), so that the general solutions remain of
the form (4.36), (4.43), and (4.44). The boundary conditions on the interface yield similar formulas to (4.45), (4.46), and (4.59). Consequently, the

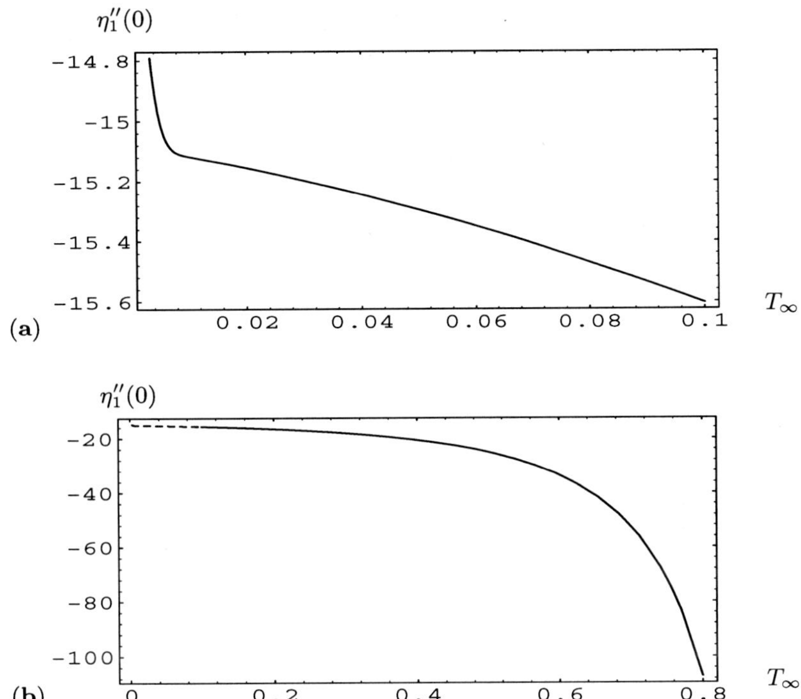

Fig. 4.3. The variations of $\eta_1''(0)$ with the parameters T_∞: **(a)** for the small under-cooling temperature regime $0 < T_\infty < 0.1$; **(b)** for the large undercooling temperature regime $0.1 < T_\infty < 0.8$

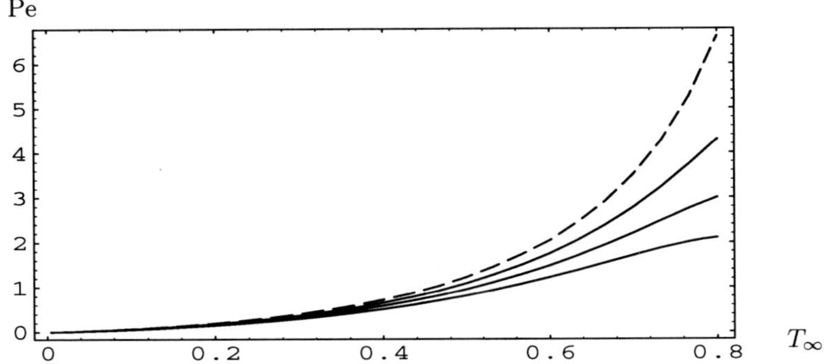

Fig. 4.4. The variations of the Peclet number $\mathrm{Pe} = \ell_t/\ell_T$ with the parameters T_∞ and ε for the cases: $\varepsilon = 0.0, 0.1, 0.15$ and 0.2 from top to bottom. The dashed line is the Ivantsov solution with $\varepsilon = 0$

Table 4.1. The values of $\eta_1''(0)$

T_∞	η_0^2	$\eta_1''(0)/2$	T_∞	η_0^2	$\eta_1''(0)/2$
-0.9806	100	-5402.5	-0.1297	0.09	-7.883
-0.9629	50	-1452.9	-0.1116	0.08	-7.846
-0.8521	10	-94.732	-0.1018	0.07	-7.809
-0.8399	9	-81.371	-0.0915	0.06	-7.773
-0.8254	8	-69.025	-0.0803	0.05	-7.737
-0.8079	7	-57.699	-0.06845	0.04	-7.699
-0.7863	6	-47.396	-0.05539	0.03	-7.662
-0.7588	5	-38.122	-0.04079	0.02	-7.624
-0.7227	4	-29.885	-0.02375	0.01	-7.585
-0.6724	3	-22.695	-0.02184	0.009	-7.581
-0.5963	2	-16.569	-0.01987	0.008	-7.577
-0.4614	1	-11.529	-0.01785	0.007	-7.573
-0.4413	0.9	-11.086	-0.01575	0.006	-7.568
-0.4191	0.8	-10.654	-0.01358	0.005	-7.562
-0.3945	0.7	-10.233			
-0.3668	0.6	-9.822			
-0.3352	0.5	-9.423			
-0.2987	0.4	-9.034			
-0.2552	0.3	-8.654			
-0.2014	0.2	-8.280			
-0.1297	0.1	-7.919			

coefficients in the general solutions can be determined in the same manner as was just carried out for the first-order approximation. The conclusions drawn from the first-order approximation solution that the solution satisfying the smooth tip conditions can be analytically extended to the complex ξ-plane, and that it algebraically decays as $\xi \to \infty$, will also be valid for the higher-order approximate solutions $\{\eta_2(\xi); \eta_3(\xi)\dots\}$.

Several remarks should be made here.

1. The solutions $T_1(\xi, \eta), T_{S1}(\xi, \eta)$ and $\eta_1(\xi)$ were determined without a root condition. Hence, the (RPE) solution we obtained must actually be the regular asymptotic expansion for many different steady solutions specified by a variety of root conditions.
2. The function $\mathcal{K}_0(\xi)$ is an analytic function of ξ in the whole complex ξ-plane, except for $\xi = \pm i$. Therefore, it is expected that the solutions $\{\eta_1(\xi), T_1(\xi, \eta)\}$ are also analytic functions of ξ in the complex ξ-plane, with isolated singular points at $\xi = \pm i$.

4.2.3 The Asymptotic Behavior of the Regular Perturbation Expansion Solution as $\xi \to \infty$

The Laguerre polynomial expansion obtained in the last section is valid for $0 \le \xi < \infty$. However, numerically, it can only be used in the region near the tip ($\xi \le \xi_{\max}$). Its partial summation with any large number of terms starts to rapidly oscillate as $\xi \gg 1$ (see Fig. 4.1). To describe the asymptotic behavior of the solution in the limit $\xi \to \infty$, one needs to find a different asymptotic form.

As we have deduced, the solution is an analytic function at $\xi = \infty$. Hence, it can be expanded as a Taylor series

$$T_1(\xi, \eta) = \frac{A_1(\eta)}{\xi} + \frac{A_2(\eta)}{\xi^2} + \cdots$$

$$T_{S1}(\xi, \eta) = \frac{A_{S1}(\eta)}{\xi} + \frac{A_{S2}(\eta)}{\xi^2} + \cdots \tag{4.69}$$

$$\eta_1(\xi) = \frac{C_1}{\xi} + \frac{C_2}{\xi^2} + \cdots .$$

The first terms in the Taylor series are subject to the system:

$$\left\{ \frac{d^2}{d\eta^2} + \left(\frac{1}{\eta} + \eta_0^2 \eta \right) \frac{d}{d\eta} + \eta_0^2 \right\} \begin{pmatrix} A_1 \\ A_{S1} \end{pmatrix} = 0 \tag{4.70}$$

with the boundary conditions at $\eta = 1$:

$$A_1(1) = A_{S1}(1) + C_1 , \tag{4.71}$$

$$A_{S1}(1) = -1 , \tag{4.72}$$

$$\frac{d}{d\eta}(A_1 - A_{S1}) + (1 + \eta_0^2)C_1 = 0 . \tag{4.73}$$

Equation (4.70) is of the same type as (4.26). Letting

$$\tau = \frac{\eta_0^2 \eta^2}{2} \qquad \text{and} \qquad A_1(\eta) = \frac{e^{-\frac{\tau}{2}}}{\tau^{\frac{1}{2}}} Z(\tau), \tag{4.74}$$

equation (4.70) is transformed to (4.38) with

$$\lambda = \mu = 0. \tag{4.75}$$

Thus, we obtain the solutions

$$A_1(\eta) = \sqrt{\pi} \hat{A}_1 e^{-\frac{\eta_0^2 \eta^2}{2}} M\left(\frac{1}{2}, 1, \frac{\eta_0^2 \eta^2}{2} \right) = \hat{A}_1 e^{-\frac{\eta_0^2 \eta^2}{4}} K_0\left(\frac{\eta_0^2 \eta^2}{4} \right) \tag{4.76}$$

$$A_{S1}(\eta) = \sqrt{\pi} \hat{A}_{S1} e^{-\frac{\eta_0^2 \eta^2}{2}} U\left(\frac{1}{2}, 1, \frac{\eta_0^2 \eta^2}{2} \right) = \hat{A}_{S1} e^{-\frac{\eta_0^2 \eta^2}{4}} I_0\left(\frac{\eta_0^2 \eta^2}{4} \right). \tag{4.77}$$

The boundary conditions (4.71)–(4.73) uniquely determine the three unknown constants $\{\hat{A}_1, \hat{A}_{S1}, C_1\}$. The results are as follows:

$$\hat{A}_1 = \frac{I_1 + (1 - a_0)I_0}{K_1 + (1 - a_0)K_0} \hat{A}_{S1} ,$$ (4.78)

$$\hat{A}_{S1} = -\frac{1}{I_0} e^{\frac{\eta_0^2}{4}} ,$$ (4.79)

$$C_1 = \frac{I_0 K_1 - K_0 I_1}{I_0\left\{(1 - a_0)K_0 + K_1\right\}} ,$$ (4.80)

where

$$K_{0,1} = K_{0,1}\left(\frac{\eta_0^2}{4}\right)$$

$$I_{0,1} = I_{0,1}\left(\frac{\eta_0^2}{4}\right)$$ (4.81)

$$a_0 = 2 + \frac{2}{\eta_0^2} ,$$

and the functions $K_\nu(x)$ and $I_\nu(x)$ are νth-order modified Bessel functions.

The above procedure can be continued to higher-order approximations with no difficulty. The asymptotic solution for $\xi \to \infty$ is shown in Fig. 4.1.

4.3 Two-Dimensional, Steady Needle Crystal Growth

We now consider two-dimensional needle-like crystal growth. The approach that was developed for the three-dimensional case can also be applied here. The results for the two cases are very similar. Hence, the description here will be briefer. We shall simply find the analytical form of the solutions, without giving the numerical results.

4.3.1 Mathematical Formulation
of Two-Dimensional Needle Growth

As in the three-dimensional case, when $\varepsilon \geq 0$, we set the dendrite tip at $\xi = 0, \eta = 1$. The dimensionless governing equation is:

$$\left(\frac{\partial^2 T}{\partial \xi^2} + \frac{\partial^2 T}{\partial \eta^2}\right) = \eta_0^2 \left(\xi \frac{\partial T}{\partial \xi} - \eta \frac{\partial T}{\partial \eta}\right) .$$ (4.82)

The boundary conditions are:

1. The up-stream far-field condition

$$T \to T_\infty \,, \qquad \text{as } \eta \to \infty \,. \tag{4.83}$$

2. The regularity condition

$$T_S = O(1) \,, \qquad \text{as } \eta \to 0 \,. \tag{4.84}$$

3. The interface conditions at $\eta = \eta_s(\xi)$:
    (i) the thermodynamic equilibrium condition

$$T = T_S \,, \tag{4.85}$$

(ii) the Gibbs–Thomson condition

$$T_S = -\eta_0^2 \varepsilon^2 \mathcal{K} \left\{ \frac{d}{d\xi}, \frac{d^2}{d\xi^2} \right\} \eta_s \,, \tag{4.86}$$

(iii) the heat balance condition

$$\frac{\partial}{\partial \eta}(T - T_S) - \eta_s' \frac{\partial}{\partial \xi}(T - T_S) + \eta_0^2 (\xi \eta_s)' = 0 \,. \tag{4.87}$$

4. The tip smoothness conditions

$$\eta_s'(0) = 0 \,, \tag{4.88}$$

$$\eta_s(0) = 1 \,. \tag{4.89}$$

4.3.2 The Regular Perturbation Expansion Solution as $\varepsilon \to 0$

As $\varepsilon \to 0$, we make the following regular perturbation expansion (RPE):

$$T = T(\xi, \eta) = T_*(\eta) + \varepsilon^2 \eta_0^2 T_1(\xi, \eta) + \cdots$$

$$T_S = T_S(\xi, \eta) = T_{S*} + \varepsilon^2 \eta_0^2 T_{S1}(\xi, \eta) + \cdots \tag{4.90}$$

$$\eta_s(\xi) = 1 + \varepsilon^2 \eta_1(\xi) + \cdots .$$

The first-order approximation is then governed by the system:

$$L\{T_1\} = \left\{ \frac{\partial^2}{\partial \xi^2} + \frac{\partial^2}{\partial \eta^2} - \eta_0^2 \xi \frac{\partial}{\partial \xi} + \eta_0^2 \eta \frac{\partial}{\partial \eta} \right\} T_1 = 0 \tag{4.91}$$

with the boundary conditions:

1. As $\eta \to \infty$,

$$T_1 \to 0 \ \text{ (exponentially)}. \tag{4.92}$$

2. As $\xi \to \infty$,

$$T_1 \to 0 \quad \text{(algebraically)}. \tag{4.93}$$

3. As $\eta \to 0$,

$$T_{S1} \text{ regular.} \tag{4.94}$$

4. At $\eta = 1$

$$T_1 = T_{S1} + \eta_1 \,, \tag{4.95}$$

$$T_{S1} = -\mathcal{K}_0(\xi) = -\frac{1}{\left(1 + \xi^2\right)^{\frac{3}{2}}} \,, \tag{4.96}$$

$$\frac{\partial}{\partial \eta}\left(T_1 - T_{S1}\right) + \left(1 + \eta_0^2\right)\eta_1 + \xi\frac{d\eta_1}{d\xi} = 0 \,. \tag{4.97}$$

5. The tip smoothness condition at $\xi = 0$,

$$\eta_1'(0) = 0 \,, \tag{4.98}$$

$$\eta_1(0) = 0 \,. \tag{4.99}$$

Let

$$T_1(\xi, \eta) = X(\xi)Y(\eta), \tag{4.100}$$

where

$$X'' - \eta_0^2\xi X' + \eta_0^2\lambda_1^{\,2}X = 0 \tag{4.101}$$

$$Y'' + \eta_0^2\eta Y' - \eta_0^2\lambda_1^{\,2}Y = 0 \,. \tag{4.102}$$

By letting

$$\sigma = \frac{\eta_0^2\xi^2}{2}, \quad X = X(\sigma), \tag{4.103}$$

equation (4.101) is transformed into the Kummer equation

$$\sigma X''(\sigma) + \left(\frac{1}{2} - \sigma\right)X'(\sigma) + \frac{\lambda_1^2}{2}X = 0, \tag{4.104}$$

whose fundamental solutions are

$$X(\sigma) = \begin{cases} M\left(-\frac{\lambda_1^2}{2}, \frac{1}{2}, \sigma\right) & \text{(regular at } \xi = \sigma = 0); \\[2mm] U\left(-\frac{\lambda_1^2}{2}, \frac{1}{2}, \sigma\right) & \text{(with a logarithmic singularity} \\[1mm] & \text{at } \xi = \sigma = 0). \end{cases} \tag{4.105}$$

Here M and U are the confluent hypergeometric functions. For the same reasons as given for the three-dimensional case, we must choose

$$X(\sigma) = M\left(-\frac{\lambda_1^2}{2}, 1, \sigma\right) \tag{4.106}$$

and

$$\frac{\lambda_1^2}{2} = n = 0, 1, 2, \ldots . \tag{4.107}$$

Thus,

$$X(\sigma) = M\left(-n, \frac{1}{2}, \sigma\right) = H_{2n}(\eta_0 \xi), \tag{4.108}$$

where H_n is the Hermite polynomial. To solve (4.102) in the solid phase region we let

$$\hat{\tau} = -\frac{\eta_0^2 \eta^2}{2}. \tag{4.109}$$

Thus (4.102) is transformed into

$$\hat{\tau} Y''(\hat{\tau}) + (1 - \hat{\tau}) Y'(\hat{\tau}) + n Y = 0. \tag{4.110}$$

The solution regular at $\hat{\tau} = \eta = 0$ is

$$Y(\hat{\tau}) = M_n\left(-n, \frac{1}{2}, -\frac{\eta_0^2 \eta^2}{2}\right) = H_{2n}(i\eta_0 \eta). \tag{4.111}$$

Therefore, the temperature in the solid phase is

$$T_{s1}(\xi, \eta) = \sum_{n=0}^{\infty} \alpha_n H_{2n}(\eta_0 \xi) H_{2n}(i\eta_0 \eta). \tag{4.112}$$

In the liquid phase region, we set

$$\tau = \frac{\eta_0^2 \eta^2}{2}, \quad Y(\eta) = \frac{1}{\tau^{\frac{1}{4}}} e^{-\frac{\tau}{2}} Z(\tau). \tag{4.113}$$

Equation (4.102) can now be transformed into the Whittaker equation

$$Z''(\tau) + \left(-\frac{1}{4} + \frac{\lambda}{\tau} + \frac{\frac{1}{4} - \mu^2}{\tau^2}\right) Z(\tau) = 0 \tag{4.114}$$

with

$$\lambda = -\left(n + \frac{1}{4}\right), \quad \mu = \frac{1}{4}. \tag{4.115}$$

As it is known that $Y(\eta)$ vanishes exponentially when $\eta \to \infty$, it is follows that

$$Y(\eta) = e^{-\frac{\eta_0^2 \eta^2}{2}} \eta \, U\left(n+1, \ \frac{3}{2}, \ \frac{\eta_0^2 \eta^2}{2}\right). \qquad (4.116)$$

Therefore, the temperature in the liquid is

$$T_1(\xi, \eta) = \sum_{n=0}^{\infty} \beta_n H_{2n}(\eta_0 \xi) \frac{e^{-\frac{\eta_0^2 \eta^2}{2}} \eta \, U\left(n+1, \ \frac{3}{2}, \ \frac{\eta_0^2 \eta^2}{2}\right)}{e^{-\frac{\eta_0^2}{2}} U\left(n+1, \ \frac{3}{2}, \ \frac{\eta_0^2}{2}\right)}. \qquad (4.117)$$

We use the following Hermite series for the function $\eta_1(\xi)$:

$$\eta_1(\xi) = \sum_{n=0}^{\infty} \gamma_n H_{2n}(\eta_0 \xi). \qquad (4.118)$$

We now need to determine the coefficients $\{\alpha_n, \beta_n, \gamma_n \ (n = 0, 1, 2 \cdots)\}$, in terms of the boundary conditions (4.95)–(4.97).

It follows from the boundary condition (4.95) that

$$\beta_n = \alpha_n + \gamma_n. \qquad (4.119)$$

Furthermore, the boundary condition (4.96) gives

$$\sum_{n=0}^{\infty} \alpha_n H_{2n}(\eta_0 \xi) = -\mathcal{K}_0(\xi) = -\frac{1}{(1+\xi^2)^{\frac{3}{2}}}. \qquad (4.120)$$

Due to the orthogonality of the set of functions $H_n(x)$ $(n = 0, 1, 3, \cdots)$, the coefficients α_n can be determined by the integrals:

$$\alpha_n = -\frac{\eta_0^3}{2^n \pi^{\frac{1}{2}} n!} \hat{\alpha}_n$$

$$\hat{\alpha}_n = \int_{\infty}^{\infty} \frac{e^{-x^2} H_{2n}(x)}{\left(x^3 + \eta_0^2\right)^{\frac{3}{2}}} dx. \qquad (4.121)$$

Finally, from (4.97), we get

$$(a_n - b_n) + \left(1 + \eta_0^2 + 2n\right)\gamma_n + 8(n+1)\left(n + \tfrac{1}{2}\right)\gamma_{n+1} = 0 \qquad (4.122)$$

$$(n = 0, 1, 2, 3, \cdots),$$

where

$$a_n = 2\beta_n A_n, \qquad b_n = 2\alpha_n B_n, \qquad (4.123)$$

and

$$A_n = n - \frac{U\left(n, \frac{3}{2}, \frac{\eta_0^2}{2}\right)}{U\left(n+1, \frac{3}{2}, \frac{\eta_0^2}{2}\right)}, \tag{4.124}$$

$$B_n = 1 - 2(2n-1)\frac{H_{2n-2}(i\eta_0)}{H_{2n}(i\eta_0)}. \tag{4.125}$$

From the above, we find that

$$\gamma_{n+1} = -g_n\gamma_n + \frac{\eta_0^3 f_n}{2^n \pi^{\frac{1}{2}} n!}\hat{\alpha}_n \quad (n = 0, 1, 2\ldots), \tag{4.126}$$

where

$$g_n = \frac{1 + 2n + \eta_0^2 + 2A_n}{8(n+1)(n+\frac{1}{2})}, \quad f_n = \frac{A_n - B_n}{4(n+1)(n+\frac{1}{2})}. \tag{4.127}$$

We may set

$$\gamma_n = \frac{\eta_0^3}{2^n \pi^{\frac{1}{2}} n!}\hat{\gamma}_n. \tag{4.128}$$

Thus, we rewrite (4.126) as

$$\hat{\gamma}_{n+1} = -2(n+1)g_n\hat{\gamma}_n + 2(n+1)f_n\hat{\alpha}_n \quad (n = 0, 1, 2\ldots). \tag{4.129}$$

For any given value of $\hat{\gamma}_0$, equation (4.129) generates the sequence

$$\{\hat{\gamma}_0, \hat{\gamma}_1, \hat{\gamma}_2, \cdots \hat{\gamma}_n \cdots\}.$$

Thus, the function $\eta_1(\hat{\gamma}_0, \xi)$ can be evaluated as

$$\eta_1(\hat{\gamma}_0, \xi) = \sum_{n=0}^{\infty} \gamma_n H_{2n}(\eta_0\xi). \tag{4.130}$$

The value of $\hat{\gamma}_0$ is then uniquely determined by the tip condition (4.99), namely

$$\eta_1(\hat{\gamma}_0, 0) = \sum_{n=0}^{\infty} \frac{\eta_0^3}{2^n \pi^{\frac{1}{2}} n!}\hat{\gamma}_n H_{2n}(0) = 0. \tag{4.131}$$

Note that

$$H_{2n}(0) = (-1)^n 2^n (2n-1)!!. \tag{4.132}$$

Thus,

$$\eta_1(\hat{\gamma}_0, 0) = \sum_{n=0}^{\infty} (-1)^n \frac{\eta_0^3}{\pi^{\frac{1}{2}}} \frac{(2n-1)!!}{n!}\hat{\gamma}_n = 0. \tag{4.133}$$

The dimensionless tip radius Pe is calculated by the formula

$$\frac{\mathcal{K}(0)}{\eta_0^2} = \frac{1}{\mathrm{Pe}}\,.$$

(4.134)

Finally, we obtain

$$\mathrm{Pe} = \frac{\ell_{\mathrm{t}}}{\ell_{\mathrm{T}}} = \frac{\eta_0^2}{1 - \varepsilon^2 \eta_1''(0)}\,.$$

(4.135)

4.3.3 Asymptotic Behavior
of the Regular Perturbation Expansion Solution as $\xi \to \infty$

In the far-field, the function $\mathcal{K}_0(\xi) \sim \frac{1}{\xi^3}$ as $\xi \to \infty$. Thus, we make the expansions

$$T_1(\xi, \eta) = \frac{A_1(\eta)}{\xi^3} + \frac{A_2(\eta)}{\xi^4} + \cdots$$

$$T_{\mathrm{S}1}(\xi, \eta) = \frac{A_{\mathrm{S}1}(\eta)}{\xi^3} + \frac{A_{\mathrm{S}2}(\eta)}{\xi^4} + \cdots$$

(4.136)

$$\eta_1(\xi) = \frac{C_1}{\xi^3} + \frac{C_2}{\xi^4} + \cdots .$$

The first-order approximate system is

$$\left\{\frac{d^2}{d\eta^2} + \eta_0^2 \eta \frac{d}{d\eta} + 3\eta_0^2\right\} \left(\begin{array}{c} A_1 \\ A_{\mathrm{S}1} \end{array}\right) = 0\,,$$

(4.137)

and, at $\eta = 1$,

$$A_1(1) = A_{\mathrm{S}1}(1) + C_1\,;$$

(4.138)

$$A_{\mathrm{S}1}(1) = -1\,;$$

(4.139)

$$\frac{d}{d\eta}\left(A_1 - A_{\mathrm{S}1}\right) + \left(\eta_0^2 - 2\right)C_1 = 0\,.$$

(4.140)

Similar to the way of treating (4.102) in the liquid region, we let

$$\tau = \frac{\eta_0^2 \eta^2}{2}\,; \quad \left(\begin{array}{c} A_{\mathrm{S}1}(\eta) \\ A_1(\eta) \end{array}\right) = \mathrm{e}^{-\frac{\tau}{2}} \tau^{-\frac{1}{4}} Z(\tau)\,.$$

(4.141)

Then (4.137) is transformed into the Whittaker equation with

$$\lambda = \frac{5}{4}\,, \qquad \mu = \frac{1}{4}\,.$$

(4.142)

Applying the boundary conditions

$$A_1(\eta) \to 0, \quad \text{as} \quad \eta \to \infty \quad \text{(exponentially)}, \tag{4.143}$$

$$A_{S1}(\eta) \quad \text{is regular at} \quad \eta = 0 , \tag{4.144}$$

we obtain

$$A_1(\eta) = \hat{A}_1 \eta e^{-\frac{\eta_0^2 \eta^2}{2}} U\left(-\frac{1}{2}, \frac{3}{2}, \frac{\eta_0^2 \eta^2}{2}\right), \tag{4.145}$$

$$A_{S1}(\eta) = \hat{A}_{S1} \eta e^{-\frac{\eta_0^2 \eta^2}{2}} M\left(-\frac{1}{2}, \frac{3}{2}, \frac{\eta_0^2 \eta^2}{2}\right). \tag{4.146}$$

From (4.138)–(4.140), one can uniquely determine the three unknown constants $\{\hat{A}_1, \hat{A}_{S1}, C_1\}$ to be

$$\begin{cases} \hat{A}_{S1} = -\dfrac{1}{M} e^{\frac{\eta_0^2}{2}} \\[2mm] \hat{A}_1 = \dfrac{M - M'}{U - U'} \hat{A}_{S1} \\[2mm] C_1 = \dfrac{U\hat{A}_1 - M\hat{A}_{S1}}{e^{-\frac{\eta_0^2}{2}}} , \end{cases} \tag{4.147}$$

where

$$\begin{aligned} M &= M\left(-\tfrac{1}{2}, \tfrac{3}{2}, \tfrac{\eta_0^2}{2}\right) \\ M' &= \tfrac{d}{d\eta} M\left(-\tfrac{1}{2}, \tfrac{3}{2}, \tfrac{\eta_0^2 \eta^2}{2}\right)\Big|_{\eta=1} = -\tfrac{1}{3} M\left(\tfrac{1}{2}, \tfrac{5}{2}, \tfrac{\eta_0^2}{2}\right) \\ U &= U\left(-\tfrac{1}{2}, \tfrac{3}{2}, \tfrac{\eta_0^2}{2}\right) \\ U' &= \tfrac{d}{d\eta} U\left(-\tfrac{1}{2}, \tfrac{3}{2}, \tfrac{\eta_0^2 \eta^2}{2}\right)\Big|_{\eta=1} = -\tfrac{1}{2} U\left(\tfrac{1}{2}, \tfrac{5}{2}, \tfrac{\eta_0^2}{2}\right) . \end{aligned} \tag{4.148}$$

4.4 Summary and Discussion

In the previous sections, we obtained the regular perturbation expansion (RPE) for both three-dimensional and two-dimensional steady needle growth. Using

$$\mathcal{R}_N = 1 + \sum_{n=1}^{N} \varepsilon^{2n} \eta_n \tag{4.149}$$

to denote the summation of the first N terms of the RPE, it can be concluded that the $2N^{\text{th}}$-order asymptotic approximation solution, $\mathcal{R}_N$ $(N = 1, 2, 3, \cdots)$ has the following properties:

1. it has a regular tip, satisfying the smooth tip condition;
2. it can be extended to the region $0 \leq \xi < \infty$. Furthermore, as $\xi \to \infty$, it approaches the Ivantsov solution, $\eta_s = 1$.

However, the RPE obtained above is not a convergent Taylor series in ε. For any fixed $0 < \varepsilon \ll 1$, the RPE diverges at any point (ξ, η). The behavior of $\mathcal{R}_N$ as $\xi \to \infty$, does not give any information about the behavior of the true steady-state solution $\eta_s(\xi)$ at $\xi \gg 1$. In other words, the fact that $\eta_n(\infty) = 0$ $(n = 1, 2, 3, \cdots)$, or, $\mathcal{R}_N(\infty) = 1$ $(N = 1, 2, 3, \cdots)$ does not imply that for given $\varepsilon \neq 0$, the exact steady needle crystal growth solution itself also satisfies the limit

$$\lim_{\xi \to \infty} \eta_s(\xi, \varepsilon) = 1.$$

This situation can be easily illustrated by the following example. Suppose the exact solution is

$$\eta_s(\xi, \varepsilon) = 1 + e^{\xi - \frac{1}{\varepsilon}} + \varepsilon \frac{\xi^2 e^{-\xi}}{1 - \varepsilon e^{-\xi}}. \tag{4.150}$$

As $\varepsilon \to 0$, the η_s has the following RPE:

$$\eta_s \sim \left\{ 1 + \varepsilon \eta_1 + \varepsilon^2 \eta_2 + \cdots \right\}$$
$$= 1 + \varepsilon \xi^2 e^{-\xi} \left\{ 1 + \varepsilon e^{-\xi} + \varepsilon^2 e^{-2\xi} + \cdots \right\}. \tag{4.151}$$

Thus,

$$\begin{aligned} \eta_1'(0) &= \eta_2'(0) &= \cdots = 0 \\ \eta_1(\infty) &= \eta_2(\infty) &= \cdots = 0. \end{aligned} \tag{4.152}$$

However, for any fixed $\varepsilon > 0$,

$$\eta_s(\infty) = \infty; \quad \eta_s'(0) = e^{-\frac{1}{\varepsilon}}. \tag{4.153}$$

This example shows that as an asymptotic expansion solution, the RPE may miss some exponentially small component of the true solution as $\varepsilon \to 0$. Such an exponentially small component may be significant for describing the solution behavior with a fixed $\varepsilon > 0$ at the far field $\xi \to \infty$, as well as at the tip. This possibility relates to some very subtle mathematical issues and has caused some controversy concerning the problems of dendrite growth and other related subjects.

References

4.1 M. Abramovitz and I. A. Stegun (Eds.), 'Handbook of Mathematical Functions', (Dover, New York 1964).

4.2 G. P. Ivantsov, "Temperature Field around a Spheroidal, Cylindrical and Acicular Crystal Growing in a Supercooled Melt", Dokl. Akad. Nauk, SSSR. **58**, No. 4, pp. 567–569, (1947).

4.3 G. Horvay and J. W. Cahn, "Dendritic and Spheroidal Growth", Acta Metall. **9**, pp. 695–705, (1961).

4.4 D. E. Temkin, "Growth Rate of the Needle-Crystal Formed in a Supercooled Melt", Dokl. Akad. Nauk. SSSR. **132**, pp. 1307–1310, (1960).

4.5 G. F. Bolling and W. A. J. Tiller, "Growth from the Melt. III. Dendrite Growth", J. Appl. Phys. **32**, No. 12, pp. 2587–2605, (1961).

4.6 R. Trivedi, "Growth of Dendritic Needles from a Supercooled Melt", Acta Metall. **18**, pp. 287–296, (1970).

4.7 M. E. Glicksman, R. J. Schaefer, and J. D. Ayers, "High-Confidence Measurement of Solid/Liquid Surface Energy in a Pure Material", Phil. Mag. **32**, pp. 725–743, (1975).

4.8 M. E. Glicksman, R. J. Schaefer, and J. D. Ayers, "Dendrite Growth — A Test of Theory", Metall. Trans. **7A**, pp. 1747–1759, (1976).

4.9 J. S. Langer and H. Müller-Krumbhaar, "Theory of Dendritic Growth — I. Elements of a Stability Analysis; II. Instabilities in the Limit of Vanishing Surface Tension; III. Effects of Surface Tension", Acta Metall. **26**, pp. 1681–1708, (1978).

4.10 J. S. Langer, "Instability and Pattern Formation in Crystal Growth", Rev. Mod. Phys. **52**, 1–28, (1980).

4.11 J. J. Xu, "Global Asymptotic Solution for Axisymmetric Dendrite Growth with Small Undercooling", in 'Structure and Dynamics of Partially Solidified System, Ed. by D.E. Loper NATO ASI Series E. No. 125, (1987), pp. 97–109.

4.12 J. J. Xu, "Asymptotic Theory of Steady Axisymmetric Needle-like Crystal Growth", Studies in Applied Mathematics, 82, pp. 71–91, (1990).

5. The Steady State for Dendrite Growth with Nonzero Surface Tension

In the last chapter we derived the steady asymptotic solution which we called the RPE solution. However, so far we have not specified the steady state solution itself. How to specify the steady state solution in dendrite growth is an important subject which must be approached with great caution. In the literature, many researchers have looked for the classic, steady needle solution for dendrite growth. Such efforts have not been successful for the case of isotropic surface tension.

The so-called classic, steady needle crystal solution, like Ivantsov's solution, has both a smooth tip and an infinitely long smooth, nonoscillating tail. Given the fact that the system with $\varepsilon = 0$ allows a steady needle solution, what about nonzero surface tension? The question of whether or not the system with $\varepsilon \neq 0$ still allows a steady needle crystal solution is not trivial. It involves the subtle mathematical issue of how to catch the exponentially small terms missed by the regular asymptotic expansion. This issue is sometimes called *asymptotics beyond all orders*. At the early stage of research on dendrite growth, most researchers thought that when a small isotropic surface tension is included, the steady needle solution would still persist with a small perturbation from the Ivantsov needle solution in the whole infinite region. It was with this idea that Nash and Glicksman formulated the needle crystal growth problem. It is now recognized that the above idea is incorrect.

What is wrong with this idea? What theoretical difficulty has occurred? How do we overcome this difficulty? In this chapter, we attempt to discuss these problems.

5.1 The Nash–Glicksman Problem and the Classic Needle Crystal Solution

The classic, steady needle crystal growth problem was first mathematically formulated by Nash and Glicksman in 1974 [5.1]. Nash and Glicksman took the isotropic surface tension into account, neglecting the anisotropy. Furthermore, as a boundary condition in the far field for any fixed $\varepsilon > 0$, they assumed that the needle crystal solution must approach the Ivantsov solution as $\xi \to \infty$.

In the mathematical formulation of three-dimensional, axially symmetric, steady needle crystal growth proposed by Nash and Glicksman, the temperature field is described by the heat equation in infinite space

$$\left(\frac{\partial^2 T}{\partial \xi^2} + \frac{\partial^2 T}{\partial \eta^2} + \frac{1}{\xi}\frac{\partial T}{\partial \xi} + \frac{1}{\eta}\frac{\partial T}{\partial \eta}\right) = \eta_0^2 \left(\xi\frac{\partial T}{\partial \xi} - \eta\frac{\partial T}{\partial \eta}\right) \tag{5.1}$$

$$(0 \le \xi < \infty; \quad 0 \le \eta < \infty).$$

The boundary conditions are:

1. The up-stream far-field condition

$$T \to T_\infty = \frac{(T_\infty)_D - T_{M0}}{\Delta H/(c_p \rho)} < 0 \quad \text{as} \quad \eta \to \infty. \tag{5.2}$$

2. The regularity condition:

$$T_S = O(1) \quad \text{as} \quad \eta \to 0. \tag{5.3}$$

3. The interface conditions at $\eta = \eta_s(\xi)$:
   (i) the thermodynamic equilibrium condition

$$T = T_S, \tag{5.4}$$

   (ii) the Gibbs–Thomson condition

$$T_S = -\varepsilon^2 \eta_0^2 \mathcal{K}\left\{\frac{d}{d\xi}, \frac{d^2}{d\xi^2}\right\} \eta_s, \tag{5.5}$$

   (iii) the heat balance condition

$$\frac{\partial}{\partial \eta}(T - T_S) - \eta_s'\frac{\partial}{\partial \xi}(T - T_S) + \eta_0^2(\xi\eta_s)' = 0. \tag{5.6}$$

4. The tip smoothness condition: at the tip of dendrite,

$$\frac{\partial}{\partial \xi}\{T, T_S, \eta_s\}(0) = 0, \quad \eta_s(0) = 1. \tag{5.7}$$

5. The far-field condition

$$\{T, T_S, \eta_s\} = \{T_*, T_{S*}, \eta_*\} \quad \text{as} \quad \xi \to \infty. \tag{5.8}$$

Obviously, the far-field condition (5.8) imposed by Nash and Glicksman is not easy to justify. In practice, the situation at the root region is rather complicated. A realistic growing dendrite is always finite. It connects with neighboring dendrites at its root. On the other hand, although an isolated dendrite at the later stage of evolution may be considered sufficiently long compared with the thermal length scale, it never really is infinitely long. Therefore, the Nash–Glicksman far-field condition certainly has greatly simplified reality. Nash and Glicksman defined the solution of the above problem as "the steady state" of dendrite growth. This definition was also accepted by many other researchers. However, it was later suggested by a number of investigators, and rigorously proved by Segur and Kruskal in terms of a simplified local model equation, that such a Nash–Glicksman problem did not have a solution [5.2]!

The result of Segur and Kruskal is quite understandable. It implies that for the problem of steady dendrite growth with surface tension, the solution cannot always be monotonically extended to infinity, as these dynamical systems have an intrinsic mathematical singularity there. It turns out that for any fixed (ξ, η), the realistic steady dendrite growth solution indeed approaches the Ivantsov solution as $\varepsilon \to 0$; however, for a fixed $\varepsilon > 0$, as $\xi \to \infty$ the realistic steady growth solution may be far away from the Ivantsov solution; it may have a nonsmooth tail which is oscillatory with a rapidly increasing amplitude.

The subsequent theoretical difficulty, caused by the above definition of the steady state, is evident. Because the system does not allow such a steady state for every point in the parameter space, the regular perturbation expansion solutions (RPE) for the steady state of needle crystal growth derived in the last chapter is meaningless. Moreover, the stability analysis of the steady state over the parameter space is also not applicable.

Further investigations during the past decades have been progressing in two different directions. Some investigators continue to adopt the Nash–Glicksman formulation, looking for the classic, steady needle crystal solution, but include an additional physical effect, the anisotropy of surface tension in the problem. This approach leads to the so-called microscopic solvability condition (MSC) theory [5.3]–[5.8]. They find that in order for the system to have a classic, needle crystal solution, a small amount of anisotropy of surface tension must be taken into account.

The other direction modifies the Nash–Glicksman formulation and redefines the steady state for the system. This alternative approach appeared to be very promising and led to the interfacial wave theory [5.9]. It is certain that the nonexistence of the mathematical solution for the classic steady needle growth does not imply that a dendrite growth system with isotropic surface tension does not permit a physically acceptable, nonclassic steady needle solution.

The key is how to formulate a physical problem as a proper mathematical problem. From the physical point of view, a mathematically infinitely large system is a simplification of a sufficiently large realistic system, whereas a mathematical steady state is also a simplification of a sufficiently slowly time-evolving state. Hence, in order to guarantee the existence of a mathematical solution, one may modify the classic Nash–Glicksman problem in these two respects: consider the system to be finite and/or take a small time dependence into account. More specifically, we can formulate the following two problems:

1. Assume that the needle steadily grows with a constant, long but finite, length, $L_0 = \frac{C}{\varepsilon^\nu}$, $(C = O(1), \nu > 0)$, and at the root $\xi = L_0$, as a boundary condition, the solution is assumed to be close to the Ivantsov solution, say, with an error of $O(\varepsilon)$. We call this problem the *nonclassic steady needle growth problem*.

2. Assume the needle is at the later stage of growth as $t \geq t_0 \gg 1$, so that it is 'nearly' steady and evolves with a slow time variable defined as $\tau = \varepsilon(t - t_0)$. In this case, the total length of the needle may be expressed in the form:

$$L(\tau, \varepsilon) = \frac{1}{\varepsilon^\nu}\left(C + \int_0^\tau \bar{U}(\tau')d\tau'\right),\tag{5.9}$$

where $\bar{U}(\tau, \varepsilon)$ is the growth velocity of the needle's tip. At the root which is now moving, the solution is also assumed to be close to the Ivantsov solution. For the sake of convenience, we shall call this problem *the needle crystal formation problem*.

Obviously, the Ivantsov solution is the special solution to both of the above problems as $\varepsilon \to 0$. Moreover, one can see that given $\varepsilon > 0$, the solution for the Nash–Glicksman problem exists if and only if the needle crystal formation problem has a steady, limiting solution as $\tau \to \infty$.

In the next section, we shall use a simple local model to further demonstrate these ideas.

5.2 The Geometric Model and Solutions of the Needle Crystal Formation Problem

5.2.1 Geometric Model of Dendrite Growth

In the literature, the geometric model equation

$$\varepsilon^2\theta'''(s) + \theta'(s) = \cos\theta \quad (0 \leq s < \infty),\tag{5.10}$$

with the boundary conditions

$$\theta(0) = 0, \quad \theta(\infty) = \tfrac{\pi}{2},\tag{5.11}$$

has often been used for the investigation of free dendrite growth. In (5.10), the variable s represents the dimensionless arc-length of the interface starting from the tip, while θ is the angle between the local tangent to the interface and the vertical line. The parameter ε represents the effect of surface tension.

It is evident that for the case of zero surface tension ($\varepsilon = 0$), equation (5.10) reduces to the first-order equation:

$$\theta'(s) = \cos\theta.\tag{5.12}$$

This allows an analytical solution:

$$\theta = \theta_*(s) = -\pi/2 + 2\tan^{-1}(e^s),$$ (5.13)

which is symmetrical, i.e.,

$$\theta(-s) = -\theta(s),$$ (5.14)

and satisfies the boundary condition (5.11).

This solution is a needle-like solution which, like the Ivantsov solution, has the following properties:

1. it has a smooth tip,
2. it is symmetrical, and has a smooth and infinitely long tail.

5.2.2 The Segur–Kruskal Problem

For the case of nonzero surface tension ($\varepsilon \neq 0$), whether the system still allows a symmetrical needle solution satisfying the boundary conditions (5.11) is not a trivial problem. As $\varepsilon \to 0$, for any fixed s, ($0 \leq s < \infty$), one can derive the following formal regular perturbation expansion (RPE) solution for (5.10):

$$\theta(s, \varepsilon) \sim \theta_*(s) + \varepsilon^2\theta_1(s) + \varepsilon^4\theta_2(s) + \cdots$$ (5.15)

By substituting (5.15) into (5.10), one can derive

$$\begin{aligned}
\frac{d\theta_*}{ds} &= \cos\theta_* \\
\frac{d\theta_1}{ds} &= -(\sin\theta_*)\theta_1 - \frac{d^3\theta_*}{ds^3} \\
\frac{d\theta_2}{ds} &= -(\sin\theta_*)\theta_2 - \frac{d^3\theta_1}{ds^3} - \frac{\cos\theta_*}{2}\theta_1^2
\end{aligned}$$ (5.16)

$$\vdots$$

Each order of solution $\theta_n(s)$ is uniquely determined by the boundary condition at the tip:

$$\theta_n(0) = 0 \quad (n = 0, 1, 2, \cdots).$$ (5.17)

It can be proved that all the solutions $\theta_n(s)$ ($n = 0, 1, 2, \cdots$) vanish at the far field, as $s \to \infty$, i. e.,

$$\theta_n(\infty) = 0 \quad (n = 0, 1, 2, \cdots).$$ (5.18)

Therefore, if one defines

$$\mathcal{R}_N = \sum_{n=1}^{n=N} \theta_n$$ (5.19)

as the partial summation of the formal RPE solution, the function $\theta(s,\varepsilon) = \theta_*(s) + \mathcal{R}_N(s,\varepsilon)$ must satisfy the boundary conditions (5.11) and (5.14) exactly. It satisfies (5.10) up to any order approximation $O(\varepsilon^{2N})$. However, up to this point, it is still uncertain whether this formal solution is mathematically meaningful or not. Evidently, if the true solution for the above problem does not exist, the above formal RPE solution would have no mathematical meaning. During the last several years, the problem of the existence of a solution to (5.10) with the conditions (5.11) and (5.14) has been extensively studied by Segur and Kruskal in terms of asymptotic methods, which we call 'the Segur–Kruskal problem'.

Segur and Kruskal first proved that this problem has no solution for any given $0 \leq \varepsilon \ll 1$ (refer to [5.2]). The system may allow the unique solution satisfying the boundary conditions (5.11). This solution, however, cannot satisfy the symmetrical condition (5.14), which is equivalent to the condition:

$$\theta''(0) = 0. \tag{5.20}$$

In fact, Segur and Kruskal proved that

$$\theta''(0^+) = \frac{\Gamma_0}{4\varepsilon^{\frac{5}{2}}} e^{-\frac{\pi}{2\varepsilon}}, \tag{5.21}$$

where the constant $\Gamma_0 = O(1)$. The results obtained by Kruskal and Segur have had an important impact on the applied mathematics community. They raise issues such as exponential asymptotics or asymptotics 'beyond all orders' of solutions to a nonlinear equation. Their work motivated a series of research efforts to develop new approaches, either numerical or analytical, exploring exponentially small factors for similar nonlinear problems arising in different applied areas, which may be missed by the formal RPE solutions [5.7].

Moreover, Segur and Kruskal found that, in order for the system to allow a needle solution, one must include an additional parameter in the system, so that (5.10) is modified to

$$\varepsilon^2 \theta'''(s) + \theta'(s) = \frac{\cos\theta}{1 + \alpha\cos(4\theta)} \quad (0 \leq s < \infty). \tag{5.22}$$

Here, α can be interpreted as the surface tension anisotropy.

Segur and Kruskal's results were soon applied by some investigators to the original, nonlocal, dendrite growth system, and interpreted as indicating the nonexistence of 'the steady state' for a system of dendrite growth with isotropic surface tension and that such a steady state is allowable only when a small amount of anisotropy of surface tension is included in the system. This is the basis of the so-called microscopic solvability condition (MSC) theory, which influenced the physics community for a long time. As pointed out in the last section, in interpreting the physical implication of Segur and Kruskal's results, one needs to recognize the connection and difference between the given physical phenomenon and the resulting simplified mathematical formulation.

The nonexistence of the true solution for the Segur–Kruskal problem does not imply that the system has no physically acceptable steady state solution. To ensure the existence of a mathematical solution, one may modify the Segur–Kruskal problem in two ways, either by assuming that the needle growth is steady and the length of needle is long but finite, of $O(\frac{1}{\varepsilon})$; or, by assuming that the needle growth is 'nearly' steady with an increasingly long length as the time $t \to \infty$.

In the next sections we shall formulate these modified problems.

5.2.3 Nonclassic Steady Needle Growth Problem

Consider the following modified Segur–Kruskal problem:

$$\varepsilon^2 \theta'''(s) + \theta'(s) = \cos\theta \quad \left(0 \le s < L_0(\varepsilon)\right), \tag{5.23}$$

with the boundary conditions:

(i) The tip conditions

$$\theta(0) = \theta''(0) = 0. \tag{5.24}$$

(ii) The root condition

$$\theta(L_0) = \theta_*(L_0) + f(\varepsilon). \tag{5.25}$$

In the above, we assume that $L_0(\varepsilon)$ represents the total length of the dendrite's stem, satisfying $\lim_{\varepsilon \to 0} L(\varepsilon) \to \infty$, $f(\varepsilon)$ is an analytic function of ε and $f(0) = 0$, while $\theta_*(s)$ is 'the Ivantsov solution' for the reduced problem:

$$\theta'(s) = \cos\theta \quad (0 \le s < \infty), \quad \theta(0) = 0. \tag{5.26}$$

For any $0 < \varepsilon \ll 1$, we attempt to find a solution which is a monotonic function in the region $\left(0 \le s \le L_0(\varepsilon)\right)$. It can be proven that a solution to the problem (5.23)–(5.25) exists for some properly chosen $L_0(\varepsilon)$ and can be extended by reflection to the region $\left(0 \ge s \ge -L_0(\varepsilon)\right)$, with no singularity at $s = 0$. We call such a solution a 'nonclassic steady needle solution', and define it as the steady state of the system. In order to prove the existence of the solution to the above two-point boundary value problem, one may use the standard shooting method. In doing so, let

$$\theta'(0, \varepsilon) = \beta(\varepsilon). \tag{5.27}$$

Since the system has no singularity in the entire interval $(0 \le s < \infty)$, the solution to the initial value problem of the ordinary differential equation (5.23) with the initial conditions (5.24) and (5.27) is guaranteed. Thus, in order to have a solution to the modified Segur–Kruskal problem, one needs to further show that the function $\beta(\varepsilon)$ can be determined in terms of the root condition (5.25). This monograph, however, is not concerned with the rigorous proof of the existence of solution. Rather, in what follows, we shall explore the behavior of the solutions and derive their asymptotic expansion form.

The outer expansion. As $\varepsilon \to 0$, for any fixed s, $(0 \le s < L_0)$, one can derive the following outer expansion solution:

$$\theta(s, \varepsilon) \sim \theta_*(s) + \varepsilon^2 \theta_1(s) + \varepsilon^4 \theta_2(s) + \cdots . \tag{5.28}$$

This outer expansion solution is just the regular perturbation expansion (RPE) solution for (5.15). This solution satisfies the boundary conditions (5.24) at the tip $s = 0$. However, it does not satisfy the root condition (5.25), nor the general initial condition (5.27). In order to find the uniformly valid asymptotic solution, normally one should use the matched asymptotic expansion method. In other words, look for the inner solution near the root region, match the inner solution and the outer solution in the intermediate region, and construct the composite solution.

The composite solution. Let us define

$$\tilde{\theta}(s, \varepsilon) = \theta(s, \varepsilon) - \theta_*(s) . \tag{5.29}$$

The Nth-order composite solution $\theta^{(N)}(s, \varepsilon)$ should be a general asymptotic expansion of the solution $\tilde{\theta}(s, \varepsilon)$ in the region $\big(0 \le s \le L_0(\varepsilon)\big)$ (refer to [5.10]). Namely, as $\varepsilon \to 0$,

$$\frac{|\tilde{\theta}(s, \varepsilon) - \theta^{(N)}(s, \varepsilon)|}{\varepsilon^N} \to 0 , \tag{5.30}$$

for any $0 \le s = s(\varepsilon) \le L_0(\varepsilon)$. Note that s is not fixed in the limit $\varepsilon \to 0$. The variable s, during the limit process, can vary with ε arbitrarily in the definition region of solution: $\big(0, L_0(\varepsilon)\big)$. Evidently, the summation of the first N terms of the RPE,

$$\mathcal{R}_N = \varepsilon^2 \theta_1(s) + \varepsilon^4 \theta_2(s) + \cdots + \varepsilon^{2N} \theta_N(s) \tag{5.31}$$

is the $2N$th-order asymptotic approximation of the solution $\tilde{\theta}(s, \varepsilon)$ in the ordinary asymptotic sense. But it is not the Nth-order asymptotic approximation in the general asymptotic sense, as it does not satisfy the root condition. Of course, it should be noted that, given a solution $\tilde{\theta}(s, \varepsilon)$, its general asymptotic expansion solution is not uniquely determined. Adding some exponentially small terms to a general asymptotic expansion solution will still result in a general asymptotic expansion solution.

To find the aforementioned composite solution for the problem under consideration, a better method is to apply the *multiple variables expansion*(MVE) technique.

In doing so, we define a stretched fast variable s_+ in the form

$$s_+ = \frac{1}{\varepsilon} \int_{L_0}^{s} k(s', \varepsilon) ds' \tag{5.32}$$

and assume that the perturbation function $\tilde{\theta}(s, \varepsilon)$ can be considered as a function of the two variables s and s_+, and as $\varepsilon \to 0$ it can be expanded in a multiple variable expansion form. According to MVE method, the variables $(s; s_+)$ are formally treated as independent variables. Hence, the derivatives in (5.23) must be replaced by

$$
\begin{aligned}
\frac{d}{ds} &= \frac{k}{\varepsilon}\frac{\partial}{\partial s_+} + \frac{\partial}{\partial s} \\[2mm]
\frac{d^2}{ds^2} &= \frac{2k}{\varepsilon}\frac{\partial^2}{\partial s\partial s_+} + \frac{k'}{\varepsilon}\frac{\partial}{\partial s_+} + \frac{k^2}{\varepsilon^2}\frac{\partial^2}{\partial s_+^2} + \frac{\partial^2}{\partial s^2} \\[2mm]
\frac{d^3}{ds^3} &= \frac{1}{\varepsilon}\left(3k'\frac{\partial^2}{\partial s\partial s_+} + 3k\frac{\partial^3}{\partial s^2\partial s_+} + k''\frac{\partial}{\partial s_+} \right) \\[2mm]
&\quad + \frac{1}{\varepsilon^2}\left(3kk'\frac{\partial^2}{\partial s_+^2} + 3k^2\frac{\partial^3}{\partial s\partial s_+^2} \right) + \frac{k^3}{\varepsilon^3}\frac{\partial^3}{s_+^3} + \frac{\partial^3}{\partial s^3}\,.
\end{aligned}
\tag{5.33}
$$

Equation (5.23) is then converted into multiple variables form:

$$
\begin{aligned}
&\frac{k^3}{\varepsilon}\frac{\partial^3\tilde\theta}{\partial s_+^3} + \frac{k}{\varepsilon}\frac{\partial\tilde\theta}{\partial s_+} + 3kk'\frac{\partial^2\tilde\theta}{\partial s_+^2} + 3k^2\frac{\partial^3\tilde\theta}{\partial s\partial s_+^2} + \frac{\partial\tilde\theta}{\partial s} \\[2mm]
&\quad + \varepsilon\left(3k'\frac{\partial^2\tilde\theta}{\partial s\partial s_+} + 3k\frac{\partial^3\tilde\theta}{\partial s^2\partial s_+} + k''\frac{\partial\tilde\theta}{\partial s_+} \right) + \varepsilon^2\frac{\partial^3\tilde\theta}{\partial s^3} \\[2mm]
&\quad = \cos\theta_*\left(\cos\tilde\theta - 1 \right) - \sin\theta_*\sin\tilde\theta - \varepsilon^2\theta_*'''(s)
\end{aligned}
$$

$$
= F(\tilde\theta, s)\,.
\tag{5.34}
$$

The boundary conditions are:

(1)

$$
\begin{cases}
\tilde\theta = 0, \\[2mm]
\dfrac{d^2\tilde\theta}{ds^2} = 0,
\end{cases}
\qquad \text{as}\quad s = 0\,.
\tag{5.35}
$$

(2) The additional initial condition,

$$
\frac{d\tilde\theta}{ds} = \beta(\varepsilon) - 1, \quad \text{as}\quad s = 0\,,
\tag{5.36}
$$

or (2)$'$ the boundary condition

$$
\tilde\theta = f(\varepsilon), \quad \text{at}\quad s = L_0.
\tag{5.37}
$$

To find this general asymptotic approximate solution $\tilde\theta^{(2N)}$, we assume that

$$
\tilde\theta^{(2N)}(s, \varepsilon) = \mathcal{R}_N + \mathcal{S}_N,
\tag{5.38}
$$

in which $\mathcal{S}_N$ is called the part of the singular perturbation expansion (SPE) which will be the dominant term near the root. We expand $\mathcal{S}_N$ in the following form:

$$\mathcal{S}_N = \alpha_0(\varepsilon)\Big\{\tilde{h}_0(s, s_+) + \varepsilon\tilde{h}_1(s, s_+) + \cdots \varepsilon^{2N}\tilde{h}_{2N}(s, s_+)\Big\}$$

$$+ \cdots \quad (\text{as } \varepsilon \to 0), \tag{5.39}$$

where the asymptotic sequences, $\alpha_0(\varepsilon) \ll \alpha_1(\varepsilon) \ll \cdots$, are to be determined. By substituting (5.38) and (5.39) into (5.34), one can successively derive each order approximation. Here, we are only interested in the leading order approximation. The results are summarized as follows:

(1) $O(\varepsilon^0)$: The equation is

$$k_0^3\frac{\partial^3 \tilde{h}_0}{\partial s_+^3} + k_0\frac{\partial \tilde{h}_0}{\partial s_+} = 0 \ . \tag{5.40}$$

Letting

$$\tilde{h}_0(s, s_+) = A(s)H(s_+) = A(s)e^{is_+}, \tag{5.41}$$

it follows that

$$-k_0^3 + k_0 = 0. \tag{5.42}$$

Here, one gets three roots:

$$k_0^{(1)} = 0; \quad k_0^{(2)} = 1; \quad k_0^{(3)} = -1. \tag{5.43}$$

Hence, the general solution of (5.40) is

$$\tilde{h}_0(s, s_+) = a_0(s)H_1(s_+) + b_0(s)H_2(s_+) + c_0(s)H_3(s_+) , \tag{5.44}$$

where

$$H_i(s_+) = e^{is_+} = e^{\frac{i}{\varepsilon}\int_0^s k^{(i)}ds} \quad (i = 1, 2, 3). \tag{5.45}$$

(2) $O(\varepsilon)$: We set $\tilde{h}_0(s, s_+) = A_0(s)H(s_+)$, where $A_0(s)H(s_+)$ represents any one of the three solutions $\{a_0(s)H_1(s_+); b_0(s)H_2(s_+); c_0(s)H_3(s_+)\}$. The equation for $\tilde{h}_1$ is then

$$k_0^3\frac{\partial^3 \tilde{h}_1}{\partial s_+^3} + k_0\frac{\partial \tilde{h}_1}{\partial s_+} = \Big[(3k_0^2 - 1)A_0'(s) + A_0(s)\Big(-\sin\theta_* +$$

$$+i3k_0^2 k_1 - ik_1 + 3k_0 k_0'\Big)\Big]H(s_+) . \tag{5.46}$$

To eliminate the secular terms, one must set

$$(3k_0^2 - 1)A_0'(s) - A_0(s)\left(-\sin\theta_* + i3k_0^2 k_1 - ik_1 + 3k_0 k_0'\right) = 0. \quad (5.47)$$

Letting

$$ik_1(1 - 3k_0^2) + \sin\theta_* - 3k_0 k_0' = 0 \quad (5.48)$$

or

$$k_1 = \frac{i\sin\theta_*(s)}{(1 - 3k_0^2)}, \quad (5.49)$$

we have

$$k_1^{(1)} = i\sin\theta_*(s); \quad k_1^{(2)} = -\frac{i}{2}\sin\theta_*(s); \quad k_1^{(3)} = -\frac{i}{2}\sin\theta_*(s) \quad (5.50)$$

and

$$A_0(s) = \hat{A} = \text{const.} \quad (5.51)$$

Thus, we derive three fundamental solutions:

$$H_1(s) = e^{-\int_0^s \sin\theta_*(s)ds}$$

$$H_2(s) = e^{\frac{is}{\varepsilon} + \frac{1}{2}\int_0^s \sin\theta_*(s)ds} \quad (5.52)$$

$$H_3(s) = e^{-\frac{is}{\varepsilon} + \frac{1}{2}\int_0^s \sin\theta_*(s)ds}.$$

The general solution (5.44) becomes

$$. \tilde{h}_0(s, s_+) = \hat{a}_0 e^{-\int_0^s \sin\theta_*(s)ds} + \hat{b}_0 e^{\frac{is}{\varepsilon} + \frac{1}{2}\int_0^s \sin\theta_*(s)ds}$$

$$+ \hat{c}_0 e^{-\frac{is}{\varepsilon} + \frac{1}{2}\int_0^s \sin\theta_*(s)ds}. \quad (5.53)$$

The above procedure can be continued to higher-order approximations. The coefficients $(\hat{a}_0, \hat{b}_0, \hat{c}_0)$ and the pre-factor $\alpha_0(\varepsilon)$ can be determined by the initial condition (5.35) with (5.36) or the boundary condition (5.37). It should be pointed out that of the three fundamental solutions (5.52), the functions $H_2(s)$ and $H_3(s)$ are dominant in the far field, as $\varepsilon \to 0$, while the function $H_1(s)$ is subdominant. It will be seen that to satisfy the root condition (5.37), the solution $\tilde{h}_0(s, s_+)$ must be asymptotically approximated by the dominant functions H_2 and H_3, as $\varepsilon \to 0$. The coefficients $(\hat{a}_0, \hat{b}_0, \hat{c}_0)$ will therefore be constant over the entire region $0 \le s \le L_0$. In contrast, if the solution $\tilde{h}_0(s, s_+)$ were approximated by the subdominant function H_1 in the far field as $\varepsilon \to 0$, the coefficients $(\hat{a}_0, \hat{b}_0, \hat{c}_0)$ might be different constants in different sections of the interval $0 \le s < \infty$ due to the so-called Stokes phenomenon. This is the subtle issue which one encounters when studying the Segur–Kruskal problem. A more detailed description of the Stokes phenomenon will be given in Chap. 6.

To leading order, we have

$$\begin{cases} \tilde{h}_0 = 0 \\ \dfrac{\mathrm{d}^2 \tilde{h}_0}{\mathrm{d}s^2} = 0 \end{cases} \quad \text{as} \quad s = 0\,, \tag{5.54}$$

and

$$\alpha_0(\varepsilon)\Big\{\tilde{h}_0 + \varepsilon \tilde{h}_1 \cdots \Big\} + \cdots$$

$$\sim f(\varepsilon) - \Big\{\varepsilon^2 \theta_1(L) + \cdots \Big\} = \tilde{f}(\varepsilon) \quad \text{at} \quad s = L_0\,. \tag{5.55}$$

From (5.54), we derive that

$$\hat{a}_0 + \hat{b}_0 + \hat{c}_0 = 0 \tag{5.56}$$

and

$$-\hat{a}_0 + \frac{1}{4}\big(\hat{b}_0 + \hat{c}_0\big) = 0\,. \tag{5.57}$$

Hence, we have

$$\hat{a}_0 = 0 \quad \text{and} \quad \hat{b}_0 = -\hat{c}_0\,. \tag{5.58}$$

Thus, one can determine $\alpha_0(\varepsilon)$ through (5.55):

$$\alpha_0(\varepsilon)\tilde{h}_0\big(L(\varepsilon)\big) = \tilde{f}(\varepsilon)\,. \tag{5.59}$$

This results in

$$\alpha_0(\varepsilon)\hat{b}_0 \sin\left(\frac{L_0(\varepsilon)}{\varepsilon}\right) \mathrm{e}^{\frac{1}{2}\int_0^{L_0(\varepsilon)} \sin\tilde{\theta}_*(s)ds} = \tilde{f}(\varepsilon)\,. \tag{5.60}$$

Noting that

$$\int_0^s \sin\theta_*(s_1)ds_1 = \ln(\mathrm{e}^s + \mathrm{e}^{-s}) - \ln 2\,, \tag{5.61}$$

we obtain

$$\hat{b}_0 = \frac{1}{\sqrt{2}} \tag{5.62}$$

and

$$\alpha_0(\varepsilon) = \frac{\tilde{f}(\varepsilon)}{\cosh^{\frac{1}{2}} L_0(\varepsilon) \sin\left(\frac{L_0}{\varepsilon}\right)}\,. \tag{5.63}$$

It is seen that in the case $\sin\left(\frac{L_0}{\varepsilon}\right) = 0$ or $\varepsilon = \frac{L_0}{n\pi}$, $\alpha_0(\varepsilon)$ does not exist. Consequently, the system does not admit a solution to the two-point boundary value problem under consideration. However, if the root L_0 is assumed a proper function of ε, the existence of the solution is guaranteed. We choose $L_0(\varepsilon)$, such that $\sin\left(\frac{L_0}{\varepsilon}\right) = 1$. In this case,

$$L_0(\varepsilon) = \left(2M + \frac{1}{2}\right)\pi\varepsilon = \frac{C}{\varepsilon^\nu} \quad (\nu > 0), \tag{5.64}$$

where the integer M is defined as the integer part of $\frac{1}{4\pi\varepsilon^{(1+\nu)}}$; namely,

$$M = \left[\frac{1}{4\pi\varepsilon^{(1+\nu)}}\right], \tag{5.65}$$

and the constant $0 < C < 1$. We derive

$$\alpha_0(\varepsilon) = \frac{\tilde{f}(\varepsilon)}{\cosh^{\frac{1}{2}} L_0(\varepsilon)} \tag{5.66}$$

and finally obtain

$$\theta(s, \varepsilon) = \theta_*(s) + \varepsilon^2\theta_1(s) + \varepsilon^4\theta_2(s) + \cdots + \varepsilon^{2N}\theta_N(s)$$

$$+ \left[\tilde{f}(\varepsilon)\sin\left(\frac{s}{\varepsilon}\right)\sqrt{\frac{\cosh(s)}{2\cosh(L_0)}}\right] + \cdots . \tag{5.67}$$

In the above, the natural number N can be set arbitrarily large. Hence, for convenience, one may express (5.67) in the form:

$$\theta(s, \varepsilon) = \theta_*(s) + \{\text{RPE}\} + \{\text{SPE}\}, \tag{5.68}$$

we can say that the nonclassic needle crystal solution consists of three parts: (i) the Ivantsov solution; (ii) the regular perturbation expansion; and (iii) the singular perturbation expansion.

The properties of the composite solutions. The above solutions are a family of solutions containing some undefined function $f(\varepsilon)$ and free constants, such as C. These functions and constants depend on how one sets the root condition. However, it can be seen that this root condition only has very little effect on the behavior of the solution in the tip region. More precisely, one can make the following statements:

(i)   All the members of this family of solutions with different root conditions have the same regular perturbation expansion (RPE) as $\varepsilon \to 0$, with the classic needle crystal solution $\theta_*(s)$ as its leading term. Thus, the RPE part has a more profound physical significance than the exact solution itself. It remains invariant for a variety of possible root conditions.

(ii)   At any fixed point s in the region $0 \leq s < L_0$, as $\varepsilon \to 0$, correction terms, such as $|\alpha_0(\varepsilon)\tilde{h}_0(s, s_+)| = O\left(\varepsilon e^{-\frac{L(\varepsilon)}{\varepsilon}}\right)$, are transcendentally small. Such correction terms are meaningful. The root conditions (5.25) cannot be satisfied without them. Therefore, in the physical region $(0 \leq s < L_0)$ that we are most interested in, our solution will be very close to the needle solution $\theta_*(s)$ and the accompanying RPE.

(iii) On the other hand, given any $\varepsilon > 0$, one may analytically extend the solution to the nonphysical region, i. e., $s > L_0(\varepsilon)$. As $s \to \infty$, the solution will be oscillating, with a rapidly growing amplitude, since $|\alpha_0(\varepsilon)\tilde{h}_0(s, s_+)| = O\left(\sin(\frac{s}{\varepsilon})e^{\frac{s}{2}}\right)$. This is clearly consistent with the nonexistence of a classic needle solution for the Segur–Kruskal problem.

In realistic dendrite growth experiments, the physical phenomenon observed in the tip region at the later stage of growth will be determined by two classes of conditions:

(i)   The macroscopic growth conditions, which include the material properties and undercooling temperature.

(ii)   The changeable, detailed conditions of operation, the history of growth and the initial settings, such as the selection of seed and device. All these factors will be reflected in the root conditions.

For different runs of the experiments, one can set the first class of macroscopic growth conditions be the same by using the same materials and the same undercooling temperature. However, the second class of conditions, specifically the root condition for different runs will never be exactly the same. Nevertheless, according to our findings, the differences between the members in this family of solutions, in the outer regions, are transcendentally small. Such small differences are indistinguishable in experiments and these solutions may be considered in practice as identical. It is sensible, therefore, to call the entire family of nonclassic needle crystal solutions for the modified Segur–Kruskal problem 'the steady state solution'. One should note that if the system allows the classic needle solution, this classic needle solution would be a special member of the nonclassic needle solutions defined above.

5.2.4 Needle Crystal Formation Problem

Consider a more general unsteady system obtained by modifying the geometric model to include a time derivative term as follows:

$$\frac{\partial \theta}{\partial t} + \varepsilon^2 \frac{\partial^3 \theta}{\partial s^3} + \frac{\partial \theta}{\partial s} = \cos\theta \quad \left(0 \leq s < L(t, \varepsilon)\right) \tag{5.69}$$

with the boundary conditions:

(i)   The tip smoothness condition

$$\theta(0) = \theta''(0) = 0. \tag{5.70}$$

(ii) The root condition

$$\theta(L) = \theta_*(L) + O(\varepsilon)\,. \tag{5.71}$$

Assume that the total length of the needle, $L(t, \varepsilon)$ increases as its tip grows with a speed $U(t)$. So that,

$$L(t, \varepsilon) = L_0 + \int_{t_0}^{t} U(t')dt'\,. \tag{5.72}$$

In addition to the above boundary conditions one may impose various initial conditions. Hereby, we restrict ourselves in the investigation of a special class of unsteady solutions describing the needle evolution at the later stage $t \geq t_0 \gg 1$. To describe such a solution, we introduce the slow time variable

$$\tau = \varepsilon(t - t_0) \tag{5.73}$$

and assume that the speed function U can be written in the form

$$U = \bar{U}(\tau, \varepsilon)\,. \tag{5.74}$$

Thus, the governing equation is transformed to

$$\varepsilon\frac{\partial\theta}{\partial\tau} + \varepsilon^2\frac{\partial^3\theta}{\partial s^3} + \frac{\partial\theta}{\partial s} = \cos\theta \tag{5.75}$$

and the root condition is transformed to that at $s = \bar{L}(\tau, \varepsilon) = \frac{1}{\varepsilon}(C + \int_0^\tau \bar{U}(\tau')d\tau')$,

$$\theta(\bar{L}, \tau) = \theta_*(\bar{L}) + O(\varepsilon). \tag{5.76}$$

In addition, we impose the following initial condition: As $\tau = 0$,

$$\theta(s, 0, \varepsilon) = \Theta(s, \varepsilon), \tag{5.77}$$

where the initial function $\Theta(s, \varepsilon)$ is the steady solution to the above system with $\tau = 0$ in the root condition. Evidently, this function is just the steady nonclassic needle solution defined in the last section.

Note that the Ivantsov solution is the exact solution of this system in the special case $\varepsilon = 0$. As $\varepsilon \to 0$, for any fixed (s, τ) we can derive the asymptotic expansion of the solution:

$$\theta(s, \tau, \varepsilon) \sim \theta_*(s) + \varepsilon^2\theta_1(s) + \varepsilon^4\theta_2(s) + \cdots\,. \tag{5.78}$$

Therefore, one can write

$$\theta(s, \tau, \varepsilon) = \mathcal{R}_N(s, \varepsilon) + \tilde{\theta}(s, \tau, \varepsilon), \tag{5.79}$$

where $\tilde{\theta}(s, \tau, \varepsilon) \ll O(\varepsilon^{2N})$ for any large natural number N. This implies that the solution under investigation is a 'nearly' steady needle solution with a transcendentally small time-dependent tail $\tilde{\theta}(s, \tau, \varepsilon)$.

From the above discussion we see that although in some cases the classic steady needle solution for the Segur–Kruskal problem does not exist, the above-defined two kinds of solutions do exist and they both have the same regular perturbation expansion solution (RPE). We may define all these solutions as *the generalized steady state* of the system and take it as the basic state for the stability analysis. Evidently, the conventional steady state defined by the classic needle solution, if it exists, is also the generalized steady state. It would therefore be governed by the same instability mechanisms to be derived in the following chapters.

From the point of view of pattern formation, we are interested in the behavior of the 'nearly' steady solution for the needle formation problem in the limit $\tau \to \infty$. The exact solution or a uniformly valid asymptotic solution as $\varepsilon \to 0$ for the 'nearly' steady needle solutions, $\tilde{\theta}(s, \tau, \varepsilon)$, in the range of $0 \leq \tau < \infty$ are very difficult to obtain, if not impossible. But, the following three possibilities may be anticipated:

1. The solution may have a steady limit solution, hence the system tends to a fixed point in a somehow defined state space.
2. The solution may have a time-periodic limit solution, so that the system approaches a limit circle in the state space.
3. The solution may have no limit solution, and, as such, the system exhibits chaotic behavior.

Only in case (1) does a solution to the Segur–Kruskal problem exist. But the result obtained by Segur and Kruskal has ruled out this case. Achieving a good understanding of the long term behavior of the solutions of a nonlinear dynamic system is one of the basic and most difficult tasks in the broad field of nonlinear science. The present monograph certainly cannot give a full exploration of this subject. Readers who are interested in studying this general topic are referred to books such as [5.11], [5.12]. One way to tackle this problem is to study the stability of the above-defined solutions. The following chapters will proceed in this direction.

5.3 The Nonclassic Steady State of Dendritic Growth with Nonzero Surface Tension

We now turn to the original dendrite growth system. The ideas of the nonclassic steady needle solution demonstrated in the last section can be well applied to this nonlocal model.

5.3.1 The Complete Mathematical Formulation for Free Dendrite Growth

Consider the needle crystal formation problem. It describes free dendrite growth as $t \geq t_0 \gg 1$. The solution is assumed to be unsteady but is dependent on the slow time variable $\tau = \varepsilon(t - t_0)$. Accordingly, the growth speed of the needle's tip, $U(t)$, may vary slowly with time and can be written as $U = \bar{U}(\tau)$ with $\bar{U}(0) = U_0$.

The coordinate system (ξ, η) is assumed to move with the constant velocity $\bar{U}(0) = U_0$, which is used as the scale of velocity. In the moving frame, the root $\xi = \xi_{\max}(\tau, \varepsilon)$ moves backwards along the interface $\eta = 1$ with speed

$$\xi'_{\max}(\tau) = \frac{2\bar{U}(\tau)}{\eta_0^2 \xi_{\max}}.$$

Therefore,

$$\xi^2_{\max}(\tau, \varepsilon) = \frac{2(L_0 + L(\tau))}{\eta_0^2 \varepsilon},$$

where

$$L(\tau) = \int_0^\tau U(\tau')d\tau'; \quad L(0) = O(1).$$

Thus, we can write

$$\xi_{\max}(\tau, \varepsilon) = \frac{\bar{\xi}_{\max}(\tau)}{\varepsilon^{\frac{1}{2}}}.$$

We assume that $\xi_{\max} \to \infty$ as $\tau \to \infty$ or as $\varepsilon \to 0$.

The three-dimensional axially symmetric dendrite growth is then subject to the following system:

$$\left(\frac{\partial^2 T}{\partial \xi^2} + \frac{\partial^2 T}{\partial \eta^2} + \frac{1}{\xi}\frac{\partial T}{\partial \xi} + \frac{1}{\eta}\frac{\partial T}{\partial \eta} \right) = \eta_0^2 \left(\xi \frac{\partial T}{\partial \xi} - \eta \frac{\partial T}{\partial \eta} \right) + \varepsilon \eta_0^4 (\xi^2 + \eta^2) \frac{\partial T}{\partial \tau}$$

$$\left(0 \le \xi \le \xi_{\max}, \quad 0 \le \eta < \infty \right). \tag{5.80}$$

The boundary conditions are:

1. The up-stream far-field condition: as $\eta \to \infty$

$$T \to T_\infty = \frac{(T_\infty)_D - T_{M0}}{\Delta H/(c_{\mathrm{p}}\rho)} < 0. \tag{5.81}$$

2. The regularity condition: as $\eta \to 0$

$$\frac{\partial T_{\mathrm{S}}}{\partial \eta} \to 0 \quad T_{\mathrm{S}} = O(1). \tag{5.82}$$

3. The interface conditions: at $\eta = \eta_{\mathrm{s}}(\xi)$:
   (i) the thermodynamic equilibrium condition

   $$T = T_{\mathrm{S}}, \tag{5.83}$$

   (ii) the Gibbs–Thomson condition

   $$T = -\eta_0^2 \varepsilon^2 \mathcal{K} \left\{ \frac{d}{d\xi}, \frac{d^2}{d\xi^2} \right\} \eta_{\mathrm{s}}, \tag{5.84}$$

(iii) the heat balance condition

$$\left(\frac{\partial}{\partial\eta} - \eta_s'\frac{\partial}{\partial\xi}\right)(T - T_S) + \varepsilon\eta_0^4(\xi^2 + \eta^2)\frac{\partial\eta_s}{\partial\tau}$$

$$+\eta_0^2(\xi\eta_s)' = 0\,. \qquad (5.85)$$

4. The tip smoothness condition

$$\eta_s(0) = 1, \qquad \frac{\partial}{\partial\xi}\{T, T_S, \eta_s\} = 0\,. \qquad (5.86)$$

5. The root condition: at $\xi = \xi_{\max}(\tau, \varepsilon)$,

$$q \equiv \{T, T_S, \eta_s\} = \{T_*, T_{S*}, \eta_*\} + \{F(\eta, \varepsilon), G(\eta, \varepsilon), R(\varepsilon)\}$$

$$= q_* + f(\varepsilon)\,, \qquad (5.87)$$

where $f(\varepsilon) \to 0$, as $\varepsilon \to 0$.

In the root condition, we assume the functions F, G and R are sufficiently smooth functions so that the existence of solution can be guaranteed for any $\varepsilon \geq 0$. In addition to the above boundary conditions, we also assume that the initial state of the solution at $\tau = 0$ is a steady solution satisfying the above system with $\frac{\partial}{\partial\tau} = 0$.

The mathematical formulation for two-dimensional dendrite growth in the parabolic coordinate system is formally almost entirely the same as (5.80)–(5.87). The only changes are in the forms of the heat conduction equation and the curvature.

It is evident that the Ivantsov solution (4.1) is the solution to the above system for the special case $\varepsilon = 0$. For the general case $\varepsilon > 0$, for any fixed (ξ, η, τ), the solution to the above needle formation problem has the steady regular perturbation expansion discussed in Chap. 4:

$$q(\xi, \eta, \tau) \sim q_* + \varepsilon^2\bar{q}_1(\xi, \eta) + \varepsilon^4\bar{q}_2(\xi, \eta) + \cdots \quad (\varepsilon \to 0). \qquad (5.88)$$

This steady RPE solution satisfies all boundary conditions except for the unsteady root condition (5.87). Let $\mathcal{R}_N$ denote the summation of the first N terms of this regular perturbation expansion (RPE), and write

$$q = \mathcal{R}_N + \mathcal{S}_N. \qquad (5.89)$$

One can deduce that the unsteady component $\mathcal{S}(\xi, \eta, \tau, \varepsilon)$ will be smaller than $O(\varepsilon^{2N})$ for any natural number N, as $\varepsilon \to 0$. Therefore the solution q of needle crystal formation is a 'nearly' steady solution. As mentioned before, we define such a 'nearly' steady needle solution q as the nonclassic steady state of dendrite growth. This definition overcomes the difficulty encountered due to the nonexistence of a solution for the Nash–Glicksman problem. Evidently, with this definition, the steady state of dendrite growth with the inclusion of surface tension does not provide the mechanism for the selection of the tip velocity. The selection problem of dendrite growth is essentially the problem of the behavior of the solution q in the limit $\tau \to \infty$. As we pointed out in the previous section, for the dynamic system under investigation, the exact solution of the needle formation problem q, or its uniformly valid asymptotic form, for $0 \leq \tau < \infty$ is very difficult to obtain. Numerical simulation for the behavior of the solution as $\tau \to \infty$ is also very hard to perform, even with the latest, most powerful computers. Nevertheless, it will be seen in the following chapters that insofar as the long term behavior of solution has a close connection with its stability properties, the selection problem may be resolved on the basis of linear stability analysis in terms of a unified asymptotic approach.

References

5.1   G. E. Nash, and M. E. Glicksman, "Capillarity–limited Steady-State Dendritic Growth I. Theoretical Development", Acta Metall. **22**, pp. 1283–1299, (1974).

5.2   M. Kruskal and H. Segur, "Asymptotics Beyond All Orders in a Model of Crystal Growth", Stud. in Appl. Math. No. 85, pp. 129–181, (1991).

5.3   J. S. Langer, '*Lectures in the Theory of Pattern Formation*', USMG NATO AS Les Houches Session XLVI 1986 — Le hasard et la matiere/ chance and matter. Ed. by J. Souletie, J. Vannimenus and R. Stora, (Elsevier Science, Amsterdam 1986)

5.4   D. A. Kessler, J. Koplik and H. Levine, "Pattern Formation Far from Equilibrium: the free space dendritic crystal", in '*Proc. NATO A.R.W. on Patterns, Defects and Microstructures in Non-equilibrium Systems*', Austin, TX, March 1986.

5.5   P. Pelce, '*Dynamics of Curved Front*', (Academic, New York 1988).

5.6   E. A. Brener and V. I. Melnikov, "Pattern Selection in Two Dimensional Dendritic Growth", Adv. Phys. **40**, pp. 53–97, (1991).

5.7   H. Segur, S. Tanveer and H. Levine (Eds.), '*Asymptotics Beyond All Orders*', NATO ASI Series, Series B: Physics, Vol. 284, (Plenum, New York 1991).

5.8   Y. Pomeau, M. Ben Amer, "Dendrite Growth and Related Topics" in '*Solids Far From Equilibrium*', Ed. by C. Godreche, (Cambridge University Press, Cambridge, New York 1991).

5.9   J. J. Xu, "Interfacial Wave Theory of Solidification — Dendritic Pattern Formation and Selection of Tip Velocity", Phys. Rev. A15 **43**, No: 2, pp. 930–947, (1991).

5.10   J. Kevorkian, J. D. Cole, '*Multiple Scale and Singular Perturbation Methods*', Applied Mathematical Sciences, Vol. 114, (Springer, Berlin, Heidelberg 1996).

5.11   J. K. Hale, L. T. Magalhaes and W. M. Oliva '*An Introduction to Infinite Dimensional Dynamical Systems — Geometric Theory*', Series of Applied Mathematical Sciences, Vol. 47, Ed. by F. John, J. E. Marsden, L. Sirovich, (Springer, New York 1984).

5.12   S. Wiggins, '*Global Bifurcations and Chaos: Analytical Methods*', Series of Applied Mathematical Sciences, Vol. 73, Ed. by F. John, J. E. Marsden, L. Sirovich, (Springer, New York 1988).

6. Global Interfacial Wave Instability of Dendrite Growth from a Pure Melt

We now turn to study global linear stability of the steady or 'nearly' steady needle solutions q demonstrated in the last chapter, which will all be referred to as the basic states. We shall first deal with the three-dimensional axially symmetric dendrite with nonzero isotropic surface tension. The effect of anisotropy will be discussed in the next chapter. For global linear stability, one needs to investigate the evolution of infinitesimal perturbations around the basic state solutions. In fluid dynamics, there are two approaches which have been used in the study of the evolution of perturbations. The first approach is to solve an initial value problem: Assume that a given initial disturbance is introduced into the system and then consider the evolution of the initial disturbance by solving the initial value problem. The second approach is a normal mode analysis where we assume the perturbations are in the form of quasi-stationary waves and then investigate their evolution by solving an eigenvalue problem under a certain set of boundary conditions. The normal mode approach has been used to solve stability problems for various inhomogeneous dynamic systems from a broad area of physics and engineering. Famous examples include the critical layer instability theory in fluid dynamics and the density wave theory for the spiral structure of galaxies by C. C. Lin in the 1970s. See for example, [6.1] – [6.4].

We shall adopt the normal mode approach. The major aim here is to derive the so-called global mode solutions. The eigenvalue corresponding to a global mode gives both the growth rate of the amplitude of the perturbation and the frequency of the oscillation. It was first found in 1989 (e.g. [6.9], [6.12]) that in the dendritic system there exists a special simple turning point in the complex plane which plays a crucial role in understanding the dynamics of dendritic growth.

To find the global mode solutions, we shall apply the matched asymptotic expansion method. Specifically, we shall divide the entire complex plane into three regions: the outer region, the turning point region, and the tip inner region. The outer solutions in the outer region will be derived by the *multiple variables expansion* (MVE) method. These outer solutions can be interpreted as some special interfacial traveling waves, propagating along the interface of the basic state. The local dispersion relationship for these interfacial waves is obtained in the zeroth-order approximation, while the amplitude functions of the waves are determined in the first-order approximation. It is the first-order approximation that identifies the singularity in the outer solutions. Near the turning point and the leading edge of the dendrite tip, the outer solutions are invalid. Different asymptotic expansions for the exact solutions are needed. Hence we must choose proper new length scales and derive the inner solutions in the inner regions of the turning point and dendrite tip, respectively. The inner equation in the vicinity of the turning point can be reduced to the Airy equation with complex coefficients, whereas the inner solutions in the vicinity of the tip can be expressed by Hankel functions. Finally, all asymptotic expansion solutions must be matched in the intermediate regions. The global mode solutions and a quantization condition for the eigenvalues are then obtained. Given $\varepsilon > 0$, the system permits a discrete sets of complex eigenvalues $\sigma_n = (\sigma_{\rm R} - i\omega)_n$ $(n = 0, 1, 2, \cdots)$ and corresponding global wave modes [6.12]. The global instability mechanism discovered here is called the *global trapped-wave*(GTW) instability, and its presence explains the origin and persistence of the pattern formation in the solidification process. In a certain sense, the global instability mechanism discovered in dendrite growth, is similar to the so-called *over-reflection mechanism* explored in the critical layer theory of shear flow of fluid dynamics and in the density wave theory of galactic dynamics.

It will be shown that when ε equals a critical number ε_*, the system permits a uniquely determined *global neutrally stable mode*; its corresponding eigenvalue σ has a zero real part. This global neutrally stable mode will be selected at the later stage of growth. Therefore the stability criterion $\varepsilon = \varepsilon_*$ is also the selection criterion for the tip speed.

6.1 Linear Perturbed System Around the Basic State of Three-Dimensional Dendrite Growth

We write the basic state in the form:

$$T_{\rm B}(\xi, \eta, \tau, \varepsilon) = T_*(\eta) + O(\varepsilon^2)$$

$$T_{\rm SB}(\xi, \eta, \tau, \varepsilon) = O(\varepsilon^2) \tag{6.1}$$

$$\eta_{\rm B}(\xi, \tau, \varepsilon) = 1 + O(\varepsilon^2)$$

and separate the general unsteady solutions into two parts:

$$T = T_{\rm B} + \tilde{T}(\xi, \eta, t, \varepsilon)$$

$$T_{\rm S} = T_{\rm SB} + \tilde{T}_{\rm S}(\xi, \eta, t, \varepsilon) \tag{6.2}$$

$$\eta_{\rm s} = \eta_{\rm B} + \tilde{h}(\xi, t, \varepsilon)/\eta_0^2 .$$

Assume that the above perturbations around the basic state are caused by initially infinitesimal perturbations with a characteristic amplitude $\delta \ll 1$. Thus, within a sufficiently short time period, we will have $|\tilde{q}| =: \{|\tilde{T}|, |\tilde{h}|\} \ll \{|T_\mathrm{B}|, \eta_\mathrm{B}\}$. Hence, the linearization around the basic state solution in the small amplitude parameter $\delta \ll 1$ is applicable. The linearized perturbed system must be a homogeneous system. It is shown below that

$$\left(\frac{\partial^2 \tilde{T}}{\partial \xi^2} + \frac{\partial^2 \tilde{T}}{\partial \eta^2} + \frac{1}{\xi}\frac{\partial \tilde{T}}{\partial \xi} + \frac{1}{\eta}\frac{\partial \tilde{T}}{\partial \eta}\right) - \eta_0^2\left(\xi\frac{\partial \tilde{T}}{\partial \xi} - \eta\frac{\partial \tilde{T}}{\partial \eta}\right) -$$

$$-\eta_0^4(\xi^2 + \eta^2)\frac{\partial \tilde{T}}{\partial t} = 0,$$

$$\left(0 \leq \xi \leq \xi_\mathrm{max};\ 0 \leq \eta < \infty\right) \tag{6.3}$$

with the boundary conditions:

1. As $\eta \to \infty$

$$\tilde{T} \to 0. \tag{6.4}$$

2. As $\eta \to 0$

$$\tilde{T}_\mathrm{S} = O(1). \tag{6.5}$$

3. The interface conditions: making the Taylor expansions around the interface of the basic state, it follows that, for the case of isotropic surface tension, at $\eta = \eta_\mathrm{B}(\xi, \tau, \varepsilon)$:

   (i)

$$\tilde{T} - \tilde{T}_\mathrm{S} = -\left\{\frac{\partial T_\mathrm{B}}{\partial \eta} - \frac{\partial T_\mathrm{SB}}{\partial \eta}\right\}\frac{\tilde{h}}{\eta_0^2}, \tag{6.6}$$

   (ii)

$$\tilde{T}_\mathrm{S} + \frac{\partial T_\mathrm{SB}}{\partial \eta}\frac{\tilde{h}}{\eta_0^2} = \frac{\varepsilon^2}{\hat{S}(\xi)}\left\{\frac{\partial^2 \tilde{h}}{\partial \xi^2} + \frac{(1 + 2\xi^2)}{\xi\hat{S}^2(\xi)}\frac{\partial \tilde{h}}{\partial \xi} - \frac{\tilde{h}}{\hat{S}^2}\right\}, \tag{6.7}$$

   (iii)

$$\frac{\partial}{\partial \eta}\left(\tilde{T} - \tilde{T}_\mathrm{S}\right) + \eta_0^2\hat{S}^2(\xi)\frac{\partial \tilde{h}}{\partial t} + \xi\frac{\partial \tilde{h}}{\partial \xi} + \tilde{h} + \left\{\frac{\partial^2}{\partial \eta^2}(T_\mathrm{B} - T_\mathrm{SB})\right\}\frac{\tilde{h}}{\eta_0^2}$$

$$-\frac{1}{\eta_0^2}\frac{\partial \tilde{h}}{\partial \xi}\frac{\partial}{\partial \xi}(T_\mathrm{B} - T_\mathrm{SB}) - \frac{\eta_\mathrm{B}'\tilde{h}}{\eta_0^2}\frac{\partial^2}{\partial \xi \partial \eta}(T_\mathrm{B} - T_\mathrm{SB})$$

$$-\eta_\mathrm{B}'\frac{\partial}{\partial \xi}\left(\tilde{T} - \tilde{T}_\mathrm{S}\right) = 0, \tag{6.8}$$

   where

$$\hat{S}(\xi) = \sqrt{\xi^2 + \eta_\mathrm{B}^2}. \tag{6.9}$$

4. The root condition: at $\xi = \xi_{\max}(\tau, \varepsilon)$,

$$\{\tilde{T}, \tilde{T}_{\mathrm{S}}, \tilde{h}\} = 0. \tag{6.10}$$

5. The tip smoothness condition: at $\xi = 0$, $\eta = \eta_{\mathrm{B}}(0, \tau, \varepsilon)$,

$$\frac{\partial}{\partial \xi}\{\tilde{T}, \tilde{T}_{\mathrm{S}}, \tilde{h}\} = 0. \tag{6.11}$$

The above system (6.3)–(6.11) contains two parameters ε, and η_0^2. It leads to a linear eigenvalue problem when one looks for solutions of the type $\tilde{q} = \hat{q}e^{\sigma t}$. The eigenvalue σ must be a function of (η_0^2, ε). We shall solve this eigenvalue problem in two steps. We first solve the system (6.3)–(6.10) for any given $(\sigma, \eta_0^2, \varepsilon)$. In doing so, we shall apply the multiple variables expansion (MVE) method to look for a uniformly valid asymptotic solution in the limit $\varepsilon \to 0$. The solution should satisfy all boundary conditions except the tip condition (6.11). Then, we apply the tip condition (6.11) to the asymptotic solution obtained. Thus, the parameter σ must be chosen as a function of η_0^2 and ε.

When finding the asymptotic solutions $\tilde{q}(\xi, \eta, t, \varepsilon)$, it is first seen that the above system has a singularity at the tip $\xi = 0$. So one must first single out the tip region $(|\xi| \ll 1)$. Some other singularity points may be found later. The inner regions around these singular points also need to be separated. We call the remaining region the outer region, and treat the outer solution first.

6.2 Outer Solution in the Outer Region away from the Tip

In the outer region, $|\xi| \gg \varepsilon$, we shall look for the asymptotic expansion of the solutions $\tilde{q}(\xi, \eta, t, \varepsilon)$ in terms of the multiple variables expansion method, similar to our approach in Chap. 2. The dynamic system under investigation is now inhomogeneous with variable coefficients. Thus, we must now use a set of stretched fast variables (ξ_+, η_+, t_+) defined as

$$\xi_+ = \frac{1}{\varepsilon} \int_{\xi_0}^{\xi} k(\xi', \eta, \varepsilon)\mathrm{d}\xi' \tag{6.12}$$

$$\eta_+ = \frac{1}{\varepsilon} \int_{1}^{\eta} g(\xi, \eta', \varepsilon)\mathrm{d}\eta' \tag{6.13}$$

$$t_+ = \frac{t}{\eta_0^2 \varepsilon}, \tag{6.14}$$

where the lower limit of the integral, ξ_0, will be specified later. Note that, in general, the wave number functions k and g may be functions of ξ and η.

Moreover, for solutions in the solid phase one may need to adopt a different function g_s instead of g. Assume that the contours $k = k(\xi, \eta) = $ const. and $g = g(\xi, \eta) = $ const. are smooth curves in the (ξ, η)-plane. Along these contours, one has

$$d\{k(\xi, \eta)\} = 0, \quad d\{g(\xi, \eta)\} = 0. \tag{6.15}$$

Thus

$$\frac{\partial k}{\partial \eta} = m \frac{\partial k}{\partial \xi}, \quad \frac{\partial g}{\partial \eta} = s \frac{\partial g}{\partial \xi}, \tag{6.16}$$

where m and s, with the opposite sign, represent the slope $\left(\frac{d\xi}{d\eta}\right)$ of the contours $k = $ const. and $g = $ const., respectively. In general, m and s may be regular functions of both variables ξ, and η and depend on the parameter ε. One may expand these functions in the asymptotic form

$$m = m_0 + m_1 \varepsilon + \cdots \tag{6.17}$$

$$s = s_0 + s_1 \varepsilon + \cdots, \tag{6.18}$$

as $\varepsilon \to 0$. In the leading order approximation, we shall set m_0, s_0 to be constants. At the interface $\eta = 1, \eta_+ = 0$, we shall define

$$\xi_+ = \hat{\xi}_+ = \frac{1}{\varepsilon} \int_{\xi_0}^{\xi} \hat{k}(\xi', \varepsilon) d\xi' \tag{6.19}$$
$$\hat{k}(\xi, \varepsilon) = k(\xi, 1, \varepsilon).$$

In terms of the above-defined multiple variables, we make the following MVE for the perturbed states:

$$\tilde{T} = \left\{ \tilde{T}_0(\xi, \eta, \xi_+, \eta_+) + \varepsilon \tilde{T}_1(\xi, \eta, \xi_+, \eta_+) + \cdots \right\} e^{\sigma t_+}$$

$$\tilde{T}_S = \left\{ \tilde{T}_{S0}(\xi, \eta, \xi_+, \eta_+) + \varepsilon \tilde{T}_{S1}(\xi, \eta, \xi_+, \eta_+) + \cdots \right\} e^{\sigma t_+}$$

$$\tilde{h} = \left\{ \tilde{h}_0(\xi, \hat{\xi}_+) + \varepsilon \tilde{h}_1(\xi, \hat{\xi}_+) + \cdots \right\} e^{\sigma t_+}$$

$$k = k_0 + \varepsilon k_1 + \varepsilon^2 k_2 + \cdots$$

$$g = k_0 + \varepsilon g_1 + \varepsilon^2 g_2 + \cdots \tag{6.20}$$

$$g_s = k_0 + \varepsilon g_{s1} + \varepsilon^2 g_{s2} + \cdots$$

$$\hat{k} = \hat{k}_0 + \varepsilon \hat{k}_1 + \varepsilon^2 \hat{k}_2 + \cdots$$

$$\sigma = \sigma_0 + \varepsilon \sigma_1 + \varepsilon^2 \sigma_2 + \cdots.$$

Here,

$$\hat{k}_0(\xi) = k_0(\xi, 1), \quad \hat{k}_1(\xi) = k_1(\xi, 1), \quad \hat{k}_2(\xi) = k_2(\xi, 1), \cdots \tag{6.21}$$

and $\sigma = \sigma_R - i\omega \;\; (\omega \geq 0)$ is, generally, a complex number. In the first step, we set σ_0 as an arbitrary constant. As before, the fast and slow variables $(\xi, \eta, \xi_+, \eta_+, \, t_+)$ are treated formally as independent variables. Hence, one must make the following replacements for the derivatives in the system (6.3)–(6.8):

$$
\begin{aligned}
\frac{\partial}{\partial \xi} &\Longrightarrow \frac{k}{\varepsilon}\frac{\partial}{\partial \xi_+} + \frac{\partial}{\partial \xi} \\
\frac{\partial}{\partial \eta} &\Longrightarrow \frac{g}{\varepsilon}\frac{\partial}{\partial \eta_+} + \frac{\partial}{\partial \eta} \\
\frac{\partial^2}{\partial \xi^2} &\Longrightarrow \left\{\frac{k}{\varepsilon}\frac{\partial}{\partial \xi_+} + \frac{\partial}{\partial \xi}\right\}^2 \\
\frac{\partial^2}{\partial \eta^2} &\Longrightarrow \left\{\frac{g}{\varepsilon}\frac{\partial}{\partial \eta_+} + \frac{\partial}{\partial \eta}\right\}^2 .
\end{aligned}
\tag{6.22}
$$

The converted system with the multiple variables is as follows:

$$
\begin{aligned}
\left(k^2 \frac{\partial^2}{\partial \xi_+^2} + g^2 \frac{\partial^2}{\partial \eta_+^2}\right)\tilde{T} &= \varepsilon \eta_0^2 (\xi^2 + \eta^2)\frac{\partial \tilde{T}}{\partial t_+} - \varepsilon^2 \left(\frac{\partial^2}{\partial \xi^2} + \frac{\partial^2}{\partial \eta^2}\right)\tilde{T} \\
&\quad + \varepsilon \eta_0^2 \xi \left(k\frac{\partial}{\partial \xi_+} + \varepsilon\frac{\partial}{\partial \xi}\right)\tilde{T} - \varepsilon \eta_0^2 \eta \left(g\frac{\partial}{\partial \eta_+} + \varepsilon\frac{\partial}{\partial \eta}\right)\tilde{T} \\
&\quad - \frac{\varepsilon}{\xi}\left(k\frac{\partial}{\partial \xi_+} + \varepsilon\frac{\partial}{\partial \xi}\right)\tilde{T} - \frac{\varepsilon}{\eta}\left(g\frac{\partial}{\partial \eta_+} + \varepsilon\frac{\partial}{\partial \eta}\right)\tilde{T} \\
&\quad - \varepsilon \left(2k\frac{\partial^2}{\partial \xi \partial \xi_+} + 2g\frac{\partial^2}{\partial \eta \partial \eta_+} + \frac{\partial k}{\partial \xi}\frac{\partial}{\partial \xi_+} + \frac{\partial g}{\partial \eta}\frac{\partial}{\partial \eta_+}\right)\tilde{T}
\end{aligned}
\tag{6.23}
$$

and

$$
\begin{aligned}
\left(k^2 \frac{\partial^2}{\partial \xi_+^2} + g_s^2 \frac{\partial^2}{\partial \eta_+^2}\right)\tilde{T}_S &= \varepsilon \eta_0^2 \left(\xi^2 + \eta^2\right)\frac{\partial \tilde{T}_S}{\partial t_+} - \varepsilon^2 \left(\frac{\partial^2}{\partial \xi^2} + \frac{\partial^2}{\partial \eta^2}\right)\tilde{T}_S \\
&\quad + \varepsilon \eta_0^2 \xi \left(k\frac{\partial}{\partial \xi_+} + \varepsilon\frac{\partial}{\partial \xi}\right)\tilde{T}_S - \varepsilon \eta_0^2 \eta \left(g_s\frac{\partial}{\partial \eta_+} + \varepsilon\frac{\partial}{\partial \eta}\right)\tilde{T}_S \\
&\quad - \frac{\varepsilon}{\xi}\left(k\frac{\partial}{\partial \xi_+} + \varepsilon\frac{\partial}{\partial \xi}\right)\tilde{T}_S - \frac{\varepsilon}{\eta}\left(g_s\frac{\partial}{\partial \eta_+} + \varepsilon\frac{\partial}{\partial \eta}\right)\tilde{T}_S \\
&\quad - \varepsilon \left(2k\frac{\partial^2}{\partial \xi \partial \xi_+} + 2g_s\frac{\partial^2}{\partial \eta \partial \eta_+} + \frac{\partial k}{\partial \xi}\frac{\partial}{\partial \xi_+} + \frac{\partial g_s}{\partial \xi}\frac{\partial}{\partial \eta_+}\right)\tilde{T}_S .
\end{aligned}
\tag{6.24}
$$

The boundary conditions are:

1. As $\eta_+ \to \infty$

$$
\tilde{T} \to 0 .
\tag{6.25}
$$

2. As $\eta_+ \to -\infty$

$$\tilde{T}_S \to 0 \, . \tag{6.26}$$

3. At the interface: $\eta_+ = 0, \eta = 1$,

(i) the thermodynamic equilibrium condition

$$\tilde{T} = \tilde{T}_S + \tilde{h} + O(\varepsilon^2) \, , \tag{6.27}$$

(ii) the Gibbs-Thomson condition

$$\tilde{T}_S = \frac{1}{S(\xi)} \left\{ \left(k^2 \frac{\partial^2}{\partial \xi_+^2} + 2\varepsilon k \frac{\partial^2}{\partial \xi_+ \partial \xi} + \varepsilon \frac{\partial k}{\partial \xi} \frac{\partial}{\partial \xi_+} + \varepsilon^2 \frac{\partial^2}{\partial \xi^2} \right) \right.$$

$$\left. + \varepsilon \left(\frac{1}{\xi} + \frac{\xi}{S^2(\xi)} \right) \left(k \frac{\partial}{\partial \xi_+} + \varepsilon \frac{\partial}{\partial \xi} \right) - \frac{\varepsilon^2}{S^2(\xi)} \right\} \tilde{h} + O(\varepsilon^2) \, , \tag{6.28}$$

(iii) the heat balance condition

$$\left(g \frac{\partial \tilde{T}}{\partial \eta_+} + \varepsilon \frac{\partial \tilde{T}}{\partial \eta} \right) - \left(g_s \frac{\partial \tilde{T}_S}{\partial \eta_+} + \varepsilon \frac{\partial \tilde{T}_S}{\partial \eta} \right) + S^2(\xi) \frac{\partial \tilde{h}}{\partial t_+}$$

$$+ \xi \left(k \frac{\partial \tilde{h}}{\partial \xi_+} + \varepsilon \frac{\partial \tilde{h}}{\partial \xi} \right) + \varepsilon \left(2 + \eta_0^2 \right) \tilde{h} + O(\varepsilon^2) = 0 \, . \tag{6.29}$$

4. The root condition: at $\xi = \xi_{\max}$ one should have

$$\{ \tilde{T}; \, \tilde{T}_S; \, \tilde{h} \} = 0 \, . \tag{6.30}$$

By substituting (6.20) into the system (6.23)–(6.30), one can successively derive each order of approximation.

6.2.1 Zeroth-Order Approximation

As zeroth-order approximation, we derive the following quasi-steady heat conduction equation:

$$\left(\frac{\partial^2}{\partial \xi_+^2} + \frac{\partial^2}{\partial \eta_+^2} \right) \tilde{T}_0 = 0$$

$$\left(\frac{\partial^2}{\partial \xi_+^2} + \frac{\partial^2}{\partial \eta_+^2} \right) \tilde{T}_{S0} = 0 \tag{6.31}$$

with the following boundary conditions:

1. As $\eta_+ \to \infty$

$$\tilde{T}_0 \to 0 \, . \tag{6.32}$$

2. As $\eta_+ \to -\infty$

$$\tilde{T}_{S0} \to 0 \,. \tag{6.33}$$

3. At the interface: $\eta_+ = 0$, $\eta = 1$,

$$\tilde{T}_0 = \tilde{T}_{S0} + \tilde{h}_0 \tag{6.34}$$

$$\tilde{T}_{S0} = \frac{\hat{k}_0^2}{S(\xi)} \frac{\partial^2 \tilde{h}_0}{\partial \hat{\xi}_+^2} \tag{6.35}$$

$$\hat{k}_0 \frac{\partial}{\partial \eta_+}(\tilde{T}_0 - \tilde{T}_{S0}) + \sigma_0 S^2(\xi)\tilde{h}_0 + \hat{k}_0 \xi \frac{\partial \tilde{h}_0}{\partial \hat{\xi}_+} = 0 \,; \tag{6.36}$$

moreover, we have

4. In the far-field, $\xi \to \infty$,

$$\left\{ \tilde{T}_0; \ \tilde{T}_{S0}; \ \tilde{h}_0 \right\} = 0 \,. \tag{6.37}$$

The above system has the following normal mode solutions:

$$\tilde{T}_0 = A_0(\xi, \eta) \exp\left\{ i\xi_+ - \eta_+ \right\}$$

$$\tilde{T}_{S0} = A_{S0}(\xi, \eta) \exp\left\{ i\xi_+ + \eta_+ \right\} \tag{6.38}$$

$$\tilde{h}_0 = \hat{D}_0 \exp\left\{ i\hat{\xi}_+ \right\} \,,$$

where the coefficient $\hat{D}_0$ is set as a constant. Setting

$$\begin{cases} \hat{A}_0(\xi) = A_0(\xi, 1) \\ \hat{A}_{S0}(\xi) = A_{S0}(\xi, 1) \,, \end{cases} \tag{6.39}$$

then from (6.34)–(6.36) we derive the following system of homogeneous equations:

$$\begin{cases} \hat{A}_0 - \hat{A}_{S0} - \hat{D}_0 = 0 \\ \hat{A}_{S0} + \frac{\hat{k}_0^2}{S}\hat{D}_0 = 0 \\ -\hat{k}_0(\hat{A}_0 + \hat{A}_{S0}) + (\sigma_0 S^2 + i\xi\hat{k}_0)\hat{D}_0 = 0 \,. \end{cases} \tag{6.40}$$

Obviously, for a nontrivial solution, one must have

$$\Delta = \det \begin{pmatrix} 1 & -1 & -1 \\ 0 & 1 & \hat{k}_0^2/S \\ -\hat{k}_0 & -\hat{k}_0 & \sigma_0 S^2 + i\xi\hat{k}_0 \end{pmatrix} = 0 \,, \tag{6.41}$$

which gives the local dispersion relation

$$\sigma_0 = \Sigma(\xi, \hat{k}_0) = \frac{\hat{k}_0}{S^2}\left(1 - \frac{2\hat{k}_0^2}{S}\right) - \frac{i\xi}{S^2}\hat{k}_0.$$

(6.42)

Then, one solves

$$\begin{cases} \hat{A}_0 + \hat{A}_{S0} = \left(1 - \frac{2\hat{k}_0^2}{S}\right)\hat{D}_0 \\ \hat{A}_0 - \hat{A}_{S0} = \hat{D}_0. \end{cases}$$

(6.43)

Remember that in the above, for simplicity, we have assumed that the thermal characteristic constants for both phases are the same (i.e., the so-called symmetrical model). As a matter of fact, for the more general case where, except for the thermal diffusivities, all other thermal constants are assumed to be the same for both liquid and solid phase, the parameter

$$\hat{\alpha} = \frac{\kappa_{TS}}{\kappa_T} \geq 0$$

(6.44)

is arbitrary, one can easily obtain the following similar local dispersion relation:

$$\sigma_0 = \Sigma(\xi, \hat{k}_0) = \frac{\hat{k}_0}{S^2}\left(1 - \frac{(1+\hat{\alpha})\hat{k}_0^2}{S}\right) - \frac{i\xi}{S^2}\hat{k}_0 .$$

(6.45)

The local dispersion formulas (6.42) and (6.45) obtained above are the generalization of the well-known Mullins–Sekerka formula in the unidirectional solidification. For any fixed parameter σ_0, one can solve for three wave numbers as functions of ξ (Fig. 6.1). Namely

$$\begin{cases} \hat{k}_0^{(1)}(\xi) = M(\xi)\cos\left\{\frac{1}{3}\cos^{-1}\left(\frac{\sigma_0}{N(\xi)}\right)\right\} & \text{(short-wave branch)} \\ \hat{k}_0^{(2)}(\xi) = M(\xi)\cos\left\{\frac{1}{3}\cos^{-1}\left(\frac{\sigma_0}{N(\xi)}\right) + \frac{2\pi}{3}\right\} \\ \hat{k}_0^{(3)}(\xi) = M(\xi)\cos\left\{\frac{1}{3}\cos^{-1}\left(\frac{\sigma_0}{N(\xi)}\right) + \frac{4\pi}{3}\right\} & \text{(long-wave branch)}, \end{cases}$$

(6.46)

where

$$\begin{cases} M(\xi) = \sqrt{\frac{2S(\xi)}{3}}\,(1 - i\xi)^{\frac{1}{2}} \\ N(\xi) = -\frac{M(\xi)}{3S^2(\xi)}\,(1 - i\xi). \end{cases}$$

(6.47)

Given σ_0, the system allows three fundamental wave solutions corresponding to the wave number functions $\{\hat{k}_0^{(1)}, \hat{k}_0^{(2)}, \hat{k}_0^{(3)}\}$:

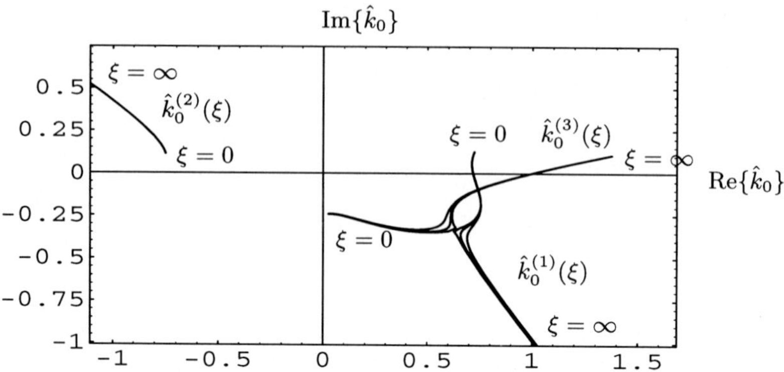

Fig. 6.1. The variations of the wave number functions $\{\hat{k}_0^{(1)}(\xi), \hat{k}_0^{(2)}(\xi), \hat{k}_0^{(3)}(\xi)\}$ in the complex $\hat{k}$-plane with ξ for given $\sigma_0 = (0.035, -0.265)$; $(0.035, -0.269)$; $(0.035, -0.270)$

H_1 wave (short wave branch)   with larger $\Re\{\hat{k}_0^{(1)}\} > 0$,

H_2 wave (no physical meaning)   with $\Re\{\hat{k}_0^{(2)}\} < 0$,

H_3 wave (long wave branch)   with smaller $\Re\{\hat{k}_0^{(3)}\} > 0$.

Among these solutions, the H_2 wave solution must be ruled out because the real part of its wave-number is negative; hence, as $\eta \to \infty$, its corresponding perturbed temperature field $\tilde{T}$ will grow exponentially, violating the boundary condition (6.32).

As a result, in the zeroth-order approximation, the general H wave solution is

$$\tilde{h} \approx D_1 \mathrm{e}^{\sigma_0 t_+ + \mathrm{i}\hat{\xi}_+^{(1)}} + D_3 \mathrm{e}^{\sigma_0 t_+ + \mathrm{i}\hat{\xi}_+^{(3)}}, \tag{6.48}$$

where

$$\hat{\xi}_+^{(1)} = \frac{1}{\varepsilon} \int_{\xi_0}^{\xi} \hat{k}_0^{(1)} \mathrm{d}\xi_1$$

$$\hat{\xi}_+^{(3)} = \frac{1}{\varepsilon} \int_{\xi_0}^{\xi} \hat{k}_0^{(3)} \mathrm{d}\xi_1 \tag{6.49}$$

and the coefficients $\{D_1, D_3\}$ are arbitrary constants independent of ε which are to be determined.

6.2.2 First-Order Approximation

The first-order approximation solution will determine the amplitude functions $A_0(\xi, \eta)$, $A_{S0}(\xi, \eta)$, the functions $k_0(\xi, \eta)$, $k_1(\xi, 1)$, and m_0, σ_1. Our major concern is σ_1. The equation for the first-order approximation can be obtained from (6.23) as follows:

$$k_0^2\left(\frac{\partial^2}{\partial\xi_+^2} + \frac{\partial^2}{\partial\eta_+^2}\right)\tilde{T}_1 = a_0 e^{\mathrm{i}\xi_+ - \eta_+}$$

$$k_0^2\left(\frac{\partial^2}{\partial\xi_+^2} + \frac{\partial^2}{\partial\eta_+^2}\right)\tilde{T}_{S1} = a_{S0}e^{\mathrm{i}\xi_+ + \eta_+}\,,$$

(6.50)

where

$$\left\{\begin{aligned}
a_0 &= 2k_0\left(\frac{\partial A_0}{\partial\eta} - \mathrm{i}\frac{\partial A_0}{\partial\xi}\right) + A_0\Big\{\sigma_0\eta_0^2(\xi^2+\eta^2) + k_0\eta_0^2(\mathrm{i}\xi + \eta)\\
&\quad + \frac{k_0}{\eta} - \mathrm{i}\frac{k_0}{\xi} - \mathrm{i}\frac{\partial k_0}{\partial\xi} + \frac{\partial k_0}{\partial\eta} + 2k_0(k_1 - g_1)\Big\}\\
a_{S0} &= -2k_0\left(\frac{\partial A_{S0}}{\partial\eta} + \mathrm{i}\frac{\partial A_{S0}}{\partial\xi}\right) + A_{S0}\Big\{\sigma_0\eta_0^2(\xi^2+\eta^2) + k_0\eta_0^2(\mathrm{i}\xi - \eta)\\
&\quad - \frac{k_0}{\eta} - \mathrm{i}\frac{k_0}{\xi} - \mathrm{i}\frac{\partial k_0}{\partial\xi} - \frac{\partial k_0}{\partial\eta} + 2k_0(k_1 - g_{s1})\Big\}\,.
\end{aligned}\right.$$

(6.51)

To ensure the uniform validity of the expansions as $\xi \to \infty$, one must eliminate the secular terms on the right-hand side of (6.50). We thus set

$$a_0 = a_{S0} = 0\,.$$

(6.52)

For simplicity, one may take $g_1 = g_{s1} = k_1$. From (6.52), it follows that

$$\left(\frac{\partial}{\partial\eta} - \mathrm{i}\frac{\partial}{\partial\xi}\right)\ln\Psi(\xi,\eta) = -\frac{\eta_0^2}{2}(\eta + \mathrm{i}\xi) - \frac{\sigma_0\eta_0^2}{2k_0}(\xi^2 + \eta^2)$$

$$\left(\frac{\partial}{\partial\eta} + \mathrm{i}\frac{\partial}{\partial\xi}\right)\ln\Phi(\xi,\eta) = -\frac{\eta_0^2}{2}(\eta - \mathrm{i}\xi) + \frac{\sigma_0\eta_0^2}{2k_0}(\xi^2 + \eta^2)\,,$$

(6.53)

where

$$\Psi(\xi,\eta) = A_0(\xi,\eta)k_0^{\frac{1}{2}}\xi^{\frac{1}{2}}\eta^{\frac{1}{2}}$$

$$\Phi(\xi,\eta) = A_{S0}(\xi,\eta)k_0^{\frac{1}{2}}\xi^{\frac{1}{2}}\eta^{\frac{1}{2}}\,.$$

(6.54)

Note that (6.16) and (6.54) are both first-order hyperbolic equations. Once m_0 is known, the wave number function $k_0(\xi,\eta)$ and the functions $A_0(\xi,\eta)$, $A_{S0}(\xi,\eta)$ can be solved in the ξ–η plane as an initial value problem given the initial values on the curve $\eta = 1$.

By adding a_0 and a_{S0}, one also derives

$$\begin{aligned}
\hat{D}_0 Q_0 &= \frac{\partial}{\partial\eta}(A_0 - A_{S0})\bigg|_{\eta=1}\\
&= \mathrm{i}\frac{\partial}{\partial\xi}(\hat{A}_0 + \hat{A}_{S0}) - (\hat{A}_0 + \hat{A}_{S0})\left[\frac{\sigma_0\eta_0^2}{2\hat{k}_0}S^2(\xi) + \frac{\mathrm{i}}{2}\left(\xi\eta_0^2 - \frac{1}{\xi}\right.\right.\\
&\quad \left.\left. - \frac{\partial\ln\hat{k}_0}{\partial\xi}\right)\right] - (\hat{A}_0 - \hat{A}_{S0})\left[\frac{1+\eta_0^2}{2} + \frac{1}{2\hat{k}_0}\frac{\partial k_0}{\partial\eta}\bigg|_{\eta=1}\right]
\end{aligned}$$

(6.55)

where

$$\left.\frac{\partial k_0}{\partial \eta}\right|_{\eta=1} = m_0 \hat{k}'_0 \,. \tag{6.56}$$

This formula will be needed later to solve for $k_1(\xi, 1)$. In terms of conditions (6.52), one obtains the first-order approximate solutions:

$$\tilde{T}_1 = A_1(\xi, \eta) \exp\left\{i\xi_+ - \eta_+\right\}$$

$$\tilde{T}_{S1} = A_{S1}(\xi, \eta) \exp\left\{i\xi_+ + \eta_+\right\} \tag{6.57}$$

$$\tilde{h}_1 = \hat{D}_1 \exp\left\{i\hat{\xi}_+\right\}.$$

Setting

$$\begin{cases} \hat{A}_1(\xi) = A_1(\xi, 1) \\ \hat{A}_{S1}(\xi) = A_{S1}(\xi, 1), \end{cases} \tag{6.58}$$

from (6.27)–(6.29) one can derive the first-order approximate interface boundary conditions, and find that

$$\begin{cases} \hat{A}_1 - \hat{A}_{S1} - \hat{D}_1 = 0 \\ \hat{A}_{S1} + \frac{\hat{k}_0^2}{S}\hat{D}_1 = I_2 \hat{D}_0 \\ -\hat{k}_0(\hat{A}_1 + \hat{A}_{S1}) + (\sigma_0 S^2 + i\xi\hat{k}_0)\hat{D}_1 = I_3 \hat{D}_0, \end{cases} \tag{6.59}$$

where we define

$$I_2 = \frac{1}{S}\left\{ i\hat{k}'_0 + i\hat{k}_0\left(\frac{1}{\xi} + \frac{\xi}{S^2}\right) - 2\hat{k}_0\hat{k}_1 \right\} \tag{6.60}$$

and

$$I_3 \hat{D}_0 = \hat{k}_1(\hat{A}_0 + \hat{A}_{S0}) - (\sigma_1 S^2 + i\xi\hat{k}_1)\hat{D}_0$$

$$- Q_0\hat{D}_0 - (2 + \eta_0^2)\hat{D}_0\,. \tag{6.61}$$

Substituting (6.43) and (6.55) in (6.60) and (6.61) for $\hat{A}_0, \hat{A}_{S0}$ and Q_0, we obtain

$$I_3 = -S^2\sigma_1 + \hat{k}_1\left(1 - i\xi - \frac{2\hat{k}_0^2}{S}\right) - \left(1 + \frac{\eta_0^2}{2}\right) - \frac{i}{2\xi}\left(1 - i\xi - \frac{2\hat{k}_0^2}{S}\right)$$

$$+ \frac{\eta_0^2}{2}\left(1 - \frac{2\hat{k}_0^2}{S}\right)^2 - i\frac{2\xi\hat{k}_0^2}{S^3} + i\frac{5\hat{k}_0\hat{k}'_0}{S} + \frac{m_0 - i}{2}\frac{\hat{k}'_0}{\hat{k}_0}\,. \tag{6.62}$$

The determinant Δ of the coefficient matrix of the above inhomogeneous system is zero. The necessary condition for the existence of a nontrivial solution for $\{\hat{A}_1, \hat{A}_{S1}, \hat{D}_1\}$ is:

$$\det \begin{pmatrix} 1 & -1 & 0 \\ 0 & 1 & I_2 \\ -\hat{k}_0 & -\hat{k}_0 & I_3 \end{pmatrix} = 0, \tag{6.63}$$

which leads to the solvability condition:

$$I_3 + 2\hat{k}_0 I_2 = 0. \tag{6.64}$$

From (6.64), we obtain

$$\hat{k}_1 \left(1 - \mathrm{i}\xi - \frac{6\hat{k}_0^2}{S} \right) = S^2 \sigma_1 + \left(1 + \frac{\eta_0^2}{2} \right) + \frac{\mathrm{i}}{2\xi} \left(1 - \mathrm{i}\xi - \frac{6\hat{k}_0^2}{S} \right)$$
$$- \frac{\eta_0^2}{2} \left(1 - \frac{2\hat{k}_0^2}{S} \right)^2 - \frac{m_0 - \mathrm{i}}{2} \frac{k_0'}{k_0} - \mathrm{i}\frac{7\hat{k}_0 \hat{k}_0'}{S} \tag{6.65}$$

or

$$\hat{k}_1 = \frac{\mathrm{i}}{2\xi} + \frac{R_1(\xi)}{F(\xi)} + \frac{R_2(\xi)}{F(\xi)} \frac{\hat{k}_0'}{\hat{k}_0}, \tag{6.66}$$

where we denote

$$F(\xi) = 1 - \mathrm{i}\xi - \frac{6\hat{k}_0^2}{S}$$

$$R_1(\xi) = S^2 \sigma_1 + \left(1 + \frac{\eta_0^2}{2} \right) - \frac{\eta_0^2}{2} \left(1 - \frac{2\hat{k}_0^2}{S} \right)^2 \tag{6.67}$$

$$R_2(\xi) = -\left(\frac{m_0 - \mathrm{i}}{2} + \mathrm{i}\frac{7\hat{k}_0^2}{S} \right).$$

From the dispersion formula (6.42), one finds that

$$F(\xi) = S^2 \left(\frac{\partial \Sigma}{\partial \hat{k}_0} \right). \tag{6.68}$$

Therefore, it is seen that at the root ξ_c of the equation:

$$\left(\frac{\partial \Sigma}{\partial \hat{k}_0} \right) = 0 \quad (\text{or} \quad F(\xi) = 0), \tag{6.69}$$

the solution $\hat{k}_1$, as well as the functions A_0 and A_{S0}, has a singularity. Note that given σ_0, from the local dispersion formula (6.42) we have

$$\frac{\hat{k}_0'(\xi)}{\hat{k}_0(\xi)} = \frac{R_0(\xi)}{F(\xi)}$$
$$R_0(\xi) = \mathrm{i} + \frac{2\xi(1 - \mathrm{i}\xi)}{S^2} - \frac{6\xi\hat{k}_0^2}{S^3}. \tag{6.70}$$

As $\xi \to \xi_c$, we have $F(\xi) \to 0$ and

$$R_0(\xi_c) = \frac{1}{\xi_c - \mathrm{i}}.$$

Moreover,

$$F'(\xi) = -\mathrm{i} + \frac{6\xi\hat{k}_0^2}{S^3} - \frac{12\hat{k}_0^2}{S}\left(\frac{\hat{k}_0'}{\hat{k}_0}\right) = -\mathrm{i} + \frac{6\xi\hat{k}_0^2}{S^3} - \frac{12\hat{k}_0^2}{S}\frac{R_0(\xi)}{F(\xi)}. \tag{6.71}$$

Therefore, as $\xi \approx \xi_c$, we have $F(\xi)F'(\xi) = O(1)$. From here, it follows that

$$F(\xi) \sim A(\xi - \xi_c)^{\frac{1}{2}}$$

$$A = \mathrm{i}\hat{k}_0\sqrt{\frac{24R_0}{S}}\Bigg|_{\xi=\xi_c} = \mathrm{i}\frac{2^{\frac{3}{2}}3^{\frac{1}{2}}\hat{k}_0}{(\xi_c - \mathrm{i})^{\frac{3}{4}}(\xi_c + \mathrm{i})^{\frac{1}{4}}}. \tag{6.72}$$

Based on these results, we derive that as $\xi \to \xi_c$,

$$\hat{k}_1 \sim \frac{R_1(\xi_c)}{A(\xi - \xi_c)^{\frac{1}{2}}} + \frac{b_1}{(\xi - \xi_c)} + O(1), \tag{6.73}$$

where

$$b_1 = \frac{S(\xi_c)}{24\hat{k}_0^2(\xi_c)}\left(\frac{m_0 - \mathrm{i}}{2} + \mathrm{i}\frac{7\hat{k}_0^2(\xi_c)}{S(\xi_c)}\right). \tag{6.74}$$

Therefore, as $\xi \to \xi_c$,

$$e^{\mathrm{i}\int_{\xi_c}^{\xi}\hat{k}_1(\xi_1)\mathrm{d}\xi_1} \approx C_1(\xi - \xi_c)^{\mathrm{i}b_1}$$

$$\times\left\{1 + \mathrm{i}\frac{2R_1(\xi_c)}{A}(\xi - \xi_c)^{\frac{1}{2}} + a_1(\xi - \xi_c) + \cdots\right\}, \tag{6.75}$$

where a_1 and C_1 are both constants.

6.2.3 Singular Point ξ_c of the Outer Solution

Note that in writing the general solution (6.48), we have actually extended our solution to the complex ξ-plane by analytic continuation. From (6.73) we see that the MVE solution (6.20) is not valid at the point ξ_c, so it is not uniformly valid in the whole complex ξ-plane. The singular point ξ_c was not seen in the zeroth-order approximation but it is now clearly seen in the first approximation. The existence of the singular point ξ_c plays a crucial role in pattern formation and selection in dendrite growth. This singularity was unknown by all previous researchers in the field until its discovery in 1989; see, e.g., [6.10], [6.12].

The location of the singular point ξ_c in the complex ξ-plane depends on the value of σ_0. As a matter of fact, combining (6.69) and (6.42), one finds that ξ_c is also a root of the equation:

$$\sigma_0 = \sqrt{\frac{2}{27}}\left(\frac{1 - \mathrm{i}\xi_c}{S}\right)^{3/2}_{\xi=\xi_c} = e^{-\mathrm{i}\frac{3\pi}{4}}\sqrt{\frac{2}{27}}\left(\frac{\xi_c + \mathrm{i}}{\xi_c - \mathrm{i}}\right)^{3/4}, \tag{6.76}$$

or

$$\sigma_0 = \sqrt{\frac{2}{27}} \left(\frac{r_2}{r_1}\right)^{3/4} e^{i\frac{3}{4}(\theta_2 - \theta_1 - \pi)} , \tag{6.77}$$

where

$$\xi_c + i = r_2 e^{i\theta_2}; \quad \xi_c - i = r_1 e^{i\theta_1} .$$

Evidently, if ξ_c is on the real axis, $|\sigma_0| = \sqrt{\frac{2}{27}}$, so that the eigenvalue σ_0 will be on a circle in the complex σ_0-plane. This will be illustrated further in the next section.

In the region away from ξ_c the outer solution (6.48) is valid. Thus the root condition can be applied to the outer solution. From Fig. 6.1, it is noted that for fixed $\varepsilon > 0$, as $\xi \gg 1$, we have $\text{Im}\{\hat{k}_0^{(1)}\} < 0$, while $\text{Im}\{\hat{k}_0^{(3)}\} > 0$. Hence, as $\xi \to \xi_{\max}(\varepsilon)$, the solution H_1 increases exponentially whereas H_3 decreases exponentially. Thus in order to satisfy the root condition in the far-field, the following asymptotic form must hold:

$$\tilde{h}_0 \sim D_3' H_3(\xi, \varepsilon) \quad \text{as } \xi \to \infty \ \ (\varepsilon > 0). \tag{6.78}$$

As a consequence, in the far-field one also has

$$\tilde{h}_0 \sim D_3' H_3(\xi, \varepsilon) \quad \text{as } \varepsilon \to 0. \tag{6.79}$$

This asymptotic condition in the far-field is called the *radiation condition*. Here, it is worthwhile to point out that there is an important difference between the solution (6.48) for dendrite growth discussed here and the solution (5.44) for the nonclassic steady needle growth problem discussed in Sect. 5.2.3. In contrast to the solution (5.44), which in the far-field is approximated by a dominant function, the solution (6.48), in the far-field, is approximated by a subdominant function. As a consequence, for a fixed $\varepsilon > 0$, the asymptotic form (6.79) of the solution $\tilde{h}$ cannot be applied to the entire complex ξ-plane as a good approximation for the solution $\tilde{h}$, due to the so-called *Stokes phenomenon*.

There are different ways to derive and demonstrate the Stokes phenomenon for an ODE system. One way is through the WKB method and contour integral, another way is through the resurgent function and Borel–Laplace transformation. For a detailed discussion, the reader is referred to [6.5]–[6.7]. In what follows, we attempt to give a brief elucidation.

Let us first choose the singular point ξ_c as the lower limit, ξ_0, of the integral in the definition of ξ_+. Subsequently, the fundamental solutions are defined as

$$H_1(\xi, \varepsilon) = \exp\left(\frac{i}{\varepsilon} \int_{\xi_c}^{\xi} \hat{k}_0^{(1)} d\xi'\right)$$

$$H_3(\xi, \varepsilon) = \exp\left(\frac{i}{\varepsilon} \int_{\xi_c}^{\xi} \hat{k}_0^{(3)} d\xi'\right) . \tag{6.80}$$

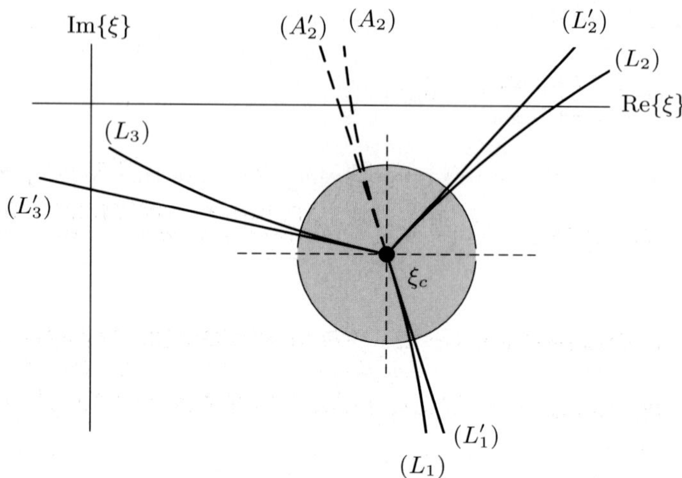

Fig. 6.2. A sketch of the structure of Stokes lines for the system of dendrite growth

The asymptotic solution (6.48) in the complex ξ-plane is an exponential function of $\frac{1}{\varepsilon}$. As $\varepsilon \to 0$, the two terms on the right-hand side are of different orders of magnitude in the complex ξ-plane, except at some isolated lines known as the Stokes lines. The Stokes lines are defined by the integral

$$\mathrm{Im}\left\{ \int_{\xi_c}^{\xi} \left(\hat{k}_0^{(1)} - \hat{k}_0^{(3)} \right) \mathrm{d}\xi' \right\} = 0 \,. \tag{6.81}$$

On the other hand, the anti-Stokes lines are defined by

$$\mathrm{Re}\left\{ \int_{\xi_c}^{\xi} \left(\hat{k}_0^{(1)} - \hat{k}_0^{(3)} \right) \mathrm{d}\xi' \right\} = 0 \,. \tag{6.82}$$

A sketch of the Stokes lines (L_1), (L_2), (L_3) and anti-Stokes lines (A_1), (A_2), (A_3) of our system are shown in Fig. 6.2. The anti-Stokes line (A_2) divides the entire complex ξ-plane into the sector (S_1) and sector (S_2) . The root of dendrite, $\xi = \xi_{\max}(\varepsilon)$, belongs to sector (S_2), while the tip of the dendrite, $\xi = 0$, belongs to sector (S_1).

Note that when ξ is located at the right side of (L_1), $H_1(\xi, \varepsilon) \gg H_3(\xi, \varepsilon)$ exponentially as $\varepsilon \to 0$. As a result, the function $H_1(\xi, \varepsilon)$ is dominant while the function $H_3(\xi, \varepsilon)$ is subdominant. When ξ is located at the left side of (L_1), $H_3(\xi, \varepsilon) \gg H_1(\xi, \varepsilon)$ exponentially as $\varepsilon \to 0$. In this case, the function $H_3(\xi, \varepsilon)$ is dominant while the function $H_1(\xi, \varepsilon)$ is subdominant. Directly on the Stokes lines, $H_1(\xi, \varepsilon) = O\big(H_3(\xi, \varepsilon)\big)$, the two functions have the same order of magnitudes as $\varepsilon \to 0$. When ξ moves across the Stokes line, the other of these two functions becomes dominant.

In general, according to the asymptotic theory, if a dominant function $F(\xi, \varepsilon)$ is an asymptotic approximation of the function $\tilde{h}$, then with any addition of the subdominant function $G(\xi, \varepsilon)$, the function $F + G$ must also be an asymptotic approximation of $\tilde{h}$. As a result, in different sectors, as a uniformly valid asymptotic approximation to the exact solution for our problem, the form (6.48) may have a pair of different constant coefficients $\{D_1, D_3\}$. This is the Stokes phenomenon.

Now let us consider the asymptotic solution (6.79) in sector (S_1). Since in this sector, $H_3(\xi, \varepsilon) \gg H_1(\xi, \varepsilon)$ the asymptotic solution of $\tilde{h}$ may change to:

$$\tilde{h} \sim D_3' H_3 + D_1 H_1 \quad \left[\text{as } \xi \in (S_1)\right], \tag{6.83}$$

with the inclusion of an additional mode H_1. In the following, we shall denote the coefficients of the solution (6.48) in (S_1) by $\{D_1, D_3\}$ and by $\{D_1', D_3'\}$ in (S_2). The determination of the relation between these two pairs of coefficients is called the connection problem. The connection problem in our case is actually the determination of the constant D_1.

In view of the above, the MVE asymptotic solution (6.48) may have a discontinuity along the anti-Stokes line (A_2). In other words, along (A_2) there is a critical layer with a thickness of $O(\varepsilon)$ where the MVE solution is not applicable. In Sect. 6.3, we shall study the inner solution near the singular point ξ_c. The connection condition between the coefficients $\{D_1, D_3 = D_3'\}$ in sector (S_1) and $\{D_1' = 0, D_3'\}$ in sector (S_2) will be derived by matching the outer solution (6.48) with the inner solution in the intermediate region. It can be proved and can also be seen by the asymptotic theory of Airy functions that, for the outer asymptotic form of the solution $\tilde{h}$, the jump of the pair of coefficients (D_1, D_3) will not immediately occur upon crossing the Stokes line. It occurs only when ξ enters a different sector by crossing the next anti-Stokes line. This means that as ξ crosses the Stokes line (L_2), moving from the far-field towards the tip along the real ξ axis, regardless of the dominance of H_3 and the subdominance of H_1, the asymptotic form (6.79) of the solution $\tilde{h}$ still holds. It is only when ξ crosses the next anti-Stokes line (A_2) and enters the sector (S_1) that the asymptotic form of the solution $\tilde{h}$ changes to (6.83).

Up to this point, one can see that the solution for dendrite growth is comprised of three major parts: (1) the Ivantsov solution, (2) the steady regular perturbation expansion (RPE) solution, and (3) the unsteady singular perturbation expansion (SPE) solution.

The interface shape of the dendrite is accordingly described as

$$\eta_s(\xi, t) = \eta_B(\xi) + \left(\tilde{h}_0(\xi) e^{\frac{\sigma t}{\varepsilon \eta_0^2}} + \cdots\right)$$

$$= 1 + \varepsilon^2 \left(\bar{h}_0(\xi) + \varepsilon^2 \bar{h}_1(\xi) + \cdots\right) + \left(\tilde{h}_0(\xi) e^{\frac{\sigma t}{\varepsilon \eta_0^2}} + \cdots\right). \tag{6.84}$$

6.3 The Inner Solutions near the Singular Point ξ_c

As previously indicated, the MVE solution (6.48) is not valid at the singular point ξ_c. This implies that the solutions in the vicinity of ξ_c: $|\xi - \xi_c| \ll 1$; $|\eta - 1| \ll 1$, no longer have a multiple scale structure. To derive the inner solutions we must start with the perturbed system (6.3)–(6.8) and construct a different asymptotic expansion. For this purpose, we introduce the inner variables:

$$\xi_* = \frac{\xi - \xi_c}{\varepsilon^\alpha}$$
$$\eta_* = \frac{\eta - 1}{\varepsilon^\alpha},$$

(6.85)

where α is to be determined. In terms of the inner variables, equation (6.3) for perturbed states, can be expressed in the form

$$\left(\frac{\partial^2}{\partial \xi_*^2} + \frac{\partial^2}{\partial \eta_*^2} \right) \tilde{T} = \left[\varepsilon^{2\alpha-1} \sigma \eta_0^2 (\xi^2 + \eta^2) + \varepsilon^\alpha \eta_0^2 \left(\xi \frac{\partial}{\partial \xi_*} - \eta \frac{\partial}{\partial \eta_*} \right) \right.$$
$$\left. - \varepsilon^\alpha \left(\frac{1}{\xi} \frac{\partial}{\partial \xi_*} + \frac{1}{\eta} \frac{\partial}{\partial \eta_*} \right) \right] \tilde{T} .$$

(6.86)

When $\alpha > \frac{1}{2}$ (which can be verified later), the above equation can be reduced, as $\varepsilon \to 0$, to

$$\left(\frac{\partial^2}{\partial \xi_*^2} + \frac{\partial^2}{\partial \eta_*^2} \right) \tilde{T} \approx 0 .$$

(6.87)

With the inner variables, the interface shape function is

$$\eta_{\mathrm{s}}(\xi, t) = 1 + \frac{\tilde{h}}{\eta_0^2} = 1 + \varepsilon^\alpha \eta_{*\mathrm{s}}.$$

(6.88)

Writing

$$\eta_{*\mathrm{s}} = \frac{\hat{h}(\xi_*, t)}{\eta_0^2},$$

(6.89)

we have

$$\tilde{h}(\xi, t) = \varepsilon^\alpha \hat{h}(\xi_*, t).$$

(6.90)

Accordingly, we put

$$\tilde{T}(\xi, \eta, t) = \varepsilon^\alpha \hat{T}(\xi_*, \eta_*, t); \quad \tilde{T}_{\mathrm{S}}(\xi, \eta, t) = \varepsilon^\alpha \hat{T}_{\mathrm{S}}(\xi_*, \eta_*, t).$$

(6.91)

The boundary conditions (6.4)–(6.8) are transformed to the following:

1. As $\eta_* \to \infty$

$$\hat{T} \to 0 \,. \tag{6.92}$$

2. As $\eta_* \to -\infty$

$$\hat{T}_{\rm S} \to 0 \,. \tag{6.93}$$

3. At the interface $\eta_* = 0$,
   (i)

$$\hat{T} = \hat{T}_{\rm S} + \hat{h} + \text{ (higher-order terms)}, \tag{6.94}$$

   (ii)

$$\hat{T}_{\rm S} = \frac{\varepsilon^{2-2\alpha}}{S(\xi)} \frac{\partial^2 \hat{h}}{\partial \xi_*^2} + \text{ (higher-order terms)}, \tag{6.95}$$

   (iii)

$$\varepsilon^{1-\alpha} \frac{\partial}{\partial \eta_*} \left(\hat{T} - \hat{T}_{\rm S} \right) + \sigma S^2(\xi)\hat{h} + \varepsilon^{1-\alpha}\xi \frac{\partial \hat{h}}{\partial \xi_*}$$

$$= \text{ (higher-order terms)} \,. \tag{6.96}$$

We seek the mode solutions and make the inner expansions:

$$\hat{T}(\xi_*, \eta_*, t) = \left[\nu_0(\varepsilon)\hat{T}_0(\xi_*, \eta_*) + \nu_1(\varepsilon)\hat{T}_1(\xi_*, \eta_*) + \dots \right] e^{\frac{\sigma t}{\varepsilon \eta_0^2}}$$

$$\hat{T}_{\rm S}(\xi_*, \eta_*, t) = \left[\nu_0(\varepsilon)\hat{T}_{\rm S0}(\xi_*, \eta_*) + \nu_1(\varepsilon)\hat{T}_{\rm S1}(\xi_*, \eta_*) + \dots \right] e^{\frac{\sigma t}{\varepsilon \eta_0^2}} \tag{6.97}$$

$$\hat{h} = \left[\nu_0(\varepsilon)\hat{h}_0 + \nu_1(\varepsilon)\hat{h}_1 + \cdots \right] e^{\frac{\sigma t}{\varepsilon \eta_0^2}} \,.$$

Letting $\varepsilon \to 0$, the above inner equations can be further simplified into the following third-order complex ODE for the unsteady interface perturbations. As a matter of fact, from (6.87) it can be shown that

1. In the liquid phase region:

$$\left(i\frac{\partial}{\partial \xi_*} - \frac{\partial}{\partial \eta_*} \right) \hat{T}_0 = 0 \,. \tag{6.98}$$

2. In the solid phase region:

$$\left(i\frac{\partial}{\partial \xi_*} + \frac{\partial}{\partial \eta_*} \right) \hat{T}_{\rm S0} = 0 \,. \tag{6.99}$$

Consequently, at the interface we have

$$\frac{\partial}{\partial \eta_*}\left(\hat{T}_0 - \hat{T}_{S0}\right) = \mathrm{i}\frac{\partial}{\partial \xi_*}\left(\hat{T}_0 + \hat{T}_{S0}\right). \tag{6.100}$$

On the other hand, from the boundary conditions (6.94) and (6.95), one can obtain

$$\left(\hat{T}_0 + \hat{T}_{S0}\right) = \hat{h}_0 + \frac{2\varepsilon^{2-2\alpha}}{S}\frac{\mathrm{d}^2\hat{h}_0}{\mathrm{d}\xi_*^2}. \tag{6.101}$$

Combining (6.96) with (6.101) and (6.102), we derive the mode equation for the perturbed interface as follows:

$$\mathrm{i}\frac{2\varepsilon^{3-3\alpha}}{S}\frac{\mathrm{d}^3\hat{h}_0}{\mathrm{d}\xi_*^3} + \varepsilon^{1-\alpha}(\xi + \mathrm{i})\frac{\mathrm{d}\hat{h}_0}{\mathrm{d}\xi_*} + \sigma S^2\hat{h}_0 = 0. \tag{6.102}$$

Returning to the outer variables ξ, this equation may be written in the form

$$\mathrm{i}\frac{2\varepsilon^3}{S}\frac{\mathrm{d}^3\tilde{h}_0}{\mathrm{d}\xi^3} + \varepsilon(\xi + \mathrm{i})\frac{\mathrm{d}\tilde{h}_0}{\mathrm{d}\xi} + \sigma_0 S^2\tilde{h}_0 = 0. \tag{6.103}$$

It should be pointed out that one may apply (6.104) to the entire outer region and derive its asymptotic solutions by the WKB method. By doing so, one can regain the local dispersion relation (6.42). Hence, in the leading order approximation, the WKB solution obtained for the ODE system (6.104) is exactly the same as the outer solution (6.48) obtained for the original PDE system. The difference between the solutions of the two systems is a small quantity of higher-order. It is for this reason that one may use (6.104) as a simplified model equation for discussing the instability mechanism of dendrite growth. However, the present book is not only concerned with the leading order, but also with the first-order approximate solutions. Therefore, this simplified approach to developing the interfacial wave theory is not adopted in this monograph.

To solve (6.103), we need to apply the transformation first introduced in [6.10] and transform the solution $\tilde{h}_0$ into a new unknown function $W_0(\xi)$,

$$\tilde{h}_0 = W_0(\xi)\exp\left[\frac{\mathrm{i}}{\varepsilon}\int_{\xi_c}^{\xi} k_c(\xi_1)\,\mathrm{d}\xi_1\right], \tag{6.104}$$

where the reference wave number function $k_c(\xi)$ is to be chosen later. It will be seen that

$$\mathrm{Re}\{\hat{k}_0^{(3)}\} < \mathrm{Re}\{k_c\} < \mathrm{Re}\{\hat{k}_0^{(1)}\}. \tag{6.105}$$

In the outer region we have

$$H_1 = W_0^{(+)}(\xi)\exp\left[\frac{\mathrm{i}}{\varepsilon}\int_{\xi_c}^{\xi} k_c(\xi_1)\,\mathrm{d}\xi_1\right] \tag{6.106}$$

and

$$H_3 = W_0^{(-)}(\xi) \exp\left[\frac{i}{\varepsilon} \int_{\xi_c}^{\xi} k_c(\xi_1)\, d\xi_1\right].$$ (6.107)

The general outer solutions for W_0 may be written in the form:

$$W_0 = D_1 W_0^{(+)} + D_3 W_0^{(-)}.$$ (6.108)

The wave number function for the $W_0^{(+)}$ wave is $\hat{k}_0^{(+)} = \hat{k}_0^{(1)} = k_0^{(1)} - k_c$. Since $\mathrm{Re}\{\hat{k}_0^{(+)}\} > 0$, the phase velocity of the $W_0^{(+)}$ wave is positive and, as such, it is an outgoing wave. On the other hand, the wave number function for the $W^{(-)}$ wave is $\hat{k}_0^{(-)} = \hat{k}_0^{(3)} = k_0^{(3)} - k_c$, and $\mathrm{Re}\{\hat{k}_0^{(-)}\} < 0$. Hence, the $W_0^{(-)}$ wave is an incoming wave with negative phase velocity. One can draw the following diagram of the relation between the H waves and the W waves:

H_1 wave $\qquad\qquad\Longleftrightarrow\; W^{(+)}$

(short-wave branch) $\qquad$ (outgoing wave);

H_3 wave $\qquad\qquad\Longleftrightarrow\; W^{(-)}$

(long-wave branch) $\qquad$ (incoming wave).

In accordance with the above, in the inner region we can set

$$\hat{h}_0 = \hat{W}_0(\xi_*) \exp\left\{\frac{i}{\varepsilon} \int_{\xi_c}^{\xi} k_c(\xi_1)\, d\xi_1\right\}.$$ (6.109)

Hence one derives

$$\frac{d\hat{h}_0}{d\xi} = \left(\frac{d\hat{W}_0}{d\xi} + \frac{ik_c}{\varepsilon}\hat{W}_0\right) \exp\left[\frac{i}{\varepsilon} \int_{\xi_c}^{\xi} k_c(\xi_1)\, d\xi_1\right]$$

$$\frac{d^2\hat{h}_0}{d\xi^2} = \left[\frac{d^2\hat{W}_0}{d\xi^2} + \frac{2ik_c}{\varepsilon}\frac{d\hat{W}_0}{d\xi} + \left(\frac{ik_c'}{\varepsilon} - \frac{k_c^2}{\varepsilon^2}\right)\hat{W}_0\right]$$

$$\times \exp\left\{\frac{i}{\varepsilon} \int_{\xi_c}^{\xi} k_c(\xi_1)\, d\xi_1\right\}$$

$$\approx \left[\frac{d^2\hat{W}_0}{d\xi^2} + \frac{2ik_c}{\varepsilon}\frac{d\hat{W}_0}{d\xi} - \frac{k_c^2}{\varepsilon^2}\hat{W}_0\right] \exp\left\{\frac{i}{\varepsilon} \int_{\xi_c}^{\xi} k_c(\xi_1)\, d\xi_1\right\}$$

$$\frac{d^3\hat{h}_0}{d\xi^3} = \left[\frac{d^3\hat{W}_0}{d\xi^3} + \frac{3ik_c}{\varepsilon}\frac{d^2\hat{W}_0}{d\xi^2} + \left(\frac{3ik_c'}{\varepsilon} - \frac{3k_c^2}{\varepsilon^2}\right)\frac{d\hat{W}_0}{d\xi}\right.$$

$$\left. - \left(\frac{ik_c^3}{\varepsilon^3} + \frac{3k_c k_c'}{\varepsilon^2} - \frac{ik_c''}{\varepsilon}\right)\hat{W}_0\right] \exp\left\{\frac{i}{\varepsilon} \int_{\xi_c}^{\xi} k_c(\xi_1)\, d\xi_1\right\}$$

$$\approx \left[\frac{d^3\hat{W}_0}{d\xi^3} + \frac{3ik_c}{\varepsilon}\frac{d^2\hat{W}_0}{d\xi^2} - \frac{3k_c^2}{\varepsilon^2}\frac{d\hat{W}_0}{d\xi} - \frac{ik_c^3}{\varepsilon^3}\hat{W}_0\right]$$ (6.110)

$$\times \exp\left\{\frac{i}{\varepsilon} \int_{\xi_c}^{\xi} k_c(\xi_1)\, d\xi_1\right\}.$$

Equation (6.104) is now changed to

$$\varepsilon^3 \Omega_3 \frac{\mathrm{d}^3 \hat{W}_0}{\mathrm{d}\xi^3} + \mathrm{i}\varepsilon^2 \Omega_2 \frac{\mathrm{d}^2 \hat{W}_0}{\mathrm{d}\xi^2} - \varepsilon \Omega_1 \frac{\mathrm{d}\hat{W}_0}{\mathrm{d}\xi} + \mathrm{i}S^2 \big\{ \sigma - \Sigma_c(\xi) \big\} \hat{W}_0 = 0 , \quad (6.111)$$

where

$$
\begin{aligned}
\Omega_3 &= \frac{1}{3!} \left[\frac{\partial^3 (S^2 \Sigma)}{\partial \hat{k}_0^3} \right]_{\hat{k}_0 = k_c} = -\frac{2}{S} \\
\Omega_2 &= \frac{1}{2!} \left[\frac{\partial^2 (S^2 \Sigma)}{\partial \hat{k}_0^2} \right]_{\hat{k}_0 = k_c} = -\frac{6k_c}{S} \\
\Omega_1 &= \left[\frac{\partial (S^2 \Sigma)}{\partial \hat{k}_0} \right]_{\hat{k}_0 = k_c} = -\frac{6k_c^2}{S} + (1 - \mathrm{i}\xi) \\
\Sigma_c(\xi) &= \Sigma(k_c, \xi) = \mathrm{e}^{-\mathrm{i}\frac{3\pi}{4}} \sqrt{\frac{2}{27}} \left(\frac{\xi + \mathrm{i}}{\xi - \mathrm{i}} \right)^{\frac{3}{4}} .
\end{aligned}
\tag{6.112}
$$

If we set

$$\Omega_1 = -\frac{6k_c^2}{S} + (1 - \mathrm{i}\xi) = 0 \tag{6.113}$$

or

$$k_c(\xi) = \mathrm{e}^{-\mathrm{i}\frac{\pi}{4}} \sqrt{\frac{S}{6}} \, (\xi + \mathrm{i}) \qquad (\, \mathrm{Re}\{k_c\} > 0) , \tag{6.114}$$

then (6.104) can be changed to

$$\varepsilon^3 \Omega_3 \frac{\mathrm{d}^3 \hat{W}_0}{\mathrm{d}\xi^3} + \mathrm{i}\varepsilon^2 \Omega_2 \frac{\mathrm{d}^2 \hat{W}_0}{\mathrm{d}\xi^2} + \mathrm{i}S^2 \big\{ \sigma - \Sigma_c(\xi) \big\} \hat{W}_0 = 0 . \tag{6.115}$$

Accordingly, with the inner variables (6.103) becomes

$$\varepsilon^{3-3\alpha} \Omega_3 \frac{\mathrm{d}^3 \hat{W}_0}{\mathrm{d}\xi_*^3} + \mathrm{i}\varepsilon^{2-2\alpha} \Omega_2 \frac{\mathrm{d}^2 \hat{W}_0}{\mathrm{d}\xi_*^2} + \mathrm{i}S^2 \big\{ \sigma - \Sigma_c(\xi) \big\} \hat{W}_0 = 0 . \tag{6.116}$$

It is clear now that this equation has a turning point singularity in the complex ξ-plane at the root of the equation

$$\sigma_0 - \Sigma_c(\xi) = \sigma_0 - \mathrm{e}^{-\mathrm{i}\frac{3\pi}{4}} \sqrt{\frac{2}{27}} \left(\frac{\xi + \mathrm{i}}{\xi - \mathrm{i}} \right)^{\frac{3}{4}} = 0 . \tag{6.117}$$

This singular point coincides with the singular point ξ_c, which was a root of the equations

$$\frac{\partial \Sigma(\xi, \hat{k}_0)}{\partial \hat{k}_0} = 0 , \quad F(\xi) = 0 , \quad \text{or} \quad \hat{k}_0^{(3)} = \hat{k}_0^{(1)} . \tag{6.118}$$

To obtain a single-valued, analytical solution in the complex ξ-plane, we make the branch cut along the Stokes line (L_1) as shown in Fig. 6.2. For any given $\sigma_0 = |\sigma_0|e^{i\theta}$, from (6.118) or (6.76) one can find the solution

$$\xi_c = i\left(\frac{|\sigma_0|^{\frac{4}{3}}e^{i\frac{4\theta}{3}} - \frac{2^{\frac{2}{3}}}{9}}{|\sigma_0|^{\frac{4}{3}}e^{i\frac{4\theta}{3}} + \frac{2^{\frac{2}{3}}}{9}}\right). \tag{6.119}$$

The conformal mapping from the complex σ_0-plane to the complex ξ_c-plane by the analytic function $\xi_c = \xi_c(\sigma_0)$ is illustrated in Fig. 6.3 and Fig. 6.4. The semicircle (γ) in the complex σ_0-plane corresponding to $|\sigma_0| = \sigma_{\max} = \sqrt{\frac{2}{27}} = 0.2722$, where $\sigma_{\max}$ is the maximum growth rate of the Mullins–Sekerka instability for a flat interface, plays a special role. It is seen that as $\sigma_0 \to 0$, this turning point approaches the point $\xi = -i$, while as $\sigma_0 \to \infty$, $\xi_c \to i$. Furthermore, when σ_0 is on the semicircle (γ) in the complex σ_0-plane, the corresponding critical point is on the real axis $(-\infty < \xi_c < \infty)$; when the eigenvalue σ_0 belongs to the domain outside the semicircle (γ), the critical point ξ_c under discussion will be in the upper half of the complex ξ-plane. Thus, the branch cut will intersect with the real axis $(\xi > 0)$. In this case, the system has no physically acceptable continuous solution. Thus, as a necessary condition, the eigenvalue σ must be in the domain inside the semicircle (γ), in order for the turning point ξ_c to be in the lower half of the complex ξ-plane. This condition is called the '*pattern formation condition*' in [6.12].

In the vicinity of ξ_c, one can make the Taylor expansions for all the coefficients of (6.117) as the functions of ξ. By balancing the leading terms, it is found that

$$\alpha = \frac{2}{3} \quad \text{and} \quad \xi_* = \frac{(\xi - \xi_c)}{\varepsilon^{\frac{2}{3}}}. \tag{6.120}$$

The leading order approximation of the inner equation is found to be the Airy equation:

$$\frac{d^2\hat{W}_0}{d\xi_*^2} + A^2\xi_*\hat{W}_0 = 0, \tag{6.121}$$

where

$$A = \left(\frac{S^3}{6k_c}\frac{\partial\Sigma_c}{\partial\xi}\right)^{\frac{1}{2}}_{\xi=\xi_c} = i\sqrt{\frac{1}{6}}\left(\frac{\xi_c + i}{\xi_c - i}\right)^{\frac{1}{4}} \tag{6.122}$$

$$\left(\frac{\pi}{2} < \arg\{A\} < \frac{3\pi}{4}\right).$$

This is written as the standard Airy equation

$$\frac{d^2\hat{W}_0}{d\hat{\xi}_*^2} + \hat{\xi}_*\hat{W}_0 = 0, \tag{6.123}$$

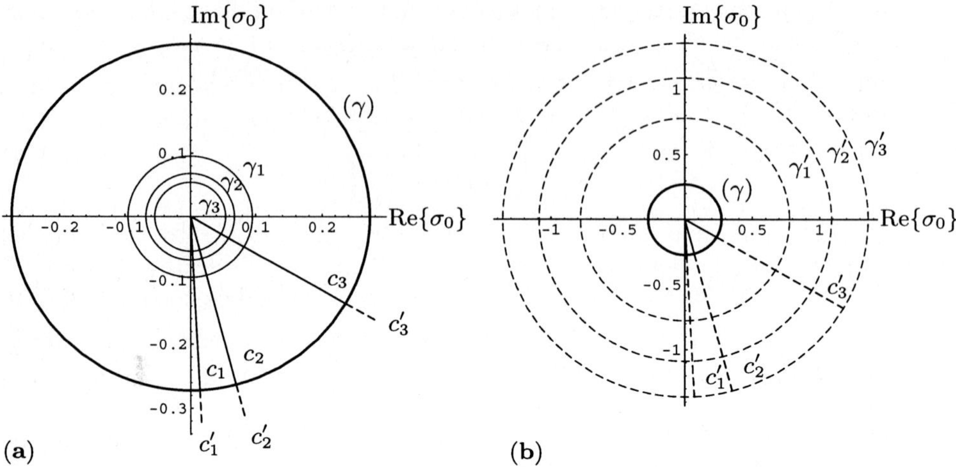

(a)   **(b)**

Fig. 6.3a,b. The rays and circles in the complex σ_0-plane, where the circle (γ) corresponds to $|\sigma_0| = 0.2722$, all the lines outside (γ) are shown as dashed lines, while all lines inside (γ) are shown as solid lines

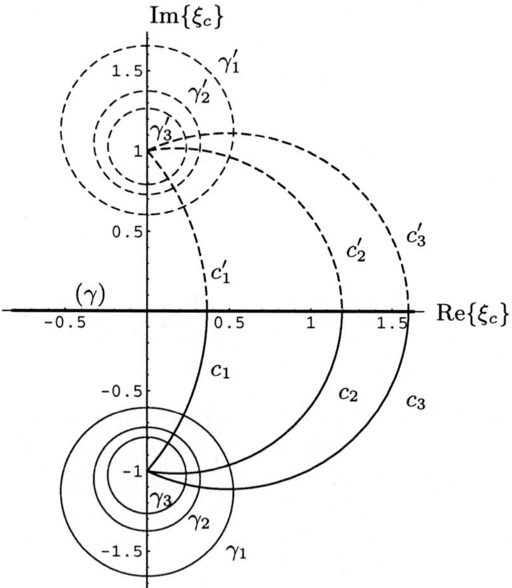

Fig. 6.4. The images in the complex ξ_c-plane under the conformal mapping $\xi_c = \xi_c(\sigma_0)$, transformed from the lines in the σ_0-plane, where dashed lines correspond to dashed lines, solid lines to solid lines

by introducing the new inner variable $\hat{\xi}_*$:

$$\hat{\xi}_* = \frac{A^{\frac{2}{3}}}{\varepsilon^{\frac{2}{3}}}(\xi - \xi_c)\,. \tag{6.124}$$

The general solution of the above Airy equation is

$$\hat{W}_0 = D_{*1}\hat{\xi}_*^{\frac{1}{2}} H_{\frac{1}{3}}^{(1)}(\zeta) + D_{*2}\hat{\xi}_*^{\frac{1}{2}} H_{\frac{1}{3}}^{(2)}(\zeta) \tag{6.125}$$

$$\left(\zeta = \frac{2}{3}\hat{\xi}_*^{\frac{3}{2}}\right)\,,$$

where $H_\nu^{(1)}(z)$ is ν^{th}-order Hankel function of the first kind, while $H_\nu^{(2)}(z)$ is the Hankel function of the second kind. In order to match this with the outer solution which satisfies the downstream far-field condition (6.79), the inner solution must be

$$\hat{W}_0 = D_*\hat{\xi}_*^{\frac{1}{2}} H_{\frac{1}{3}}^{(2)}(\zeta)\,. \tag{6.126}$$

As $\hat{\xi}_* \to \infty$, the above Airy equation has two asymptotic solutions expressible in the form

$$\hat{W}_0 \sim \begin{cases} D_* \exp\left\{i\int_0^{\hat{\xi}_*} \hat{k}_* d\xi_*\right\} \\[2mm] D_* \exp\left\{-i\int_0^{\hat{\xi}_*} \hat{k}_* d\xi_*\right\}, \end{cases} \tag{6.127}$$

where

$$\hat{k}_* = \hat{\xi}_*^{\frac{1}{2}}\,. \tag{6.128}$$

From this we find that

$$\int_0^{\hat{\xi}_*} \hat{k}_* d\xi_* = \frac{2}{3}\hat{\xi}_*^{\frac{3}{2}} = \zeta\,. \tag{6.129}$$

The Stokes lines of the asymptotic solutions (6.128) for the Airy equation are found to be the rays of $\arg(\zeta) = 0, \pi$, and 2π (or $\arg(\hat{\xi}_*) = 0, \frac{2\pi}{3}$, and $\frac{4\pi}{3}$), which we label by (L_1'), (L_2'), (L_3'), respectively. The anti-Stokes lines of the asymptotic solutions (6.128 are the rays of $\arg(\zeta) = \frac{\pi}{2}, \frac{3\pi}{2}$, and $\frac{5\pi}{2}$ (or $\arg(\hat{\xi}_*) = \frac{\pi}{3}, \pi$, and $\frac{5\pi}{3}$), which we label by (A_1'), (A_2'), (A_3'), respectively. These lines are sketched in Fig. 6.2. The Stokes and anti-Stokes lines of the inner equation (6.127) are respectively tangential lines of the Stokes and anti-Stokes lines of the outer equation determined by (6.81) and (6.82), which are also sketched in Fig. 6.2 and are labeled $(L_1), (L_2), (L_3)$ and $(A_1), (A_2), (A_3)$. As mentioned in the last section, along the anti-Stokes line (A_2), which crosses the real ξ-axis at the point ξ_c', the coefficient pair of the outer solution will have a discontinuity.

We now turn to matching the inner solution with the outer solution. In doing so, we need the asymptotic expansions of the inner solution (6.127) in the far-field, $\xi_* \to \infty$. From Bessel function theory, one can find the asymptotic expansions of (6.127) along different directions in the different sectors, and the connection of these expansions (refer to [6.8]). Namely,

$$H_\nu^{(1)}(\zeta) \sim \sqrt{\frac{2}{\pi\zeta}}\, e^{i\zeta - i(\nu/2 + 1/4)\pi}\left\{1 + O(1/\zeta)\right\}$$
$$(-\pi < \arg(\zeta) \leq 2\pi)$$

$$H_\nu^{(2)}(\zeta) \sim \sqrt{\frac{2}{\pi\zeta}}\, e^{-i\zeta + i(\nu/2 + 1/4)\pi}\left\{1 + O(1/\zeta)\right\}$$
$$(-2\pi < \arg(\zeta) \leq \pi). \tag{6.130}$$

When $\zeta_1 \in (S_1)$,

$$H_\nu^{(2)}(\zeta_1) = H_\nu^{(2)}(\zeta e^{i\pi}) = 2\cos(\nu\pi) H_\nu^{(2)}(\zeta) + e^{i\nu\pi} H_\nu^{(1)}(\zeta) \tag{6.131}$$
$$(-\pi \leq \arg(\zeta) \leq \pi).$$

In terms of these formulas, it is seen that in order to match the inner solution with the outer solution in sector (S_2) as $|\xi_*| \to \infty$, one needs to balance

$$\nu_0(\varepsilon)\hat{W}_0 \iff D'W_0^{(-)}(\xi, \varepsilon),$$

or

$$\nu_0(\varepsilon)\varepsilon^{\frac{1}{6}}\sqrt{\frac{2}{\pi}}\, e^{i(\nu/2 + 1/4)\pi} D_* \frac{e^{-i\zeta}}{(\xi - \xi_c)^{\frac{1}{4}}}$$

$$\iff D'H_3(\xi, \varepsilon)\exp\left\{-\frac{i}{\varepsilon}\int_{\xi_c}^{\xi} k_c(\xi_1)\, d\xi_1\right\}.$$

Noting that as $\xi \to \xi_c$ in (S_2),

$$\hat{k}_0^{(1)} - k_c \sim \varepsilon^{1/3}\hat{k}_*, \quad \text{and} \quad \hat{k}_0^{(3)} - k_c \sim -\varepsilon^{1/3}\hat{k}_*, \tag{6.132}$$

it is derived that in (S_2),

$$\begin{cases} W_0^{(+)}(\xi, \varepsilon) = e^{\frac{i}{\varepsilon}\int_{\xi_c}^{\xi}\left(k_0^{(1)} - k_c\right)d\xi} \sim e^{i\zeta} \\ W_0^{(-)}(\xi, \varepsilon)e^{\frac{i}{\varepsilon}\int_{\xi_c}^{\xi}\left(k_0^{(3)} - k_c\right)d\xi} \sim e^{-i\zeta}. \end{cases} \tag{6.133}$$

Therefore, the matching condition leads to

$$D_* = D'\sqrt{\frac{2}{\pi}}\, e^{-i(\nu/2 + 1/4)\pi}, \quad \text{and} \quad \nu_0(\varepsilon) = \varepsilon^{-\frac{1}{6}}. \tag{6.134}$$

On the other hand, in sector (S_1), we have

$$\hat{W}_0(\xi_*) \sim D'\hat{\xi}_*^{-\frac{1}{4}}\left\{e^{-i\zeta}\left(1 + O(1/\zeta)\right) - ie^{i\zeta}\left(1 + O(1/\zeta)\right)\right\}, \quad (6.135)$$

as $\hat{\xi}_* \to \infty$. Thus in order to match the inner solution with the outer solution in the intermediate region in the sector (S_1) as $\varepsilon \to 0$, one needs to balance

$$\nu_0(\varepsilon)\hat{W}_0 \iff D_1 W_0^{(+)}(\xi, \varepsilon) + D_3 W_0^{(-)}(\xi, \varepsilon)$$

or

$$\frac{D'}{(\xi - \xi_c)^{\frac{1}{4}}}\left\{e^{-i\zeta} - ie^{i\zeta} + O(1/\zeta)\right\} \tag{6.136}$$

$$\iff D_1 W_0^{(+)}(\xi, \varepsilon) + D_3 W_0^{(-)}(\xi, \varepsilon).$$

It should be remarked that $\hat{\xi}_* = 0$ is the branch point of the function $\hat{k}_*(\hat{\xi}_*)$. Hence, as ξ moves from sector (S_2) to (S_1), the function $\zeta(\hat{\xi}_*) = \int_0^{\hat{\xi}_*} \hat{k}_* d\xi_*$ will be analytically extended to the function $-\zeta(\hat{\xi}_*)$. Accordingly, as $\xi \to \xi_c$ in (S_1), we have

$$W_0^{(+)}(\xi, \varepsilon) = e^{\frac{i}{\varepsilon}\int_{\xi_c}^{\xi}\left(k_0^{(1)} - k_c\right)d\xi} \sim e^{-i\zeta},$$

$$W_0^{(-)}(\xi, \varepsilon) = e^{\frac{i}{\varepsilon}\int_{\xi_c}^{\xi}\left(k_0^{(3)} - k_c\right)d\xi} \sim e^{i\zeta}. \tag{6.137}$$

Therefore, for matching, the connection condition

$$\frac{D_1}{D_3} = e^{\frac{i\pi}{2}} \tag{6.138}$$

must hold. Moreover, in order to match

$$\frac{1}{(\xi - \xi_c)^{\frac{1}{4}}} \iff e^{i\int_{\xi_c}^{\xi} \hat{k}_1 d\xi_1} \tag{6.139}$$

as $\xi \to \xi_c$, one must set

$$b_1 = \frac{i}{4} \quad \text{and} \quad R_1(\xi_c) = 0. \tag{6.140}$$

This leads to

$$\sigma_1 = \frac{1}{(1 + \xi_c^2)}\left[\frac{\eta_0^2}{2}\left(\frac{2}{3} + \frac{i\xi_c}{3}\right)^2 - \left(1 + \frac{\eta_0^2}{2}\right)\right]. \tag{6.141}$$

For the case of small undercooling with $\eta_0^2 \ll 1$, we have

$$\sigma_1 \approx -\frac{1}{(1 + \xi_c^2)}. \tag{6.142}$$

6.4 Tip Inner Solution in the Tip Region

As previously indicated, the MVE solution is also not valid in the tip inner region ($|\xi| = O(\varepsilon); |1 - \eta| = O(\varepsilon)$), for in this region, the terms containing the factor $\frac{\varepsilon}{\xi}$ on the left-hand side of the linear system (6.3)–(6.11) are no longer higher-order terms. They must all be included in the leading order approximation. Moreover, the first-order approximation solution also shows a singularity at $\xi = 0$. Thus, in the tip region, we need to rescale the variables and make different asymptotic expansions for the solution. We define the tip inner variables

$$\hat{\xi}_* = \frac{\hat{k}_* \xi}{\varepsilon}$$

$$\hat{\eta}_* = \frac{\hat{q}_*(\eta - 1)}{\varepsilon} \tag{6.143}$$

$$\hat{t}_* = \frac{t}{\eta_0^2 \varepsilon},$$

where $|\xi| \ll \varepsilon$, and $|\eta - 1| \ll 1$. The tip solution can be expressed as a function of these inner variables and expanded in the following asymptotic form as $\varepsilon \to 0$:

$$\hat{T} = \left\{ \mu_0(\varepsilon)\hat{T}_0 + \mu_1(\varepsilon)\hat{T}_1 + \cdots \right\} e^{\sigma t_*}$$

$$\hat{T}_{\mathrm{S}} = \left\{ \mu_0(\varepsilon)\hat{T}_{\mathrm{S}0} + \mu_1(\varepsilon)\hat{T}_{\mathrm{S}1} + \cdots \right\} e^{\sigma t_*}$$

$$\hat{h} = \left\{ \mu_0(\varepsilon)\hat{h}_0 + \mu_1(\varepsilon)\hat{h}_1 + \cdots \right\} e^{\sigma t_*} \tag{6.144}$$

$$\hat{k}_* = \hat{k}_{*0} + \varepsilon\hat{k}_{*1} + \cdots$$

$$\hat{q}_* = \hat{k}_{*0} + \varepsilon\hat{q}_{*1} + \cdots .$$

At the zeroth order in the tip region, the system (6.3)–(6.8) can be reduced to

$$\frac{\partial^2 \hat{T}_0}{\partial \hat{\xi}_*^2} + \frac{\partial^2 \hat{T}_0}{\partial \hat{\eta}_*^2} + \frac{1}{\hat{\xi}_*} \frac{\partial \hat{T}_0}{\partial \hat{\xi}_*} = 0$$

$$\frac{\partial^2 \hat{T}_{\mathrm{S}0}}{\partial \hat{\xi}_*^2} + \frac{\partial^2 \hat{T}_{\mathrm{S}0}}{\partial \hat{\eta}_*^2} + \frac{1}{\hat{\xi}_*} \frac{\partial \hat{T}_{\mathrm{S}0}}{\partial \hat{\xi}_*} = 0 \tag{6.145}$$

with the boundary conditions at $\hat{\eta}_* = 0$

$$\hat{T}_0 = \hat{T}_{\mathrm{S}0} + \hat{h}_0 \tag{6.146}$$

$$\hat{T}_{\mathrm{S}0} = \hat{k}_0^2 \left(\frac{\partial^2 \hat{h}_0}{\partial \hat{\xi}_*^2} + \frac{1}{\hat{\xi}_*} \frac{\partial \hat{h}_0}{\partial \hat{\xi}_*} \right) \tag{6.147}$$

$$\hat{k}_{*0} \frac{\partial}{\partial \hat{\eta}_*} (\hat{T}_0 - \hat{T}_{\mathrm{S}0}) + \sigma_0 \hat{h}_0 = 0. \tag{6.148}$$

This system admits the inner solutions:

$$\tilde{T}_0 = \tilde{a}_0 H_0^{(1)}(\hat{\xi}_*)e^{-\hat{\eta}_*}$$

$$\tilde{T}_{S0} = \tilde{a}_{S0} H_0^{(1)}(\hat{\xi}_*)e^{\hat{\eta}_*} \tag{6.149}$$

$$\tilde{h}_0 = \tilde{d}_0 H_0^{(1)}(\hat{\xi}_*)\,,$$

where $H_0^{(1)}(\hat{\xi}_*)$ is the zeroth-order Hankel function of the first kind.

From the boundary conditions (6.147)–(6.149), one obtains

$$\tilde{a}_0 = (1 - \hat{k}_{*0}^2)\tilde{d}_0\,, \qquad \tilde{a}_{S0} = -\hat{k}_{*0}^2 \tilde{d}_0\,, \tag{6.150}$$

and the local dispersion relation in the tip region

$$\sigma_0 = \hat{k}_{*0}\big(1 - 2\hat{k}_{*0}^2\big). \tag{6.151}$$

For fixed σ_0, (6.152) has three roots for $\hat{k}_0$. Comparing the local dispersion relation in the tip region (6.152) with the local dispersion relation in the outer region (6.42) one can evidently write

$$\begin{cases} \hat{k}_{*0}^{(1)} = \hat{k}_0^{(1)}(0) \\[2mm] \hat{k}_{*0}^{(2)} = \hat{k}_0^{(2)}(0) \\[2mm] \hat{k}_{*0}^{(3)} = \hat{k}_0^{(3)}(0). \end{cases} \tag{6.152}$$

The root $\hat{k}_{*0}^{(2)}$ must be ruled out due to the fact that $\mathrm{Re}\{\hat{k}_{*0}^{(2)}\} < 0$. Therefore, the general solution of $\hat{h}_0$ in the tip region is

$$\hat{h}_0 = \tilde{d}_0^{(1)} H_0^{(1)}\left(\frac{\hat{k}_{*0}^{(1)}\xi}{\varepsilon}\right) + \tilde{d}_0^{(3)} H_0^{(1)}\left(\frac{\hat{k}_{*0}^{(3)}\xi}{\varepsilon}\right). \tag{6.153}$$

The above tip inner solution must satisfy the tip smoothness condition. For the three-dimensional problem there are only axially symmetrical modes, so the tip smoothness condition is

$$\text{as } \xi \to 0, \quad \tilde{h}_0(0) < \infty, \quad \text{and} \quad \tilde{h}_0'(0) = 0\,. \tag{6.154}$$

This gives

$$\tilde{d}_0^{(1)} + \tilde{d}_0^{(3)} = 0 \tag{6.155}$$

and the tip inner solution reduces to

$$\tilde{h}_0 = \tilde{d}_0 \left\{ H_0^{(1)}\left(\frac{\hat{k}_{*0}^{(1)}\xi}{\varepsilon}\right) - H_0^{(1)}\left(\frac{\hat{k}_{*0}^{(3)}\xi}{\varepsilon}\right) \right\}. \tag{6.156}$$

As $\hat{\xi}_* \to \infty$, one has

$$\tilde{h}_0 \approx \tilde{d}_0 \sqrt{\frac{2}{\pi}} \left\{ \sqrt{\frac{\varepsilon}{\hat{k}_0^{(1)}(0)\xi}}\, e^{\frac{i\hat{k}_0^{(1)}(0)\xi}{\varepsilon}} - \sqrt{\frac{\varepsilon}{\hat{k}_0^{(3)}(0)\xi}}\, e^{\frac{i\hat{k}_0^{(3)}(0)\xi}{\varepsilon}} \right\}. \tag{6.157}$$

6.5 Global Trapped-Wave Modes and the Quantization Condition

We now turn to the second step: constructing the global eigenmodes and deriving the quantization condition for the eigenvalues by applying the tip smoothness condition (6.11) to the asymptotic solutions obtained above.

In the three-dimensional case, applying the smooth tip condition means matching the outer solution with the tip inner solution (6.158) derived in the last section. In doing so, one needs to balance

$$
\mu_0(\varepsilon)\sqrt{\frac{2\varepsilon}{\pi}}\,d_0\left\{\sqrt{\frac{1}{\hat{k}_{*0}^{(1)}\xi}}\,\mathrm{e}^{\frac{i\hat{k}_{*0}^{(1)}\xi}{\varepsilon}} - \sqrt{\frac{1}{\hat{k}_{*0}^{(3)}\xi}}\,\mathrm{e}^{\frac{i\hat{k}_{*0}^{(3)}\xi}{\varepsilon}}\right\}
$$

$$
\Longleftrightarrow \sqrt{\frac{\xi_c}{\xi}}\left(D_1\exp\left\{\frac{i}{\varepsilon}\int_{\xi_c}^{\xi}\hat{k}_0^{(1)}\mathrm{d}\xi'\right\} + D_3\exp\left\{\frac{i}{\varepsilon}\int_{\xi_c}^{\xi}\hat{k}_0^{(3)}\mathrm{d}\xi'\right\}\right).
$$

To satisfy this matching condition, the parameter σ_0 must be a proper function of ε, such that as $\varepsilon \to 0$, the two functions, $\left\{\mathrm{e}^{i\chi_1(\varepsilon)}, \mathrm{e}^{i\chi_3(\varepsilon)}\right\}$ are of the same order of magnitude. Hereby, we have defined

$$
\begin{aligned}
\chi_1(\varepsilon) &= \frac{1}{\varepsilon}\int_0^{\xi_c}\hat{k}_0^{(1)}(\xi')\mathrm{d}\xi' \\
\chi_3(\varepsilon) &= \frac{1}{\varepsilon}\int_0^{\xi_c}\hat{k}_0^{(3)}(\xi')\mathrm{d}\xi'.
\end{aligned}
\tag{6.158}
$$

In other words, the parameter σ_0 must be properly chosen, so that the turning point ξ_c and the structure of Stokes lines, as the functions of σ_0, are arranged in such a way, that the tip $\xi = 0$ is located on the Stokes line (L_3).

Thus it follows from the matching condition that

$$
\mu_0(\varepsilon) = \varepsilon^{-\frac{1}{2}},
\tag{6.159}
$$

and

$$
\frac{D_3\mathrm{e}^{-i\chi_3}}{D_1\mathrm{e}^{-i\chi_1}} = -i\mathrm{e}^{i(\chi_1-\chi_3)} = -\left(\frac{\hat{k}_0^{(1)}(0)}{\hat{k}_0^{(3)}(0)}\right)^{\frac{1}{2}}.
\tag{6.160}
$$

Equation (6.161) is the quantization condition for the eigenvalues and it may be rewritten in the form

$$
\frac{1}{\varepsilon}\int_0^{\xi_c}\left(\hat{k}_0^{(1)} - \hat{k}_0^{(3)}\right)\mathrm{d}\xi = \left(2n + 1 + \frac{1}{2} + \frac{\theta_0}{2}\right)\pi - \frac{i}{2}\ln\alpha_0
\tag{6.161}
$$

$$
(n = 0, \pm 1, \pm 2, \pm 3, \cdots),
$$

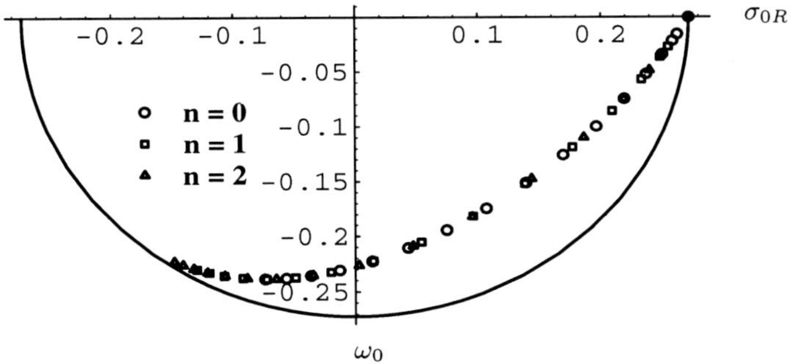

Fig. 6.5. The variation of eigenvalues σ_0 with ε in the complex σ_0-plane

where

$$\alpha_0 \, e^{i\theta_0 \pi} = \frac{\hat{k}_0^{(1)}(0)}{\hat{k}_0^{(3)}(0)} \, . \tag{6.162}$$

This quantization condition gives a discrete set of complex eigenvalues

$$\sigma_0^{(n)} \quad (n = 0, \pm 1, \pm 2, \cdots) \, ,$$

which are functions of ε. In Figs. 6.6 and 6.7, we show the variations of σ_{0R} and ω_0 of global trapped-wave (GTW) modes $n = 0, 1, 2, 3$, for the case $\hat{\alpha} = 1$. The system under consideration has no real spectrum. It is seen that the system allows a unique neutral n mode ($\sigma_R = 0$) with the eigenvalue $\sigma = -i\omega_{*n}$ when $\varepsilon = \varepsilon_{*n}$ (where $\varepsilon_{*0} = \varepsilon_* > \varepsilon_{*1} > \varepsilon_{*2} > \cdots$). Here, the critical number ε_* corresponds to the neutrally stable mode with the index $n = 0$. Obviously, when $\varepsilon > \varepsilon_*$ the system will be absolutely stable. When $\varepsilon_{*1} < \varepsilon < \varepsilon_{*0}$, the system has one growing mode and infinitely many decaying modes; when $\varepsilon_{*2} < \varepsilon < \varepsilon_{*1}$, the system has two growing modes and, in general, when $\varepsilon_{*m} < \varepsilon < \varepsilon_{*(m-1)}$, the system has m growing modes. As $\varepsilon \to 0$, the eigenvalues of these growing modes apparently tend to the limit $\sigma_0 = (0.2722, 0.0)$, which corresponds to the maximum growth rate of the Mullins–Sekerka instability We also show the variation of the eigenvalues on the complex σ_0-plane with ε in Fig. 6.5. It is very interesting to see that in the leading order approximation, all eigenvalues of the modes $n = 0, 1, 2, \cdots$ are on the same curve in the complex σ_0-plane.

In Table 6.1, we list both the zeroth- and the first-order approximate eigenvalues of the first four modes for the case $\varepsilon = 0.1, \eta_0^2 = 0.01$.

In the leading order approximation, the eigenvalues $\sigma \approx \sigma_0$ are independent of the Peclet number η_0^2. We have calculated that the global neutrally stable mode with the index $n = 0$ has the eigenvalue $\sigma = -i\omega_*^{(0)} = -0.21291i$. It corresponds to the critical number

$$\varepsilon = \varepsilon_*^{(0)} = 0.1590. \tag{6.163}$$

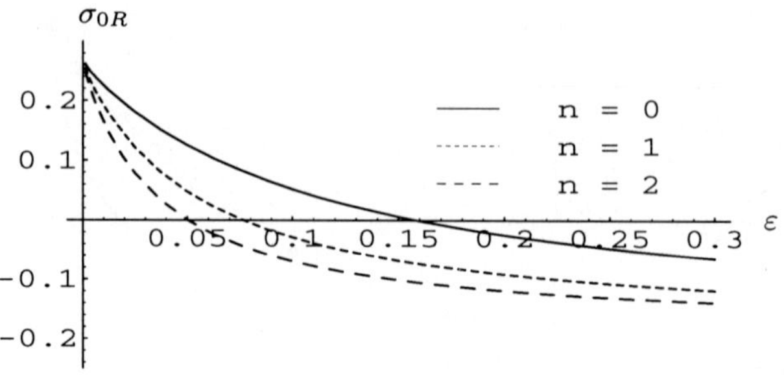

Fig. 6.6. The variations of the real part of the zeroth-order approximation of eigenvalues, σ_{0R}, of 3D, axially symmetrical GTW modes ($n = 0, 1, 2$) with ε for the case $\hat{\alpha} = 1$

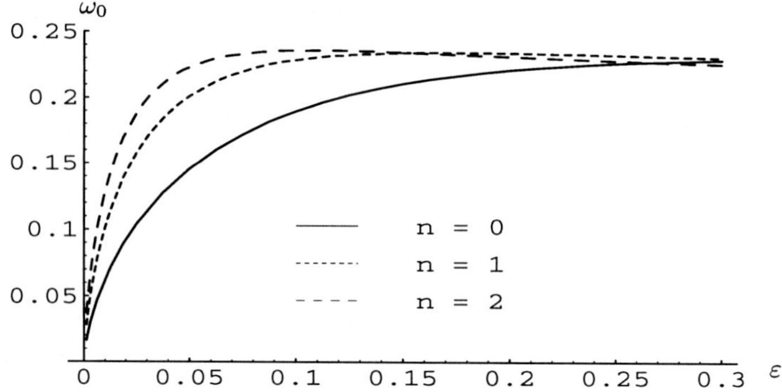

Fig. 6.7. The variations of the imaginary part of the zeroth-order approximation of eigenvalues, ω_0, of 3D, axially symmetrical GTW modes ($n = 0, 1, 2$) with ε for the case $\hat{\alpha} = 1$

In the first-order approximation, the eigenvalues $\sigma \approx \sigma_0 + \varepsilon\sigma_1$ will be a function of the Peclet number η_0^2. Consequently, the eigenvalue $\sigma = -i\omega_*^{(1)}$ of the neutral mode ($n = 0$), as well as the corresponding critical number $\varepsilon = \varepsilon_*^{(1)}$, are functions of the Peclet number $\mathrm{Pe}_0 = \eta_0^2$. For small undercooling ($\eta_0^2 \ll 1$), such a dependence is very insensitive. We find that

$$\varepsilon_*^{(1)} \approx 0.1108, \quad \omega_*^{(1)} \approx 0.2183. \tag{6.164}$$

However, for large undercooling, say, $\mathrm{Pe}_0 > 1.0$, the situation is changed. The critical number $\varepsilon_*^{(1)}$ decreases rapidly as the undercooling temperature increases (see Table 6.2).

The global mode solutions obtained above have important physical implications. A wave diagram for these global modes is sketched in Fig. 6.8. It is seen that an incident outgoing wave $W_0^{(+)}$ from the tip collides with an

Table 6.1. The eigenvalues of 3D GTW Modes ($\varepsilon = 0.1$; $\eta_0^2 = 0.01$)

n	σ_0	$\sigma = \sigma_0 + \varepsilon\sigma_1$	V_{p}
0	$(\ 0.05205, -0.1896)$	$(\ 0.01098, -0.2118)$	1.0334
1	$(-0.02853, -0.2284)$	$(-0.04601, -0.2372)$	0.9743
2	$(-0.06899, -0.2355)$	$(-0.07941, -0.2399)$	0.9259
3	$(-0.09294, -0.2357)$	$(-0.1001,\ -0.2382)$	0.8908

Table 6.2. The zeroth and first-order approximations of the critical numbers $\varepsilon_*^{(0)}$, $\varepsilon_*^{(1)}$ and the corresponding frequencies $\omega_*^{(0)}$, $\omega_*^{(1)}$ of the GTW neutral modes ($n = 0$) for various Peclet number $\mathrm{Pe}_0 = \eta_0^2$

$\mathrm{Pe}_0 = \eta_0^2$	T_∞	$\varepsilon_*^{(0)}$	$\omega_*^{(0)}$	$\varepsilon_*^{(1)}$	$\omega_*^{(1)}$	V_{p}
0.001	$-0.3514\mathrm{E}{-}2$	0.1590	0.2129	0.1108	0.2183	1.0241
0.01	$-0.2375\mathrm{E}{-}1$	0.1590	0.2129	0.1107	0.2184	1.0240
0.1	-0.1297	0.1590	0.2129	0.1094	0.2193	1.0235
1	-0.4615	0.1590	0.2129	0.09754	0.2266	1.0183
10	-0.8521	0.1590	0.2129	0.04185	0.2038	1.0027

incoming wave $W_0^{(T)}$ from the far-field at the point ξ_c' on the anti-Stokes line (A_2); the collision generates an incoming wave $W_0^{(-)}$ propagating towards to the tip region. This incoming wave $W_0^{(-)}$ is then reflected at the tip region, and again becomes an outgoing wave $W_0^{(+)}$. The waves appear trapped in the sector (S_2) between the tip point and the point ξ_c'. No wave escapes beyond the anti-Stokes line (A_2). This is the reason why we call these global modes the *Global Trapped-Wave* (GTW) modes. In the far-field, the solution $\tilde{h}(\xi, t)$ describes a long outgoing H_3 wave.

It is of interest to examine the phase velocity of the traveling wave in the far-field $\xi \to \infty$. In doing so, we express the global mode solution (6.48) in the far-field in the form

$$\tilde{h} \sim A(\xi, t)\mathrm{e}^{\frac{\mathrm{i}}{\varepsilon}\Phi(\xi,t)} \, , \tag{6.165}$$

where

$$\Phi(\xi, t) = \int_{\xi_c}^{\xi} \mathrm{Re}\{\hat{k}_0^{(3)}\}\mathrm{d}\xi - \frac{\omega_* t}{\eta_0^2}$$
$$A(\xi, t) = \exp\left\{\frac{\sigma_{\mathrm{R}} t}{\varepsilon\eta_0^2} - \frac{1}{\varepsilon}\int_{\xi_c}^{\xi} \mathrm{Im}\{\hat{k}_0^{(3)}\}\mathrm{d}\xi\right\}. \tag{6.166}$$

As $\xi \to \infty$, for any fixed eigenvalue σ, the wave number functions $\hat{k}_0^{(i)}$ ($i = 1, 2, 3$) have the asymptotic expansion

$$\hat{k}_0^{(i)}(\xi) \sim a_0^{(i)}\xi + a_1^{(i)} + \frac{a_2^{(i)}}{\xi} + \cdots . \tag{6.167}$$

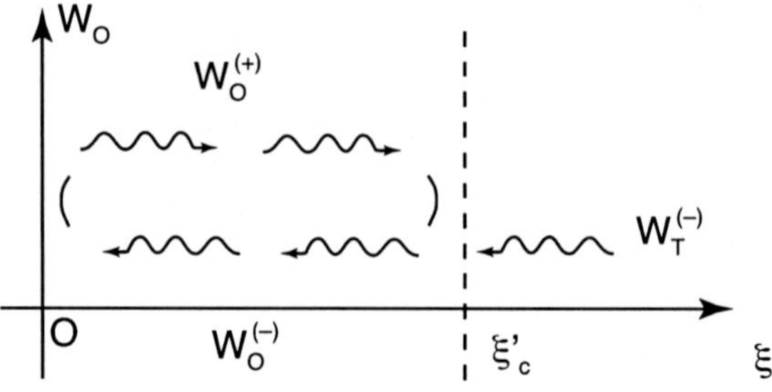

Fig. 6.8. Wave diagram of the GTW mechanism

The phase velocity of the wave along the interface $\eta = 1$ is calculated as

$$V_{\rm p} = -\left(\frac{\partial \Phi}{\partial t}\right)\left(\frac{\partial \Phi}{\partial \ell}\right) = \frac{\omega_*(1 + \xi^2)^{\frac{1}{2}}}{\mathrm{Re}\{\hat{k}_0^{(3)}(\xi)\}}, \tag{6.168}$$

and in the far-field, we have

$$V_{\rm p} \approx \frac{\omega_*}{Re\{a_0^{(3)}\}} \qquad (\xi \to \infty). \tag{6.169}$$

In the above, ℓ represents the arc length measured along the interface $\eta = 1$ starting from the tip. We have $\frac{d\ell}{d\xi} = \frac{1}{\sqrt{1+\xi^2}}$. The numerical computations show that for the GTW neutral modes the phase velocity $V_{\rm p} \approx 1.0$ (see Table 6.2). This implies that the phase velocity of the GTW modes, in the moving frame fixed at the tip, is approximately equal to the tip velocity in the laboratory frame. This result is in agreement with experimental observations.

The existence of growing GTW modes explains the origin and essence of the dendritic structure in the solidifying system. Any initial perturbation in the growth process will stimulate a spectrum of the above global modes. As $t \to \infty$, all decaying modes will vanish, while the amplitudes of the growing modes exponentially increase. Eventually, the GTW mode with the largest growth rate dominates the features of the microstructure of the dendrite. From the above analysis, one sees that to form the GTW mechanism, properly imposing the boundary conditions at both the tip and the root is important. In linear stability theory, the amplitude of a growing mode increases exponentially with time. In the real system, however, one can anticipate that when the amplitude becomes large, further increase will be suppressed by the nonlinearity and other dissipative effects that may be involved. Eventually, it appears that with the GTW mode the head of the dendrite persistently emits a long, outgoing, interfacial wave-train propagating along the interface toward the far-field with a phase velocity near unity. This forms the fantastic patterns that are observed in experiments. The selection problem for realistic dendrite growth addressed in Chap. 4 actually implies the question of what

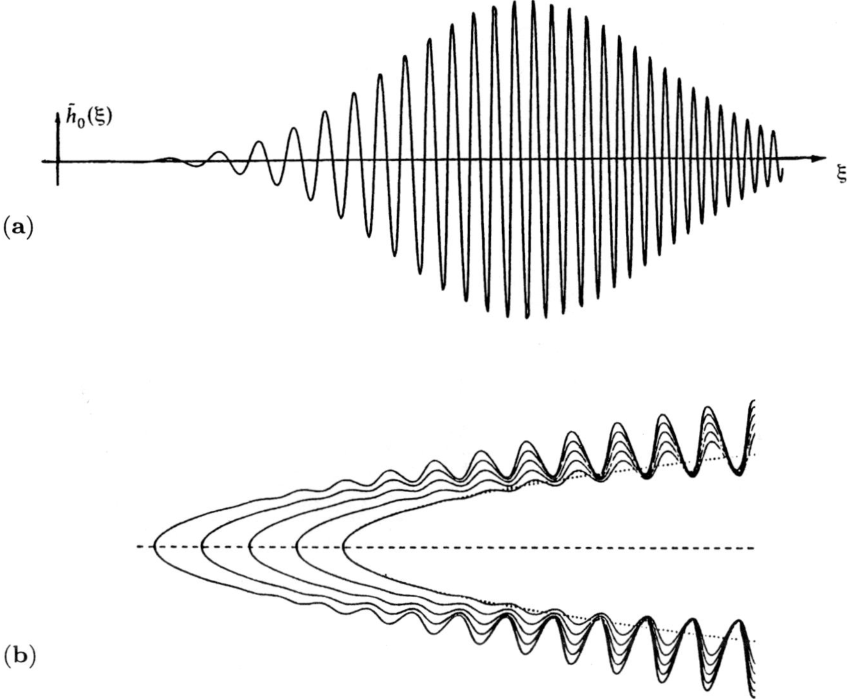

Fig. 6.9. A typical GTW neutral mode: **(a)** the graphics of the eigenfunction; **(b)** the interface shape in a time sequence

limit solution will be naturally approached by the basic state under investigation as $\tau \to \infty$? Since the spectrum of the system does not contain the zero eigenvalue ($\sigma = 0$), dendrite growth cannot approach a steady state. However, with a stationary pattern at the later stage of evolution realistic dendrite growth may approach a nonlinear periodic solution as $t \to \infty$. Indeed, it is observed in experiments that in the frame moving with the tip and at the later stage of the process, the amplitude of oscillation at any point on the interface is apparently time independent. In the scope of linear theory, a nonlinear limit circle solution will correspond to the neutral point of linear stability. Thus we deduce that as time tends to infinity, dendrite growth will approach the neutrally stable GTW mode. As a consequence, the selection condition of dendritic growth can be expressed as

$$\sigma_{\mathrm{R}}^*(\varepsilon_*) = 0 \,. \tag{6.170}$$

The critical number ε_* is directly connected with the selected dendrite's tip velocity, tip radius, as well as the oscillation frequency of the dendrite. In fact, if one uses the capillary length ℓ_{c} as the length scale, one can write the

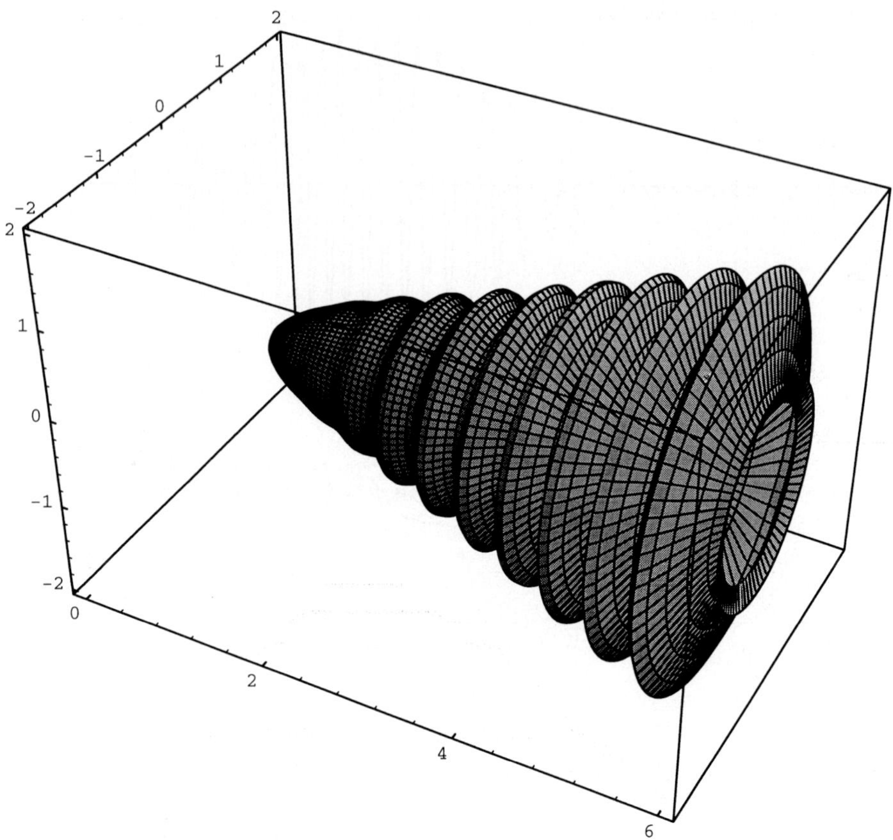

Fig. 6.10. The 3D graphics of the interface shape of a typical GTW neutral mode

dimensionless tip velocity as

$$U_{\text{tip}} = \frac{U\ell_{\text{c}}}{\kappa_{\text{T}}} = \frac{\ell_{\text{c}}}{\ell_{\text{T}}} = \varepsilon_*^2 \text{Pe}_0^2 , \qquad (6.171)$$

and the dimensionless tip radius and frequency of oscillation as

$$R_{\text{tip}} = \frac{\ell_{\text{t}}}{\ell_{\text{T}}} = \frac{\ell_{\text{t}}}{\ell_{\text{T}}} \frac{\ell_{\text{T}}}{\ell_{\text{c}}} = \frac{\text{Pe}}{\varepsilon_*^2 \eta_0^4} = \frac{\text{Pe}}{\varepsilon_*^2 \text{Pe}_0^2} \qquad (6.172)$$

$$\Omega_* = \frac{\omega_*}{\eta_0^2 \varepsilon_*} = \frac{\omega_*}{\text{Pe}_0 \varepsilon_*} . \qquad (6.173)$$

We recall that $\text{Pe}_0 = \eta_0^2$ is the Peclet number for the case of zero surface tension.

The eigenfunction of a typical selected global neutrally stable mode and its interface shape in a time sequence are shown in Fig. 6.9. The 3D graphics of dendrite growth is shown in Fig. 6.10.

6.6 Global Interfacial Wave Instability of Two-Dimensional Dendrite Growth

For completeness, we attempt to give the corresponding results for the two-dimensional case in this section. One can adopt the same procedure as in the previous sections to deal with this case.

Before proceeding, let us summarize this approach:

1. First of all solve the steady growth problem with zero surface tension ($\varepsilon = 0$). Many systems often allow an analytical solution, such as the Ivantsov solution for this special case.
2. For the general case ($\varepsilon \neq 0$), consider the nonclassic steady needle growth problem (or the needle formation problem) and treat the nonclassic steady needle solution (or the 'nearly' steady needle solutions) as the basic states. Note that the exact form of these basic states is not important. For our purpose, the important things are that these solutions exist and, in the region away from the root region, that they have a regular perturbation expansion independent of the root conditions and can thus be well approximated by the Ivantsov solution. Namely,

$$T_{\mathrm{B}}(\xi, \eta, \varepsilon) = T_*(\eta) + O(\varepsilon^2)$$

$$T_{\mathrm{SB}}(\xi, \eta, \varepsilon) = O(\varepsilon^2) \tag{6.174}$$

$$\eta_{\mathrm{B}}(\xi, \varepsilon) = 1 + O(\varepsilon).$$

3. Consider the stability of the basic state. Separate the general unsteady solutions into

$$T = T_{\mathrm{B}} + \tilde{T}(\xi, \eta, t, \varepsilon)$$

$$T_{\mathrm{S}} = T_{\mathrm{SB}} + \tilde{T}_{\mathrm{S}}(\xi, \eta, t, \varepsilon) \tag{6.175}$$

$$\eta_{\mathrm{s}} = \eta_{\mathrm{B}} + \tilde{h}(\xi, t, \varepsilon)/\eta_0^2$$

and linearize around the basic state solution in terms of the small amplitude parameter of the initially infinitesimal perturbations $\delta \ll 1$. This leads to a linear eigenvalue problem. Note that the difference between this basic solution and the Ivantsov solution is only in the second and higher-order approximations. Hence, in the linear perturbed system, for the zeroth- and the first-order approximation, one can simply replace the basic state solution by the Ivantsov solution.

4. Solve the eigenvalue problem in two steps. First, find the uniformly valid asymptotic solutions for the linear perturbed system with any fixed σ and related parameters. This system consists of the governing equations and all boundary conditions except for the smooth tip condition. Then, apply the smooth tip condition to obtain the quantization condition for the eigenvalues.

5. In finding the uniformly valid asymptotic solutions, first apply the MVE method to derive the outer solutions. This gives the local dispersion relation for the zeroth-order approximation. Some singularities of the outer solutions, which may not appear in the zeroth-order approximation but rather in the first-order approximation, will be found in the complex plane Thus, one needs to introduce the new scales and derive the inner equations in the inner regions of the singular points and then match the outer solutions to these inner solutions and apply the root condition. In deriving the asymptotic solution, the root condition (6.11) can be replaced by the asymptotic condition (6.78), the so-called radiation condition in the far-field.

6. Finally we must apply the tip conditions.

The above recipe gives rise to the uniformly valid global modes and the quantization conditions of the eigenvalues. In the remaining part of the book, we shall always follow this recipe.

Linear perturbed system. The linearized perturbed equation for two-dimensional dendrite growth is

$$\frac{\partial^2 \tilde{T}}{\partial \xi^2} + \frac{\partial^2 \tilde{T}}{\partial \eta^2} - \eta_0^4(\xi^2 + \eta^2)\frac{\partial \tilde{T}}{\partial t} + \eta_0^2\left(\xi\frac{\partial \tilde{T}}{\partial \xi} - \eta\frac{\partial \tilde{T}}{\partial \eta}\right) = 0. \quad (6.176)$$

The boundary conditions for the two-dimensional case are formally the same as for the three-dimensional case. Exceptions are the Gibbs–Thomson interface condition due to the different expression for the curvature and the tip conditions. For the reader's convenience, we list all the conditions below:

1. At $\eta \to \infty$

$$\tilde{T} \to 0. \tag{6.177}$$

2. At $\eta \to 0$

$$\tilde{T}_S = O(1). \tag{6.178}$$

3. On the interface of the Ivantsov solution, $\eta = 1$, one has
   (i) the thermodynamic equilibrium condition

$$\tilde{T} = \tilde{T}_S + \tilde{h} + O(\varepsilon^2), \tag{6.179}$$

   (ii) the Gibbs–Thomson condition

$$\tilde{T}_S = \frac{\varepsilon^2}{S(\xi)}\left\{\frac{\partial^2 \tilde{h}}{\partial \xi^2} + \frac{\xi}{S^2(\xi)}\frac{\partial \tilde{h}}{\partial \xi} - \frac{1}{S^2(\xi)}\tilde{h}\right\} + O(\varepsilon^2), \quad (6.180)$$

   (iii) the heat balance condition

$$\frac{\partial}{\partial \eta}\left(\tilde{T} - \tilde{T}_S\right) + \eta_0^2 S^2(\xi)\frac{\partial \tilde{h}}{\partial t} + \xi\frac{\partial \tilde{h}}{\partial \xi} + \varepsilon(1 + \eta_0^2)\tilde{h} + O(\varepsilon^2), (6.181)$$

where

$$S(\xi) = \sqrt{1 + \xi^2}\,. \tag{6.182}$$

4. In the far-field, as $\xi \to \infty$, the solution describes an outgoing wave $H_3(\xi)$,

$$\tilde{h} \sim C_3\, H_3(\xi)\,. \tag{6.183}$$

5. At the tip $\xi = 0$, $\eta = 1$, two-dimensional dendrite growth allows two different types of smooth tip condition:
   (i) for a symmetrical mode (S-mode)

$$\frac{\partial}{\partial \xi}\{\tilde{T},\ \tilde{T}_{\mathrm{S}},\ \tilde{h}\} = 0, \qquad \{\tilde{T},\ \tilde{T}_{\mathrm{S}},\ \tilde{h}\} < \infty; \tag{6.184}$$

   (ii) for an anti-symmetrical mode (A-mode)

$$\frac{\partial}{\partial \xi}\{\tilde{T},\ \tilde{T}_{\mathrm{S}},\ \tilde{h}\} < \infty, \qquad \{\tilde{T},\ \tilde{T}_{\mathrm{S}},\ \tilde{h}\} = 0\,. \tag{6.185}$$

The above system gives a linear eigenvalue problem which can been solved using the same approach as in the last section for the three-dimensional case. For the two-dimensional case, the problem is simpler. The present system does not have a singularity at the tip. Consequently, there is no need to look for the tip inner solution. Nevertheless, it does have the same turning point singularity at $\xi = \xi_c$ as in three-dimensional case.

Multiple variables form of the perturbed system. Define the fast and slow variables $(\xi, \eta, \xi_+, \eta_+, t_+)$ as in the three-dimensional case. In the outer region, away from the singular point ξ_c, we can make the following MVE for the perturbed state:

$$\tilde{T} = \left\{\tilde{T}_0(\xi,\eta,\xi_+,\eta_+) + \varepsilon\tilde{T}_1(\xi,\eta,\xi_+,\eta_+) + \cdots\right\} e^{\sigma t_+}$$

$$\tilde{h} = \left\{\tilde{h}_0(\xi,\xi_+) + \varepsilon\tilde{h}_1(\xi,\xi_+) + \cdots\right\} e^{\sigma t_+}$$

$$k = k_0 + \varepsilon k_1 + \varepsilon^2 k_2 + \cdots$$

$$g = k_0 + \varepsilon k_1 + \varepsilon^2 g_2 + \cdots \tag{6.186}$$

$$g_{\mathrm{s}} = k_0 + \varepsilon k_1 + \varepsilon^2 g_{\mathrm{s}2} + \cdots$$

$$\sigma = \sigma_0 + \varepsilon\sigma_1 + \varepsilon^2\sigma_2 + \cdots\,,$$

where we have set $g_0 = g_{\mathrm{s}0} = k_0$ and $g_1 = g_{\mathrm{s}1} = k_1$, as in the three-dimensional case. The converted system, with the multiple variables, is

$$\left(k^2\frac{\partial^2}{\partial\xi_+^2} + g^2\frac{\partial^2}{\partial\eta_+^2}\right)\tilde{T} = \varepsilon\eta_0^2(\xi^2 + \eta^2)\frac{\partial\tilde{T}}{\partial t_+} - \varepsilon^2\left(\frac{\partial^2}{\partial\xi^2} + \frac{\partial^2}{\partial\eta^2}\right)\tilde{T}$$

$$+ \varepsilon\eta_0^2\xi\left(k\frac{\partial}{\partial\xi_+} + \varepsilon\frac{\partial}{\partial\xi}\right)\tilde{T} - \varepsilon\eta_0^2\eta\left(g\frac{\partial}{\partial\eta_+} + \varepsilon\frac{\partial}{\partial\eta}\right)\tilde{T}$$

$$- \varepsilon\left(2k\frac{\partial^2}{\partial\xi\partial\xi_+} + 2g\frac{\partial^2}{\partial\eta\partial\eta_+} + \frac{\partial k}{\partial\xi}\frac{\partial}{\partial\xi} + \frac{\partial g}{\partial\eta}\frac{\partial}{\partial\eta}\right)\tilde{T}\,. \tag{6.187}$$

Since most of the boundary conditions in the multiple variables form are formally the same as those in the three-dimensional case, and converting the up-stream condition, tip smooth condition, and the root condition into the multiple variables form are trivial, we shall only give the interface conditions. At the interface, $\eta_+ = 0, \eta = 1,$

(i)

$$\tilde{T} = \tilde{T}_{\mathrm{S}} + \tilde{h} + O(\varepsilon^2)\,, \tag{6.188}$$

(ii)

$$\tilde{T}_{\mathrm{S}} = \frac{1}{S(\xi)}\left\{\left(k^2\frac{\partial^2}{\partial\xi_+^2} + 2\varepsilon k\frac{\partial^2}{\partial\xi_+\partial\xi} + \varepsilon\frac{\partial k}{\partial\xi}\frac{\partial}{\partial\xi_+} + \varepsilon^2\frac{\partial^2}{\partial\xi^2}\right)\right.$$

$$\left. +\frac{\varepsilon\xi}{S^2(\xi)}\left(k\frac{\partial}{\partial\xi_+} + \varepsilon\frac{\partial}{\partial\xi}\right) - \frac{\varepsilon^2}{S^2(\xi)}\right\}\tilde{h} + O(\varepsilon^2)\,, \tag{6.189}$$

(iii)

$$\left(g\frac{\partial}{\partial\eta_+} + \varepsilon\frac{\partial}{\partial\eta}\right)\tilde{T} - \left(g_{\mathrm{s}}\frac{\partial}{\partial\eta_+} + \varepsilon\frac{\partial}{\partial\eta}\right)\tilde{T}_{\mathrm{S}} + S^2(\xi)\frac{\partial\tilde{h}}{\partial t_+}$$

$$+\xi\left(k\frac{\partial}{\partial\xi_+} + \varepsilon\frac{\partial}{\partial\xi}\right)\tilde{h} + \varepsilon\left(1 + \eta_0^2\right)\tilde{h} + O(\varepsilon^2) = 0\,. \tag{6.190}$$

Zeroth-order approximation solution. By substituting (6.187) into the system (6.188)–(6.191), one can successively derive each order of approximation in the outer region. At the zeroth-order, one obtains the same normal mode solutions as in the three-dimensional case:

$$\tilde{T}_0 = A_0(\xi,\eta)\ \exp\left(i\xi_+ - \eta_+\right)$$

$$\tilde{T}_{\mathrm{S}0} = A_{\mathrm{S}0}(\xi,\eta)\ \exp\left(i\xi_+ + \eta_+\right) \tag{6.191}$$

$$\tilde{h}_0 = \hat{D}_0\ \exp\left(i\hat{\xi}_+\right)\ .$$

The coefficient $\hat{D}_0$ is an arbitrary constant; the wave number function $\hat{k}_0(\xi) = k_0(\xi, 1)$ is subject to the same form of local dispersion formula as in the three-dimensional case,

$$\sigma_0 = \Sigma(\xi, \hat{k}_0) = \frac{\hat{k}_0}{S^2}\left(1 - \frac{2\hat{k}_0^2}{S}\right) - \frac{i\xi}{S^2}\hat{k}_0\,. \tag{6.192}$$

For the more general case, $\hat{\alpha} = \frac{\kappa_{\mathrm{TS}}}{\kappa_{\mathrm{T}}} \geq 0$, one obtains the following local dispersion relation:

$$\sigma_0 = \Sigma(\xi, \hat{k}_0) = \frac{\hat{k}_0}{S^2}\left(1 - \frac{(1 + \hat{\alpha})\hat{k}_0^2}{S}\right) - \frac{i\xi}{S^2}\hat{k}_0\,. \tag{6.193}$$

The general solution in the outer region then has the same form as in the three-dimensional case.

$$\tilde{h} = D_1 \exp\left(\frac{\sigma t}{\varepsilon \eta_0^2} + \frac{\mathrm{i}}{\varepsilon}\int_{\xi_c}^{\xi} \hat{k}_0^{(1)}\,\mathrm{d}\xi_1\right)$$

$$+ D_3 \exp\left(\frac{\sigma t}{\varepsilon \eta_0^2} + \frac{\mathrm{i}}{\varepsilon}\int_{\xi_c}^{\xi} \hat{k}_0^{(3)}\,\mathrm{d}\xi_1\right), \tag{6.194}$$

where the coefficients $\{D_1, D_3\}$ are arbitrary constants to be determined. The structure of the Stokes lines is also the same as for the three-dimensional case.

First-order approximation solution of two-dimensional dendrite growth. In the first-order approximation, we have

$$k_0^2\left(\frac{\partial^2}{\partial \xi_+^2} + \frac{\partial^2}{\partial \eta_+^2}\right)\tilde{T}_1 = a_0 e^{\mathrm{i}\xi_+ - \eta_+}$$

$$k_0^2\left(\frac{\partial^2}{\partial \xi_+^2} + \frac{\partial^2}{\partial \eta_+^2}\right)\tilde{T}_{S1} = a_{S0} e^{\mathrm{i}\xi_+ + \eta_+}, \tag{6.195}$$

where

$$\begin{cases}
a_0 = 2k_0\left(\dfrac{\partial A_0}{\partial \eta} - \mathrm{i}\dfrac{\partial A_0}{\partial \xi}\right) + A_0\left\{\sigma\eta_0^2(\xi^2 + \eta^2) + k_0\eta_0^2(\mathrm{i}\xi + \eta)\right. \\
\qquad \left. + \dfrac{k_0}{\eta} - \dfrac{\mathrm{i}k_0}{\xi} - \mathrm{i}\dfrac{\partial k_0}{\partial \xi} + \dfrac{\partial k_0}{\partial \eta}\right\} \\[2mm]
a_{S0} = -2k_0\left(\dfrac{\partial A_{S0}}{\partial \eta} + \mathrm{i}\dfrac{\partial A_{S0}}{\partial \xi}\right) + A_{S0}\left\{\sigma\eta_0^2(\xi^2 + \eta^2) + k_0\eta_0^2(\mathrm{i}\xi - \eta)\right. \\
\qquad \left. - \dfrac{k_0}{\eta} - \dfrac{\mathrm{i}k_0}{\xi} - \mathrm{i}\dfrac{\partial k_0}{\partial \xi} - \dfrac{\partial k_0}{\partial \eta}\right\}.
\end{cases}$$

To eliminate the secular term on the right-hand side of (6.196), we must set

$$a_0 = a_{S0} = 0. \tag{6.196}$$

Hence, we have

$$\hat{D}_0 Q_0 = \frac{\partial}{\partial \eta}\left(A_0 - A_{S0}\right)\Big|_{\eta=1} = \mathrm{i}\frac{\partial}{\partial \xi}\left(\hat{A}_0 + \hat{A}_{S0}\right)$$

$$- (\hat{A}_0 + \hat{A}_{S0})\left[\frac{\sigma_0\eta_0^2}{2k_0}S^2(\xi) + \frac{\mathrm{i}}{2}\left(\xi\eta_0^2 - \frac{\mathrm{d}\log k_0}{\mathrm{d}\xi}\right)\right]$$

$$- (\hat{A}_0 - \hat{A}_{S0})\left(\frac{\eta_0^2}{2} + \frac{1}{2\hat{k}_0}\frac{\partial k_0}{\partial \eta}\Big|_{\eta=1}\right). \tag{6.197}$$

The mode solutions of the first-order approximation are:

$$\tilde{T}_1 = A_1(\xi, \eta) \exp\left(\mathrm{i}\xi_+ - \eta_+\right)$$

$$\tilde{T}_{S1} = A_{S1}(\xi, \eta) \exp\left(\mathrm{i}\xi_+ + \eta_+\right) \tag{6.198}$$

$$\tilde{h}_1 = \hat{D}_1 \exp\left(\mathrm{i}\hat{\xi}_+\right) .$$

Setting

$$\left\{ \begin{array}{l} \hat{k}_1(\xi) = k_1(\xi, 1) \\[2mm] \hat{A}_1(\xi) = A_1(\xi, 1) \\[2mm] \hat{A}_{S1}(\xi) = A_{S1}(\xi, 1), \end{array} \right. \tag{6.199}$$

from (6.197)–(6.200), one can derive the first-order approximate interface boundary conditions. Then, it is found that

$$\left\{ \begin{array}{l} \hat{A}_1 - \hat{A}_{S1} - \hat{D}_1 = 0 \\[3mm] \hat{A}_{S1} + \frac{\hat{k}_0^2}{S} \hat{D}_1 = I_2 \hat{D}_0 \\[3mm] -\hat{k}_0(\hat{A}_1 + \hat{A}_{S1}) + (\sigma_0 S^2 + \mathrm{i}\xi\hat{k}_0)\hat{D}_1 = I_3 \hat{D}_0, \end{array} \right. \tag{6.200}$$

where

$$I_2 = \frac{1}{S}\left(\mathrm{i}\frac{d\hat{k}_0}{d\xi} + \mathrm{i}\frac{\mathrm{i}\xi\hat{k}_0}{S^2} - 2\hat{k}_0\hat{k}_1\right) \tag{6.201}$$

$$I_3 = -S^2\sigma_1 + \hat{k}_1\left(1 - \mathrm{i}\xi - \frac{2\hat{k}_0^2}{S}\right) - \frac{\eta_0^2}{2} - \frac{\eta_0^2}{2}\left(1 - \frac{2\hat{k}_0^2}{S}\right)^2$$

$$\qquad -\mathrm{i}\frac{2\xi\hat{k}_0^2}{S^3} + \mathrm{i}\frac{5\hat{k}_0\hat{k}_0'}{S} + \frac{m_0 - \mathrm{i}}{2}\frac{\hat{k}_0'}{\hat{k}_0} . \tag{6.202}$$

The solvability condition for the above inhomogeneous system is

$$I_3 + 2\hat{k}_0 I_2 = 0. \tag{6.203}$$

Eventually, we find that

$$\hat{k}_1 = \frac{R_1(\xi)}{F(\xi)} + \frac{R_2(\xi)}{F(\xi)}\frac{\hat{k}_0'}{\hat{k}_0} , \tag{6.204}$$

where

$$F(\xi) = 1 - \mathrm{i}\xi - \frac{6\hat{k}_0^2}{S}$$

$$R_1(\xi) = S^2\sigma_1 + \left(1 + \frac{\eta_0^2}{2}\right) - \frac{\eta_0^2}{2}\left(1 - \frac{2\hat{k}_0^2}{S}\right)^2 \tag{6.205}$$

$$R_2(\xi) = \frac{m_0 - \mathrm{i}}{2} + \mathrm{i}\frac{7\hat{k}_0^2}{S} .$$

It is seen that at the root ξ_c of the equation:

$$\left(\frac{\partial \Sigma}{\partial \hat{k}_0}\right) = 0 \quad \text{or} \quad F(\xi) = 0 \tag{6.206}$$

the solution $\hat{k}_1$ has a singularity. As in the three-dimensional case, at $\xi \approx \xi_c$, we have

$$F(\xi) \sim A(\xi - \xi_c)^{\frac{1}{2}}$$

$$A = \mathrm{i}\hat{k}_0 \sqrt{\frac{24R_0}{S}}\bigg|_{\xi=\xi_c} = \mathrm{i}\frac{2^{\frac{3}{2}}3^{\frac{1}{2}}\hat{k}_0}{(\xi_c - \mathrm{i})^{\frac{3}{4}}(\xi_c + \mathrm{i})^{\frac{1}{4}}} \tag{6.207}$$

and

$$k_1 \sim \frac{R_1(\xi_c)}{A(\xi - \xi_c)^{\frac{1}{2}}} + \frac{b_1}{(\xi - \xi_c)} + O(1), \tag{6.208}$$

where

$$R_0(\xi_c) = \frac{1}{\xi_c - \mathrm{i}}$$

$$b_1 = \frac{S(\xi_c)}{24\hat{k}_0^2}\left(\frac{m_0 - \mathrm{i}}{2} + \mathrm{i}\frac{7\hat{k}_0^2(\xi_c)}{S(\xi_c)}\right). \tag{6.209}$$

It follows that as $\xi \to \xi_c$,

$$\mathrm{e}^{\mathrm{i}\int_{\xi_c}^{\xi} \hat{k}_1(\xi_1)\mathrm{d}\xi_1} \approx C_1(\xi - \xi_c)^{\mathrm{i}b_1}$$

$$\times \left\{1 + \mathrm{i}\frac{2R_1(\xi_c)}{A}(\xi - \xi_c)^{\frac{1}{2}} + a_1(\xi - \xi_c) + \cdots\right\}, \tag{6.210}$$

where a_1 and C_1 are some constants.

The global modes and quantization conditions. In the vicinity of the turning point ξ_c, two-dimensional dendrite growth has the same inner equation as that in the three-dimensional case:

$$\varepsilon^2 \frac{\mathrm{d}^2 \hat{W}_0}{\mathrm{d}\xi_*^2} + A^2(\xi - \xi_c)\hat{W}_0 = 0 \tag{6.211}$$

or

$$\frac{\mathrm{d}^2 \hat{W}_0}{\mathrm{d}\xi_*^2} + \xi_* \hat{W}_0 = 0, \tag{6.212}$$

where

$$A = \mathrm{i}\sqrt{\frac{1}{6}}\left(\frac{\xi_c + \mathrm{i}}{\xi_c - \mathrm{i}}\right)^{\frac{1}{4}} \quad \left(\frac{\pi}{2} < \arg\{A\} < \frac{3\pi}{4}\right),$$

$$\xi_* = \frac{A^{\frac{2}{3}}}{\varepsilon^{\frac{2}{3}}}(\xi - \xi_c). \tag{6.213}$$

Matching the solution of (6.213) to the outer solution (6.195) leads to the following connection conditions for the pair of coefficients of the outer solution in different sectors:

In sector (S_2):

$$D_1' = 0 \quad \text{and} \quad D_3' = D' \neq 0 \,. \tag{6.214}$$

In sector (S_1):

$$\frac{D_1}{D_3} = e^{\frac{i\pi}{2}} \,, \tag{6.215}$$

and

$$b_1 = \frac{i}{4}, \quad R_1(\xi_c) = 0 \,. \tag{6.216}$$

Thus, as in the 3D case, we find

$$\sigma_1 = \frac{1}{1 + \xi_c^2} \left\{ \frac{\eta_0^2}{2} \left(\frac{2}{3} + \frac{i\xi_c}{3} \right)^2 - \left(1 + \frac{\eta_0^2}{2} \right) \right\}, \tag{6.217}$$

and for the case of small undercooling $\eta_0^2 \ll 1$, we have

$$\sigma_1 \approx -\frac{1}{(1 + \xi_c^2)} \,. \tag{6.218}$$

The only difference between the two- and the three-dimensional case is the tip condition. Now, for the two-dimensional case, to satisfy the tip smoothness conditions, the constants D_1 and D_3 must be subject to the following conditions:

(i) For the anti-symmetrical A-modes,

$$\frac{D_3 e^{-i\chi_3}}{D_1 e^{-i\chi_1}} = -1 \,. \tag{6.219}$$

(ii) For the symmetrical, S-modes,

$$\frac{D_3 e^{-i\chi_3}}{D_1 e^{-i\chi_1}} = -\frac{\hat{k}_0^{(1)}(0)}{\hat{k}_0^{(3)}(0)} \,. \tag{6.220}$$

Thus, with the connection condition (6.216), one obtains the following quantization conditions:

(i) For the anti-symmetrical A-mode:

$$\frac{1}{\varepsilon} \int_0^{\xi_c} \left(\hat{k}_0^{(1)} - \hat{k}_0^{(3)} \right) d\xi = \left(2n + 1 + \frac{1}{2} \right) \pi \tag{6.221}$$

$$(n = 0, \pm 1, \pm 2, \pm 3, \cdots) \,.$$

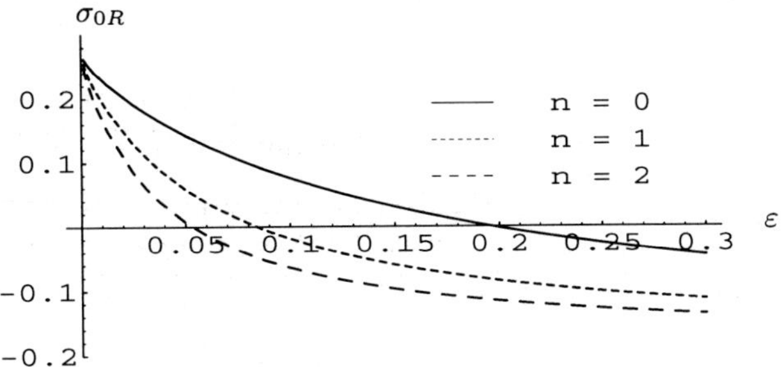

Fig. 6.11. The variations of the real part of the zeroth-order approximation of eigenvalues, σ_{0R}, of 2D, GTW A-modes $(n = 0, 1, 2)$ with ε for the case $\hat{\alpha} = 1$

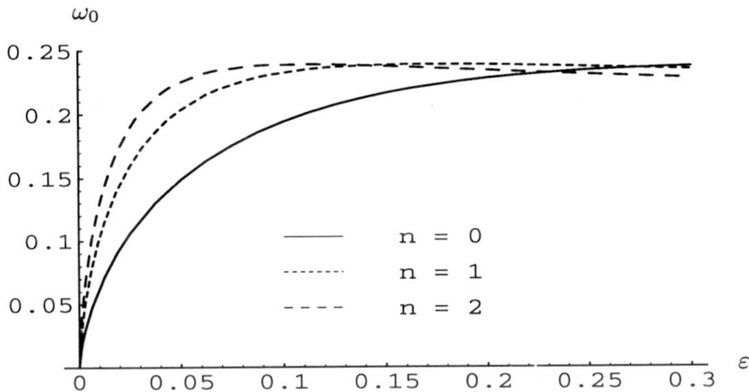

Fig. 6.12. The variations of the imaginary part of the zeroth-order approximation of eigenvalues, ω_0, of 2D, GTW A-modes $(n = 0, 1, 2)$ with ε for the case $\hat{\alpha} = 1$

(ii) For the symmetrical S-modes:

$$\frac{1}{\varepsilon} \int_0^{\xi_c} \left(k_0^{(1)} - k_0^{(3)} \right) d\xi = \left(2n + 1 + \frac{1}{2} + \theta_0 \right) \pi - i \log \alpha_0 \qquad (6.222)$$

$$\left(n = 0, \pm 1, \pm 2, \pm 3, \cdots \right),$$

where

$$\alpha_0 \, e^{i\theta_0 \pi} = \frac{\hat{k}_0^{(1)}(0)}{\hat{k}_0^{(3)}(0)} . \qquad (6.223)$$

These quantization conditions give two discrete sets of the zeroth-order approximate, complex eigenvalues σ_{0n} $(n = 0, \pm 1, \pm 2, \cdots)$.

In Figs. 6.11 and 6.12, we show for $\hat{\alpha} = 1$ the zeroth-order approximate σ_{0R} and ω_0 of GTW A-modes $(n = 0, 1, 2, 3)$ versus ε. In Figs. 6.13 and

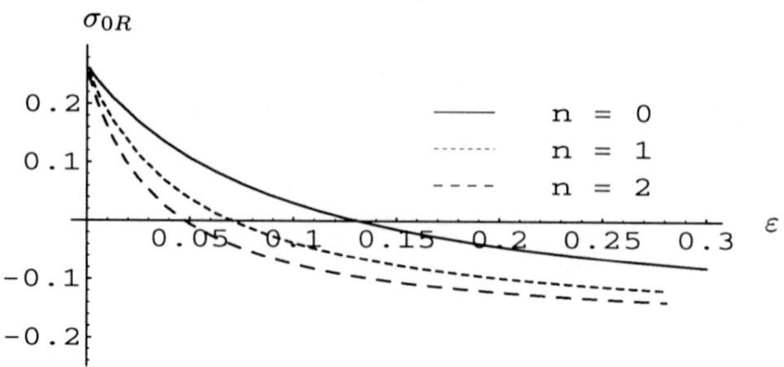

Fig. 6.13. The variations of the real part of the zeroth-order approximation of eigenvalues, σ_{0R}, of 2D, GTW S-modes ($n = 0, 1, 2$) with ε for the case $\hat{\alpha} = 1$

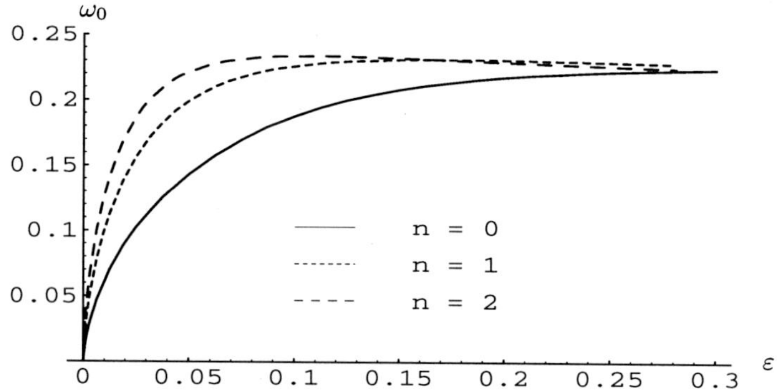

Fig. 6.14. The variations of the imaginary part of the zeroth-order approximation of eigenvalues, ω_0, of 2D, GTW S-modes ($n = 0, 1, 2$) with ε for the case $\hat{\alpha} = 1$

6.14, we show the zeroth-order approximate σ_{0R} and ω_0 of GTW S-modes ($n = 0, 1, 2, 3$) versus ε for the same case $\hat{\alpha} = 1$.

It is seen that the system allows a unique neutrally stable A-mode ($n = 0$) and a unique neutrally stable S-mode ($n = 0$); furthermore, the neutrally stable A-mode ($n = 0$) is more stable than the neutrally stable S-mode ($n = 0$).

One can also calculate the first-order approximate, complex eigenvalues $\sigma = \sigma_0 + \varepsilon\sigma_1 = \sigma_R - i\omega$ for given ε for the symmetrical modes (S-modes) and anti-symmetrical modes (A-modes) by using (6.217) and the quantization conditions (6.222) and (6.221). The results for the typical case ($\varepsilon = 0.1, \eta_0^2 = 0.01$) are listed in Tables 6.3 and 6.4.

The first-order approximations of the critical number $\varepsilon_*^{(1)}$, the corresponding frequencies $\omega_*^{(1)}$, and the phase velocity V_{p} of the interfacial wave in the far-field away from the tip, for the GTW neutral A-modes ($n = 0$) are computed with various Peclet numbers η_0^2. The numerical results for the case

Table 6.3. The eigenvalues of 2D GTW S-modes ($\varepsilon = 0.1$; $\eta_0^2 = 0.01$)

n	σ_0	$\sigma = \sigma_0 + \varepsilon\sigma_1$	V_{p}
0	(0.03028, −0.18771)	(−0.00278, −0.2115)	1.0185
1	(−0.03745, −0.2255)	(−0.05278, −0.2343)	0.9637
2	(−0.07369, −0.2331)	(−0.08321, −0.2375)	0.9189
3	(−0.09580, −0.2338)	(−0.1025, −0.2364)	0.8860

Table 6.4. The eigenvalues of 2D GTW A-modes ($\varepsilon = 0.1$; $\eta_0^2 = 0.01$)

n	σ_0	$\sigma = \sigma_0 + \varepsilon\sigma_1$	V_{p}
0	(0.07564, −0.1939)	(0.02696, −0.2124)	1.0496
1	(−0.01882, −0.2321)	(−0.03874, −0.2406)	0.9857
2	(−0.06396, −0.2383)	(−0.07538, −0.2426)	0.9333
3	(−0.08994, −0.2378)	(−0.0976, −0.2403)	0.8959

Table 6.5. The critical numbers $\varepsilon_*^{(1)}$ and the corresponding frequencies $\omega_*^{(1)}$ of 2D neutral GTW A-modes ($n = 0$) for the case $\hat{\alpha} = 1$

Pe_0	$\varepsilon_*^{(1)}$	$\omega_*^{(1)}$	V_{p}
0.001	0.1268	0.2281	1.0282
0.01	0.1267	0.2279	1.0281
0.1	0.1248	0.2261	1.0273
1.0	0.1089	0.2097	1.0209
10.0	0.0509	0.1237	1.0028

Table 6.6. The critical numbers $\varepsilon_*^{(1)}$ and the corresponding frequencies $\omega_*^{(1)}$ of 2D neutral, GTW A-modes (n=0) for the case $\hat{\alpha} = 0$

Pe_0	$\varepsilon_*^{(1)}$	$\omega_*^{(1)}$	V_{p}
0.001	0.1794	0.3226	1.0282
0.01	0.1792	0.3224	1.0281
0.1	0.1765	0.3198	1.0273
1.0	0.1540	0.2966	1.0209
10.0	0.0720	0.1750	1.0028

$\hat{\alpha} = 1$ are shown in Table 6.5, while the results for the case $\hat{\alpha} = 0$ are shown in Table 6.6.

6.7 The Comparison of Theoretical Predictions with Experimental Data

The dendrite growth theory developed above shows that dendritic growth is essentially a wave phenomenon involving the interaction of interfacial waves along the interface. This theory is therefore called the interfacial wave (IFW) theory.

The IFW theory states that when the surface tension is isotropic, in the later stage of growth, dendrite growth is not described by a steady state. Instead, it is described by a time periodic oscillatory state, the so-called global neutrally stable (GNS) state. Such a GNS state consists mathematically of three parts: (1) the Ivantsov solution, (2) the steady regular perturbation expansion (RPE) part due to the surface tension, and (3) the unsteady singular perturbation expansion (SPE) part. The interface shape of the dendrite can be expressed approximately in the form,

$$\eta_s(\xi, t) \approx 1 + \varepsilon^2 \eta_1(\xi) + \hat{h}_0(\xi) e^{\frac{\sigma t}{\eta_0^2 \varepsilon}} \qquad (\text{as } \varepsilon \to 0). \tag{6.224}$$

In the above, $\eta_1(\xi)$ is the leading term in the RPE part; while $\hat{h}_0(\xi) e^{\frac{\sigma t}{\eta_0^2 \varepsilon}}$ is the leading term in the SPE part. The parameter $\sigma = \sigma_R - i\omega$ is the eigenvalue with the expansion form

$$\sigma = \sigma_0 + \varepsilon \sigma_1 + \cdots \qquad (\text{as } \varepsilon \to 0). \tag{6.225}$$

For the selected solution, $\sigma_R = 0$.

In 1994, a series of careful experiments on free dendrite growth in pure succinonitrile (SCN) were conducted, for the first time, in the space shuttle Columbia by the research team headed by Glicksman. Under the microgravity environment, convective motion in the melt is greatly reduced [6.15], [6.16]. As expected, the new data for tip velocity and tip radius during the dendrite growth obtained by Glicksman et al. is more accurate.

Since the material SCN has a very small surface tension anisotropy, one can expect that the theoretical results for three-dimensional axially symmetric dendrite growth from a pure melt obtained in this chapter are comparable with these experimental results obtained by Glicksman et al.

According to the IFW theory, the selected tip growth velocity is determined by the critical number ε_*. Thus, with the capillary length scale ℓ_c, we can express the dimensionless tip velocity as

$$U_{\text{tip}} = \varepsilon_*^2 \text{Pe}_0^2, \tag{6.226}$$

or

$$\varepsilon_* = \frac{\sqrt{U_{\text{tip}}}}{\text{Pe}_0}. \tag{6.227}$$

The dimensionless tip radius R_{tip} can be expressed as

$$R_{\mathrm{tip}} = \frac{\ell_{\mathrm{t}}}{\ell_{\mathrm{c}}} = \frac{\mathrm{Pe}}{\varepsilon_*^2 \mathrm{Pe}_0^2}. \tag{6.228}$$

We recall that $\mathrm{Pe}_0 = \eta_0^2$ is the Peclet number for the case of zero surface tension. In the leading order approximation, $\varepsilon_* = \varepsilon_*^{(0)} = 0.1590$, while in the first-order approximation, as $\eta_0^2 \ll 1$ (which is in the range of T_∞ applied in the Glicksman's experiments), $\varepsilon_* = \varepsilon_*^{(1)} \approx 0.1108$.

The above theoretical prediction is free of adjustable parameters and can be directly tested in terms of the experimental data of Glicksman et al. It is therefore of great interest to compare quantitatively the predictions of this theory with these newest experimental data. In this section, we attempt to make such direct comparisons.

The dendrite tip velocity and tip radius. The thermodynamic data for SCN provided by Glicksman and his co-workers are listed in Table 6.7. Based on these data, one can calculate that the capillary length for SCN is $\ell_{\mathrm{c}} = 2.804 \times 10^{-7}$cm, the velocity unit is 4069.19 cm/s and the temperature unit is 23.067 K.

Table 6.7. The thermodynamic properties of SCN

mol	80.092 g/mol
ΔH	11.051 cal/g
c_{pL}	0.4791 cal g^{-1}K^{-1}
c_{pS}	0.4468 cal g^{-1}K^{-1}
$c_{\mathrm{p}} = (c_{\mathrm{pL}} + c_{\mathrm{pS}})/2$	0.4630 cal g^{-1}K^{-1}
κ_{TL}	1.127×10^{-3} m^{2}s^{-1}
κ_{TS}	1.155×10^{-3} m^{2}s^{-1}
$\kappa_{\mathrm{T}} = (\kappa_{\mathrm{TL}} + \kappa_{\mathrm{TS}})/2$	1.141×10^{-3} m^{2}s^{-1}
T_{M0}	331.233 K
γ' (Gibbs–Thomson coefficient)	6.480×10^{-6}cm K
γ	2.136×10^{-7} cal cm^{-2}

In Fig. 6.15, we show the tip velocity U_{tip} versus the undercooling temperature T_∞ and compare the experimental data with the theoretical curve determined by (6.226). In Fig. 6.16 we show the tip radius R_{tip} versus the undercooling T_∞. It is seen that the overall agreement between the theoretical curve and the experimental data is very satisfactory, especially considering that the experimental data of the tip radius has an error of about 10%. Only in the regime of small undercooling, $|\Delta T| < 0.4$ K or $|T_\infty| < 0.017$, can one see a growing deviation between the theoretical curve and the experimental data.

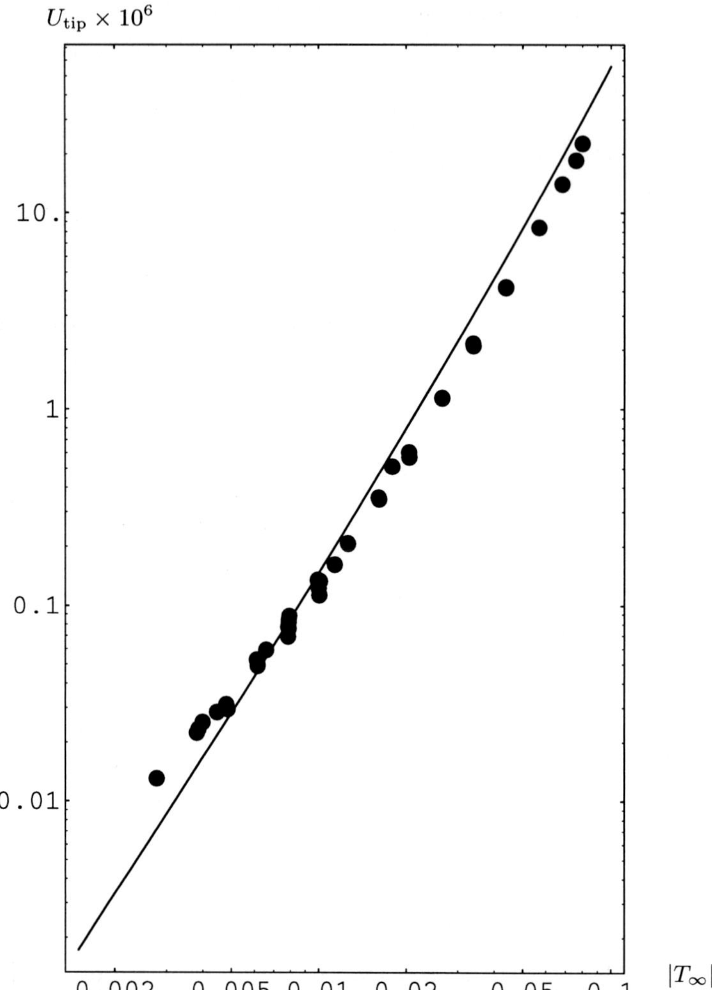

Fig. 6.15. The variation of U_{tip} with T_∞. The solid line is given by the IFW theory, while the dots are the microgravity experimental data

The critical number ε_*. To explore the deviations between the theoretical curve and the experimental data in more detail, we calculate $U_{\mathrm{tip}}^{\frac{1}{2}}/\mathrm{Pe}_0 = (\varepsilon_*)_{\mathrm{exp}}$ in terms of the data obtained from both the flight experiments and ground experiments (shown in Fig. 6.17 by two sets of the round dots, data I and data II).

It is seen that these two sets of data are well separated. The data $(\varepsilon_*)_{\mathrm{exp}}$ derived from the experiments in microgravity remain approximately constant at 0.094 within the regime of large undercooling, $|T_\infty| > 0.02$, whereas in the regime of small T_∞, starting from $|T_\infty| = 0.017$, the data $(\varepsilon_*)_{\mathrm{exp}}$ increase as

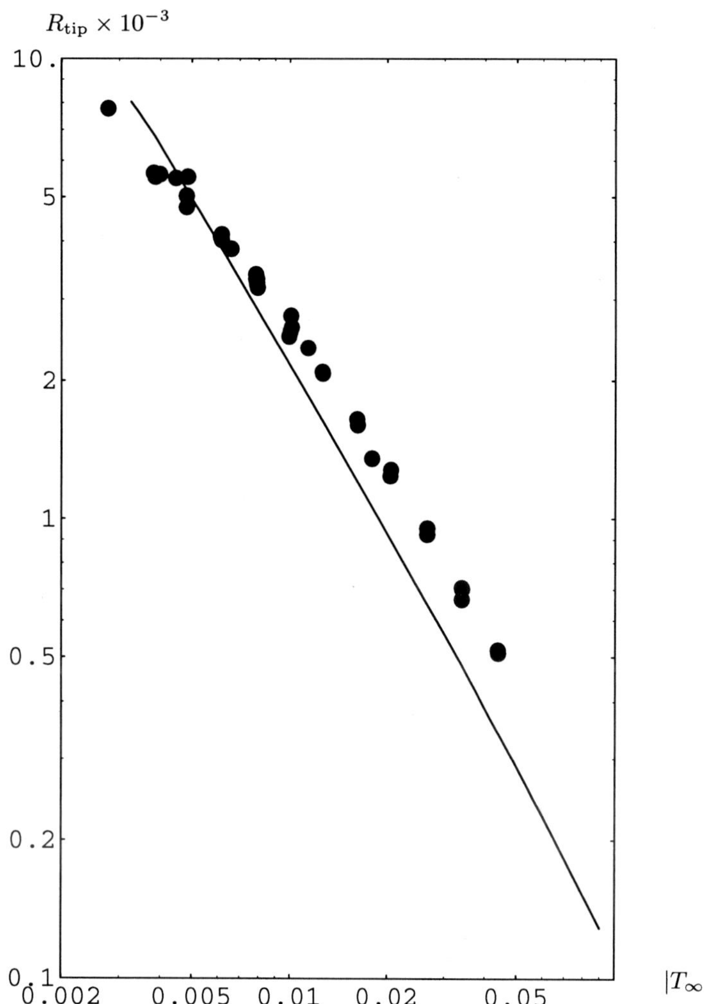

Fig. 6.16. The variation of R_tip with T_∞. The solid line is given by the IFW theory, while the dots are the microgravity experimental data

T_∞ decreases. The data derived from the ground experiments show a similar tendency, but the increase of the data $(\varepsilon_*)_\mathrm{exp}$ starts at a higher undercooling temperature $|T_\infty| = 0.1$. This fact clearly shows that the increase of $(\varepsilon_*)_\mathrm{exp}$ in the low undercooling regime is due to the convection caused by the buoyancy effect, as gravity is greatly reduced in the flight experiment but not completely eliminated. As a comparison, in Fig. 6.17 we also show the theoretical value of ε_* by the thin solid line. It is seen that in the entire undercooling regime under discussion, the theoretical curve remains flat with $\varepsilon_* \approx 0.1108$, which agrees with the experimental data in the high T_∞ regime. The quantitative

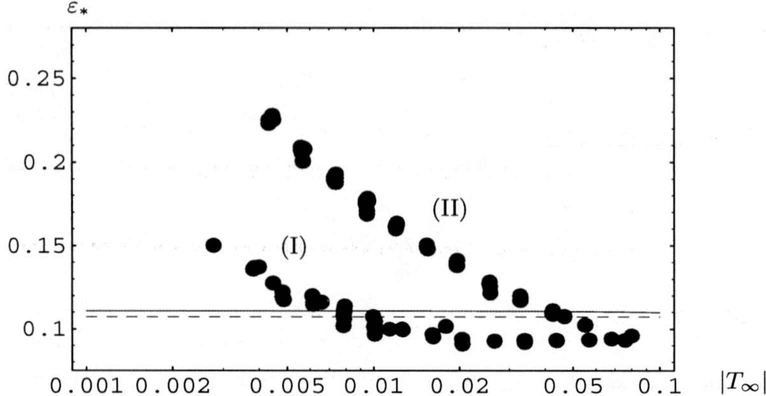

Fig. 6.17. Comparison of ε_* with $(\varepsilon_*)_{\mathrm{exp}}$ within the region of undercooling, $0.002 < |T_\infty| < 0.1$. The sets of dots (I) and (II) are respectively calculated from the microgravity experimental data and ground experimental data. The solid line is the stability criterion $\varepsilon_* = 0.01108$ predicted by IFW theory, while the dashed line is the stability criterion with the anisotropy correction $\tilde{\varepsilon}_* = 0.1073$

discrepancy between the theoretical and experimental data is of $O(\varepsilon_*^2)$, which can be attributed to the following sources.

1. In the calculation of the critical number ε_* the higher-order terms of $O(\varepsilon^2)$ in the asymptotic solution (6.20) have been neglected. As a consequence, the ε_* obtained is expected to have an error up to $O(\varepsilon_*^2)$.
2. The theory given above neglected the effects of many physical parameters such as the anisotropy of surface tension α_4, kinetic attachment, convection motion, nonlinearity, etc. These effects are presumed to be of secondary importance for the selection criterion for the materials under investigation, but they do have small quantitative effects on the numerical values of ε_*. For instance, the anisotropy for SCN is $\alpha_4 \approx 0.075$. Taking this effect into account, we can calculate $\tilde{\varepsilon}_* = 0.1073$ in terms of the formula to be derived in Chap. 7. This number is shown by the dashed line in Fig. 6.17, which displays an improvement in the agreement with experimental data ($\varepsilon_{\mathrm{exp}} \approx 0.094$), in the large undercooling regime.
3. The convection motion in the melt not only affects the stability criterion ε_*, but also significantly changes the Peclet number of growth for the case of zero surface tension by some factor, $\tilde{\mathrm{Pe}}_0$. When convection is induced by the buoyancy effect, this factor will be a function of the Grashof number Gr measuring the strength of convection, and the Prandtl number, $\mathrm{Pr} = \frac{\nu}{\kappa_{\mathrm{T}}}$. Therefore, the resultant Peclet number can be written in the form

$$\tilde{\mathrm{Pe}}_0 = \mathrm{Pe}_0 H(\mathrm{Gr}, \mathrm{Pr}). \tag{6.229}$$

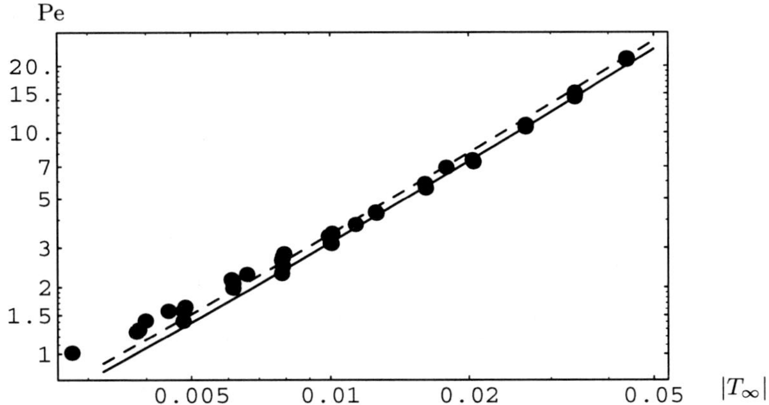

Fig. 6.18. The variation of Pe number with T_∞

Accordingly,

$$(\varepsilon_*)_{\mathrm{exp}} = \varepsilon_* H(\mathrm{Gr}, \mathrm{Pr}). \qquad (6.230)$$

How to determine the factor H and the critical number ε_* as functions of Gr and Pr theoretically is an interesting and open subject, but its discussion is beyond the scope of the present book. For discussions of the effect of convective motion on dendrite growth, the reader is referred to the articles [6.17]–[6.20] and the references therein.

The Peclet number. In Chap. 4, we obtained the formula for the Peclet number,

$$\mathrm{Pe} = \frac{\mathrm{Pe}_0}{1 - \varepsilon_*^2 \eta_1''(0)/2}. \qquad (6.231)$$

In Fig. 6.18, we show the Peclet number Pe versus T_∞, in which the dashed line represents the Ivantsov solution $\mathrm{Pe}_0 = \eta_0^2$, while the solid line represents the Peclet number including the modification of surface tension $\varepsilon = \varepsilon_*$. It is seen that the overall agreement in both figures is again very satisfactory.

Some remarks about the parameter σ_*. It is now necessary and appropriate to make some remarks about the parameter σ_*. In the literature, for historical reasons, some researchers used to employ the parameter

$$\sigma_* = \frac{2}{U_{\mathrm{tip}} R_{\mathrm{tip}}^2} \qquad (6.232)$$

as the selection criterion. As we mentioned in Chap. 1, the parameter σ_* was first introduced by Langer and Müller-Krumbhaar in the so-called marginal stability hypothesis (MSH) in 1979 (e.g. [1.13]) as the selection criterion.

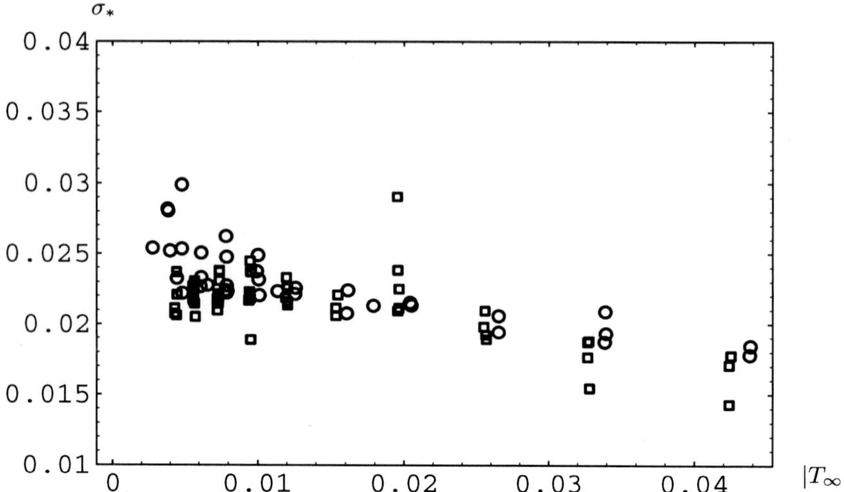

Fig. 6.19. The variation of σ_* calculated with the experimental data versus T_∞. The circles are the microgravity data, while the squares are the terrestrial experimental data

On the basis of scaling analysis, Langer and Müller-Krumbhaar proposed the parameter $\sigma = \frac{2\ell_T\ell_c}{\ell_t^2}$. They assumed that the system of dendrite growth is linearly unstable as $\sigma < \sigma_*$ and that the selected dendrite growth would be in the marginal stable state corresponding to this critical number σ_*. As a hypothesis, Langer and Müller-Krumbhaar further proposed that the critical number $\sigma_* \approx 0.0253$ is a universal constant, independent of the undercooling temperature.

Glicksman et al. calculated σ_* using the experimental data of U_{tip} and R_{tip} obtained in both their own flight experiments and the ground experiments, and plotted the result against undercooling. They found that these sets of data for σ_* both under terrestrial and microgravity conditions, are indistinguishable from one another. In Fig. 6.19, we show these data versus T_∞. In contrast to the data ε_* versus T_∞ shown in Fig. 6.17, the two sets of data, σ_*, are well mixed up. This result shows that at a higher level of precision, σ_*, unlike the parameter ε_*, is not suitable for use as a criterion for dendrite growth, since it remains insensitive to such a large change of growth conditions.

Another significant difference between the parameter σ_* introduced by MSH and the critical number ε_* derived by IFW theory is that σ_* was claimed by MSH as a universal constant independent of the undercooling, whereas ε_* is derived as a function of the undercooling T_∞. Especially in the large undercooling regime, the numerical value of ε_* is quite sensitive to the variation of T_∞. Only for small undercooling, is the critical number ε_* approximately a constant, as we have seen in the previous sections. In reality, the parameter σ_* will not be a universal constant. It is only in the regime of small undercooling that the hypothesis of constant σ_* is correct and $\sigma_* \approx 2\varepsilon_*^2$.

References

6.1   C. C. Lin, '*The Theory of Hydrodynamic Stability*', (Cambridge University Press, Cambridge 1955).

6.2   P. G. Drazin and W. H. Reid, '*Hydrodynamic Stability*', (Cambridge University Press, Cambridge 1971).

6.3   C. C. Lin, and Y. Y. Lau, "On Spiral Waves in Galaxies — A Gas Dynamic Approach", SIMA. J. Appl. Math. **29**, No. 2, pp. 352–370, (1975).

6.4   C. C. Lin, and Y. Y. Lau, "Density Wave Theory of Spiral Structure of Galaxies", Studies In Applied Mathematics, **60**, pp. 97–163, (1979).

6.5   R. B. Dingle, '*Asymptotic Expansions: Their Derivation and Interpretation*', (Academic, London 1973).

6.6   C. M. Bender and S. A. Orszag, '*Advanced Mathematical Methods for Scientists and Engineers*', (McGraw–Hill, New York 1978).

6.7   B. Steinin and V. Shatalov, '*Borel-Laplace Transform and Asymptotic Theory: Introduction to Resurgent Analysis*', (CRC–Press, 1995).

6.8   M. Abramovitz and I. A. Stegun (Eds), '*Handbook of Mathematical Functions*', (Dover, New York 1964).

6.9   J. J. Xu, "Interfacial Wave Theory for Dendritic Structure of a Growing Needle Crystal (I): Local Instability Mechanism", Phys. Rev. A **40**, No. 3, pp. 1599–1608, (1989).

6.10   J. J. Xu, "Interfacial Wave Theory for Dendritic Structure of a Growing Needle Crystal (II): Wave-Emission Mechanism at the Turning Point", Phys. Rev. A **40**, No. 3, pp. 1609–1614, (1989).

6.11   J. J. Xu, "Global Neutral Stable State and Selection Condition of Tip Growth Velocity", J. Crystal Growth, **100**, pp. 481–490, (1990).

6.12   J. J. Xu, "Interfacial Wave Theory of Solidification — Dendritic Pattern Formation and Selection of Tip Velocity", Phys. Rev. A15 **43**, No. 2, pp. 930–947, (1991).

6.13   J. J. Xu, "Two-Dimensional Dendritic Growth with Anisotropy of Surface Tension", Physics (D) **51**, pp. 579–595, (1991).

6.14   J. J. Xu, "Interfacial Wave Theory of Two-Dimensional Dendritic Growth with Anisotropy of Surface Tension", Canad. J. Physics **69**, No. 7, pp. 789–800, (1991).

6.15   M. E. Glicksman, M. B. Koss and E. A. Winsa, "Dendritic Growth Velocities in Microgravity", Phy. Rev. Letters **73**, No. 4, pp. 573–576, (1994).

6.16   M. E. Glicksman and M. B. Koss, L. T. Bushnell, J. C. Lacombe, and E. A. Winsa, "Dendritic Growth of Succinonitrile in Terrestrial and Microgravity Conditions as a Test of Theory", ISIJ International **35**, No. 6, pp. 604–610, (1995).

6.17   J. J. Xu, "Dendritic Growth From a Melt in an External Flow: Uniformly Valid Asymptotic Solution for The Steady State", J. Fluid Mech. **263**, pp. 227–243, (1994).

6.18   R. Ananth and W. N. Gill, "Dendritic Growth of an Elliptical Paraboloid with Forced Convection in the Melt", J. Fluid Mech. **208**, pp. 575–593, (1989).

6.19 R. Ananth and W. N. Gill, "Self-consistent Theory of Dendritic Growth with Convection", J. Crystal Growth **108**, pp. 173–189, (1991).

6.20 D. Canright and S. H. Davis, "Buoyancy Effect of a Growing, Isolated Dendrite", J. Crystal Growth **114**, pp. 153–185, (1991).

6.21 J. S. Langer and H. Müller-Krumbhaar, "Theory of Dendritic Growth — I. Elements of a Stability Analysis; II. Instabilities in the Limit of Vanishing Surface Tension; III. Effects of Surface Tension", Acta Metall. **26**, pp. 1681–1708, (1978).

7. The Effect of Surface Tension Anisotropy and Low-Frequency Instability on Dendrite Growth

So far we have only considered the case of isotropic surface tension. The anisotropy of surface tension was neglected. In the study of dendrite growth, the role and significance of the anisotropy of surface tension has been an issue of great importance. Experimental observations show that the effect of the anisotropy of surface tension on the morphology of the dendrite interface is certainly significant. However, its role in the stability of dendrite growth and selection of the tip velocity still remains unclear.

In this chapter we attempt to study this problem. It will be seen that the inclusion of a small anisotropy only causes a slight quantitative change for the GTW instability mechanism. However the inclusion of anisotropy invokes a new instability mechanism which we call the low-frequency (LF) instability. As a consequence, the system is now subject to two distinct instability mechanisms: the global trapped wave (GTW) instability and the low-frequency (LF) instability. When the anisotropy parameter is smaller than a critical number, the GTW instability still dominates, but when the anisotropy is sufficiently large, this new LF instability will dominate the system.

Hereby, for simplicity, we only consider two-dimensional dendrite growth, so the system described in Sect. 2.3 will be adopted. We note that, in the leading approximation, the only difference between the two- and the three-dimensional case is the tip condition. One can very easily write down the solution for three-dimensional growth from the results of the two-dimensional case.

Taking the anisotropy of surface tension into account, the only change in the mathematical formulation of the problem is the Gibbs–Thomson condition at the interface $\eta = \eta_{\mathrm{s}}(\xi, t, \varepsilon)$,

$$
T = -\frac{\Gamma}{\eta_0^2} \mathcal{K} \left\{ \frac{d}{d\xi}, \frac{\mathrm{d}^2}{\mathrm{d}\xi^2} \right\} \eta_{\mathrm{s}} .
\tag{7.1}
$$

The surface tension parameter Γ has been defined as

$$
\Gamma = \frac{\ell_{\mathrm{c}}}{\ell_{\mathrm{T}}} ,
\tag{7.2}
$$

where the capillary length is

$$
\ell_{\mathrm{c}} = \frac{\gamma c_{\mathrm{p}} T_{\mathrm{M0}}}{(\Delta H)^2} .
\tag{7.3}
$$

Now, with the inclusion of anisotropy, the coefficient of surface tension, γ, is no longer a constant but rather a function of the local orientation of the interface. In a commonly used form, γ is expressed as

$$\begin{cases} \gamma = \hat{\gamma} A_s(\theta) \\ A_s(\theta) = 1 - \alpha_m \cos(m\theta) \, , \end{cases} \tag{7.4}$$

where $\hat{\gamma}$ is the coefficient of isotropic surface tension, α_m is the anisotropy parameter, and θ is the orientation angle.

Thus we have

$$\Gamma = \hat{\Gamma} A_s(\theta) \, , \tag{7.5}$$

where

$$\hat{\Gamma} = \frac{\hat{\ell}_c}{\ell_T} \, , \quad \hat{\ell}_c = \frac{\hat{\gamma} c_p T_{M0}}{(\Delta H)^2} \, . \tag{7.6}$$

In general,

$$\tan\theta = \frac{\mathrm{d}x}{\mathrm{d}y} = \frac{\xi - \eta_s \eta_s'}{\eta_s + \xi \eta_s'} \, , \tag{7.7}$$

where $\eta = \eta_s(\xi,t)$ is the interface shape function. For the interface with $\eta_s = 1 + \tilde{\eta}_s$ and $\tilde{\eta}_s, \tilde{\eta}_s' \ll 1$, the above formula can be linearized to obtain

$$\begin{cases} \tan\theta \approx \xi - \xi\tilde{\eta}_s - S^2\tilde{\eta}_s' \, , \\ \cos^2\theta \approx \dfrac{1}{S^2}\left(1 + \dfrac{2\xi^2}{S^2}\tilde{\eta}_s + 2\xi\tilde{\eta}_s'\right) \, , \end{cases} \tag{7.8}$$

where $S = \sqrt{1 + \xi^2}$. Typically, for the four-fold anisotropy, one can set $m = 4$. By using

$$\cos 4\theta = 8\cos^4\theta - 8\cos^2\theta + 1 \, , \tag{7.9}$$

we obtain

$$A_s(\theta) = \bar{A}_s(\xi) + \tilde{A}_{s0}(\xi)\tilde{\eta}_s(\xi) + \tilde{A}_{s1}(\xi)\tilde{\eta}_s'(\xi) \, , \tag{7.10}$$

where

$$\begin{aligned} \bar{A}_s(\xi) &= \frac{f(\xi)}{S^4} \, , \\ \tilde{A}_{s0}(\xi) &= -16\alpha_4 \frac{\xi^2(1-\xi^2)}{S^6} \, , \\ \tilde{A}_{s1}(\xi) &= -16\alpha_4 \frac{\xi(1-\xi^2)}{S^4} \, , \end{aligned} \tag{7.11}$$

and

$$f(\xi) = (1 - \alpha_4)(1 + \xi^2)^2 + 8\alpha_4\xi^2 \, . \tag{7.12}$$

In this case, the interfacial stability parameter of isotropic surface tension, ε, is defined as

$$\varepsilon^2 = \frac{\hat{\Gamma}}{\eta_0^4}\,.$$ (7.13)

With the inclusion of the anisotropy of surface tension, the basic state for the case of $\varepsilon = 0$ is still the Ivantsov solution. The basic state for $\varepsilon \neq 0$ can be defined in the same way as for the case of zero anisotropy, through the 'nearly' steady needle solution and nonclassic steady needle solution. One still has the approximate expressions:

$$T_{\mathrm{B}}(\xi,\eta,\varepsilon) = T_*(\eta) + O(\varepsilon^2)$$

$$T_{\mathrm{SB}}(\xi,\eta,\varepsilon) = O(\varepsilon^2)$$ (7.14)

$$\eta_{\mathrm{B}}(\xi,\varepsilon) = 1 + O(\varepsilon^2)\,.$$

7.1 Linear Perturbed System Around the Basic State

As before, we separate the general unsteady solutions into two parts

$$T = T_{\mathrm{B}} + \tilde{T}(\xi,\eta,t,\varepsilon)$$

$$T_{\mathrm{S}} = T_{\mathrm{SB}} + \tilde{T}_{\mathrm{S}}(\xi,\eta,t,\varepsilon)$$ (7.15)

$$\eta_s = \eta_{\mathrm{B}} + \tilde{h}(\xi,t,\varepsilon)/\eta_0^2$$

and linearize around the basic state solution. The linearized perturbed system is the same as (6.176)–(6.185) derived in Chap. 6, except that the Gibbs–Thomson interface condition is now modified by the effect of anisotropy. For simplicity, we write the interface conditions simply as follows: at $\eta = 1$,

(i)

$$\tilde{T} - \tilde{T}_{\mathrm{S}} = \tilde{h} + \text{(higher-order terms)}\,,$$ (7.16)

(ii)

$$\tilde{T}_{\mathrm{S}} = \frac{\varepsilon^2 \bar{A}_s(\xi)}{S(\xi)}\left[\frac{\partial^2 \tilde{h}}{\partial \xi^2} + \frac{\xi}{S^2(\xi)}\frac{\partial \tilde{h}}{\partial \xi} - \frac{\tilde{h}}{S^2(\xi)}\right]$$

$$-\varepsilon^2 K_0(\xi)\left(\tilde{A}_{s0}\tilde{h} + \tilde{A}_{s1}\frac{\partial \tilde{h}}{\partial \xi}\right) + \text{(higher-order terms)}\,,$$ (7.17)

where $K_0(\xi) = \frac{1}{S^3}$,

(iii)

$$\frac{\partial}{\partial \eta}\left(\tilde{T} - \tilde{T}_{\mathrm{S}}\right) + \eta_0^2 S^2(\xi)\frac{\partial \tilde{h}}{\partial t} + \xi\frac{\partial \tilde{h}}{\partial \xi} + \varepsilon(1 + \eta_0^2)\tilde{h}$$

$$+\text{(higher-order terms)} = 0\,.$$ (7.18)

7.2 Multiple Variable Expansion Solution in the Outer Region

The stretched fast variables are defined as in Chap. 6. The outer solution can be derived in the same MVE form as follows

$$\tilde{T} = \left\{ \tilde{T}_0(\xi, \eta, \xi_+, \eta_+) + \varepsilon \tilde{T}_1(\xi, \eta, \xi_+, \eta_+) + \cdots \right\} e^{\sigma t_+}$$

$$\tilde{h} = \left\{ \tilde{h}_0(\xi, \xi_+) + \varepsilon \tilde{h}_1(\xi, \xi_+) + \cdots \right\} e^{\sigma t_+}$$

$$k = k_0 + \varepsilon k_1 + \varepsilon^2 k_2 + \cdots \tag{7.19}$$

$$g = k_0 + \varepsilon k_1 + \varepsilon^2 g_2 + \cdots$$

$$\sigma = \sigma_0 + \varepsilon \sigma_1 + \varepsilon^2 \sigma_2 + \cdots.$$

The converted system with the multiple variables is almost entirely the same as (6.20)–(6.29) with the corresponding tip conditions and far field condition given in Chap. 6. The only change is in the Gibbs–Thomson condition, which now reads

$$\tilde{T}_S = \frac{\bar{A}_s(\xi)}{S(\xi)} \left[\left(k^2 \frac{\partial^2 \tilde{h}}{\partial \xi_+^2} + 2\varepsilon k \frac{\partial^2 \tilde{h}}{\partial \xi_+ \partial \xi} + \varepsilon k' \frac{\partial \tilde{h}}{\partial \xi_+} + \varepsilon^2 \frac{\partial^2 \tilde{h}}{\partial \xi^2} \right) \right.$$

$$\left. + \frac{\varepsilon \xi}{S^2(\xi)} \left(k \frac{\partial \tilde{h}}{\partial \xi_+} + \varepsilon \frac{\partial \tilde{h}}{\partial \xi} \right) - \frac{\varepsilon^2}{S^2(\xi)} \tilde{h} \right] - \varepsilon^2 \tilde{A}_{s0} K_0(\xi) \tilde{h}$$

$$- \varepsilon \tilde{A}_{s1} K_0(\xi) \left(k \frac{\partial \tilde{h}}{\partial \xi_+} + \varepsilon \frac{\partial \tilde{h}}{\partial \xi} \right) + O(\varepsilon^3). \tag{7.20}$$

At the zeroth-order we have

$$\left(\frac{\partial^2}{\partial \xi_+^2} + \frac{\partial^2}{\partial \eta_+^2} \right) \tilde{T}_0 = 0$$

$$\left(\frac{\partial^2}{\partial \xi_+^2} + \frac{\partial^2}{\partial \eta_+^2} \right) \tilde{T}_{S0} = 0 \tag{7.21}$$

and the boundary conditions:

1. As $\eta_+ \to \infty$

$$\tilde{T}_0 \to 0. \tag{7.22}$$

2. As $\eta_+ \to -\infty$

$$\tilde{T}_{S0} \to 0. \tag{7.23}$$

3. At $\eta_+ = 0$ and $\eta = 1$,

$$\tilde{T}_0 = \tilde{T}_{S0} + \tilde{h}_0 \,, \tag{7.24}$$

$$\tilde{T}_{S0} = \frac{\bar{A}_s(\xi)\hat{k}_0^2}{S(\xi)} \frac{\partial^2 \tilde{h}_0}{\partial \xi_+^2} \,, \tag{7.25}$$

$$\hat{k}_0 \frac{\partial}{\partial \eta_+}(\tilde{T}_0 - \tilde{T}_{S0}) + \sigma_0 S^2(\xi)\tilde{h}_0 + \hat{k}_0 \xi \frac{\partial \tilde{h}_0}{\partial \xi_+} = 0 \,. \tag{7.26}$$

4. The root condition:

$$\left\{ \tilde{T}_0, \ \tilde{T}_{S0}, \ \tilde{h}_0 \right\} = 0 \tag{7.27}$$

as ξ and $\xi_+ \to \infty$.

5. The tip smoothness condition: at $\xi = \eta_+ = 0$,

$$\frac{\partial}{\partial \xi}\left\{ \tilde{T}_0, \ \tilde{T}_{S0}, \ \tilde{h}_0 \right\} = 0 \qquad \text{for the S-modes,} \tag{7.28}$$

$$\left\{ \tilde{T}_0, \ \tilde{T}_{S0}, \ \tilde{h}_0 \right\} = 0 \qquad \text{for the A-modes}. \tag{7.29}$$

This system has the normal mode solutions

$$\begin{aligned}
\tilde{T}_0 &= A_0(\xi,\eta) \ \exp\left\{ i\xi_+ - \eta_+ \right\} \\
\tilde{T}_{S0} &= A_{S0}(\xi,\eta) \ \exp\left\{ i\xi_+ + \eta_+ \right\} \\
\tilde{h}_0 &= \hat{D}_0 \ \exp\left\{ i\hat{\xi}_+ \right\} \,,
\end{aligned} \tag{7.30}$$

where $\hat{D}_0$ is an arbitrary constant. The local dispersion relation is

$$\sigma_0 = \Sigma(\xi, \hat{k}_0) = \frac{\hat{k}_0}{S^2}\left(1 - \frac{2\bar{A}_s(\xi)\hat{k}_0^2}{S} \right) - i\frac{\xi \hat{k}_0}{S^2} \,. \tag{7.31}$$

Remember that, until this point, we have assumed the symmetrical model. For the more general case, $\hat{\alpha} = \frac{\kappa_{TS}}{\kappa_T} \geq 0$, one obtains the local dispersion relation

$$\sigma_0 = \frac{\hat{k}_0}{S^2}\left(1 - \frac{(1 + \hat{\alpha})\bar{A}_s(\xi)\hat{k}_0^2}{S} \right) - i\frac{\xi \hat{k}_0}{S^2} \,. \tag{7.32}$$

When the surface tension anisotropy is zero ($\alpha_4 = 0$), we recover

$$\sigma_0 = \frac{\hat{k}_0}{S^2}\left(1 - \frac{(1 + \hat{\alpha})\hat{k}_0^2}{S} \right) - i\frac{\xi}{S^2}\hat{k}_0 \,. \tag{7.33}$$

Many pattern formation systems have a similar cubic dispersion formula. Thus, for convenience of discussion, we express the above local dispersion relation in a more general form:

$$\sigma = \Sigma(\xi, \hat{k}_0) = \frac{\hat{k}_0}{\Lambda_0 S^2}\left(1 - \frac{\hat{k}_0^2}{\Lambda_1 S}\right) - \mathrm{i}\frac{\xi \hat{k}_0}{S^2}\,. \tag{7.34}$$

Obviously, for the system with anisotropy, we have

$$\begin{cases} \Lambda_1 = \dfrac{1}{(1+\hat{\alpha})\bar{A}_s(\xi)} \\ \Lambda_0 = 1\,. \end{cases} \tag{7.35}$$

For any fixed eigenvalue σ_0, one can solve for three wave numbers as functions of ξ yielding results similar to Fig. 6.1, namely

$$\begin{cases} \hat{k}_0^{(1)}(\xi) = \hat{M}(\xi)\cos\left[\frac{1}{3}\cos^{-1}\left(\frac{\sigma}{\hat{N}(\xi)}\right)\right] & \text{(short-wavelength branch)} \\[2mm] \hat{k}_0^{(2)}(\xi) = \hat{M}(\xi)\cos\left[\frac{1}{3}\cos^{-1}\left(\frac{\sigma}{\hat{N}(\xi)}\right) + \frac{2\pi}{3}\right] \\[2mm] \hat{k}_0^{(3)}(\xi) = \hat{M}(\xi)\cos\left[\frac{1}{3}\cos^{-1}\left(\frac{\sigma}{\hat{N}(\xi)}\right) + \frac{4\pi}{3}\right] \text{(long-wavelength branch)}, \end{cases} \tag{7.36}$$

where we define

$$\begin{cases} \hat{M}(\xi) = \sqrt{\dfrac{4\Lambda_1 S(\xi)}{3}}\,(1 - \mathrm{i}\Lambda_0\xi)^{\frac{1}{2}} \\[4mm] \hat{N}(\xi) = -\dfrac{\hat{M}(\xi)}{3\Lambda_0 S^2(\xi)}\,(1 - \mathrm{i}\Lambda_0\xi)\,. \end{cases} \tag{7.37}$$

As we have discussed before, only the short-wavelength branch, the H_1 wave, with the larger $\mathrm{Re}\{\hat{k}_0^{(1)}\} > 0$ and the long-wavelength branch, the H_3 wave, with the smaller $\mathrm{Re}\{\hat{k}_0^{(3)}\} > 0$ are physically meaningful. The H_2 wave, with $\mathrm{Re}\{\hat{k}_0^{(2)}\} < 0$, must be ruled out. The general solution of the H wave is

$$\tilde{h}_0 = D_1 H_1(\xi) + D_3 H_3(\xi)$$

$$\tilde{h} = D_1\,\exp\left[\frac{\sigma t}{\varepsilon \eta_0^2} + \frac{\mathrm{i}}{\varepsilon}\int_{\xi_c}^{\xi}\hat{k}_0^{(1)}\mathrm{d}\xi_1\right] \tag{7.38}$$

$$+ D_3\,\exp\left[\frac{\sigma t}{\varepsilon \eta_0^2} + \frac{\mathrm{i}}{\varepsilon}\int_{\xi_c}^{\xi}\hat{k}_0^{(3)}\mathrm{d}\xi_1\right]\,.$$

The coefficients $\{D_1, D_3\}$ are arbitrary constants to be determined. As seen in Chap. 6, the above MVE solution (7.38) also has a turning point, ξ_c, in the complex ξ-plane, which appears in the first-order approximation. This singular point is the root of the equation

$$\frac{\partial \Sigma(\hat{k}_0, \xi)}{\partial \hat{k}_0} = -6\bar{A}_s \hat{k}_0^2 + (1 - \mathrm{i}\xi)S = 0, \tag{7.39}$$

or

$$\sigma_0 = \mathrm{e}^{-\mathrm{i}\frac{3\pi}{4}} \sqrt{\frac{2}{27}} \frac{(\xi + \mathrm{i})^{\frac{7}{4}}(\xi - \mathrm{i})^{\frac{1}{4}}}{\sqrt{f(\xi)}}. \tag{7.40}$$

7.3 The Inner Equation near the Singular Point ξ_c

Similar to the procedure adopted in Chap. 6, in the inner region, $|\xi - \xi_c| \ll 1$; $|\eta - 1| \ll 1$, we introduce the inner variables:

$$\begin{aligned}
\xi_* &= \frac{\xi - \xi_c}{\varepsilon^\alpha} \\
\eta_* &= \frac{\eta - 1}{\varepsilon^\alpha},
\end{aligned} \tag{7.41}$$

where α is to be determined. In addition, we denote the interface shape function by

$$\eta_{\mathrm{s}}(\xi, t) = 1 + \frac{\tilde{h}}{\eta_0^2} = 1 + \varepsilon^\alpha \eta_{*\mathrm{s}} \tag{7.42}$$

and

$$\eta_{*\mathrm{s}} = \frac{\hat{h}(\xi_*, t)}{\eta_0^2}, \tag{7.43}$$

so that we have

$$\tilde{h}(\xi, t) = \varepsilon^\alpha \hat{h}(\xi_*, t). \tag{7.44}$$

Finally, let

$$\tilde{T}(\xi, \eta, t) = \varepsilon^\alpha \hat{T}(\xi_*, \eta_*, t), \quad \text{and} \quad \tilde{T}_{\mathrm{S}}(\xi, \eta, t) = \varepsilon^\alpha \hat{T}_{\mathrm{S}}(\xi_*, \eta_*, t). \tag{7.45}$$

We look for mode solutions and make the inner expansions:

$$\begin{aligned}
\hat{T}(\xi_*, \eta_*, t) &= \left\{ \nu_0(\varepsilon)\hat{T}_0(\xi_*, \eta_*) + \nu_1(\varepsilon)\hat{T}_1(\xi_*, \eta_*) + \dots \right\} \mathrm{e}^{\frac{\sigma t}{\varepsilon \eta_0^2}} \\
\hat{T}_{\mathrm{S}}(\xi_*, \eta_*, t) &= \left\{ \nu_0(\varepsilon)\hat{T}_{\mathrm{S}0}(\xi_*, \eta_*) + \nu_1(\varepsilon)\hat{T}_{\mathrm{S}1}(\xi_*, \eta_*) + \dots \right\} \mathrm{e}^{\frac{\sigma t}{\varepsilon \eta_0^2}} \\
\hat{h} &= \left\{ \nu_0(\varepsilon)\hat{h}_0 + \nu_1(\varepsilon)\hat{h}_1 + \dots \right\} \mathrm{e}^{\frac{\sigma t}{\varepsilon \eta_0^2}}.
\end{aligned} \tag{7.46}$$

In terms of the inner variables, (6.176) becomes

$$\left(\frac{\partial^2}{\partial \xi_*^2} + \frac{\partial^2}{\partial \eta_*^2}\right)\hat{T} = \left[\varepsilon^{2\alpha-1}\sigma\eta_0^2(\xi^2 + \eta^2) + \varepsilon^\alpha\eta_0^2\left(\xi\frac{\partial}{\partial \xi_*}\right. \right.$$
$$\left. \left. - \eta\frac{\partial}{\partial \eta_*}\right)\right]\hat{T},$$
(7.47)

and the interface boundary conditions (7.16)–(7.18) are transformed to the following: at $\eta_* = 0$

(i)

$$\hat{T} = \hat{T}_{\mathrm{S}} + \hat{h} + (\text{higher-order terms});$$
(7.48)

(ii)

$$\hat{T}_{\mathrm{S}} = \frac{\varepsilon^{2-2\alpha}\bar{A}_s(\xi)}{S(\xi)}\frac{\partial^2\hat{h}}{\partial\hat{\xi}_*^2} + (\text{higher-order terms});$$
(7.49)

(iii)

$$\varepsilon^{1-\alpha}\frac{\partial}{\partial\eta_*}\left(\hat{T} - \hat{T}_{\mathrm{S}}\right) + \sigma S^2(\xi)\hat{h} + \varepsilon^{1-\alpha}\xi\frac{\partial\hat{h}}{\partial\xi_*}$$
$$= (\text{higher-order terms}).$$
(7.50)

Letting $\varepsilon \to 0$, for the leading order approximation, the above inner equation can be further simplified into the following third-order, complex, ODE for the interface perturbation $\hat{h}$:

$$\mathrm{i}\frac{2\varepsilon^{3-3\alpha}\bar{A}_s(\xi)}{S}\frac{\partial^3\hat{h}}{\partial\xi_*^3} + \varepsilon^{1-\alpha}(\xi + \mathrm{i})\frac{\partial\hat{h}}{\partial\xi_*} + \sigma S^2\hat{h}$$
$$= (\text{higher-order terms}).$$
(7.51)

This simplification holds, provided

$$\varepsilon^{2\alpha-1}\sigma \to 0.$$
(7.52)

As before, we use the transformation introduced in [6.12] and relate $\tilde{h}_0$ to a new unknown function $W_0(\xi)$, i.e.,

$$\tilde{h}_0 = W_0(\xi)\exp\left[\frac{\mathrm{i}}{\varepsilon}\int_{\xi_c}^{\xi} k_c(\xi_1)\,\mathrm{d}\xi_1\right],$$
(7.53)

where the reference wave number function $k_c(\xi)$ is to be chosen later. It will seen that

$$\mathrm{Re}\{k_3\} < \mathrm{Re}\{k_c\} < \mathrm{Re}\{k_1\};$$
(7.54)

therefore, in the outer region we have the fundamental solutions:

$$H_1 = W_0^{(+)}(\xi)\,\exp\left[\frac{\mathrm{i}}{\varepsilon}\int_{\xi_c}^{\xi} k_c(\xi_1)\,\mathrm{d}\xi_1\right]$$

$$H_3 = W_0^{(-)}(\xi)\,\exp\left[\frac{\mathrm{i}}{\varepsilon}\int_{\xi_c}^{\xi} k_c(\xi_1)\,\mathrm{d}\xi_1\right].$$

$$(7.55)$$

One can draw the same diagram of the relation between the H waves and the W waves as in Chap. 6:

$$H_1 \text{ wave} \qquad\qquad \Longleftrightarrow \quad W^{(+)}$$

(short-wavelength branch)   (outgoing wave);

$$H_3 \text{ wave} \qquad\qquad \Longleftrightarrow \quad W^{(-)}$$

(long-wavelength branch)   (incoming wave).

In accordance with the above, we set

$$\hat{h}_0 = \hat{W}_0(\xi_*)\exp\left\{\frac{\mathrm{i}}{\varepsilon}\int_{\xi_c}^{\xi} k_c(\xi_1)\,\mathrm{d}\xi_1\right\}$$

$$(7.56)$$

in the inner region. Similar to what was found in Chap. 6, by setting

$$S^2\left(\frac{\partial\Sigma}{\partial k_0}\right)_{k_0=k_c} = -\frac{6\bar{A}_s k_c^2}{S} + (1 - \mathrm{i}\xi) = 0,$$

$$(7.57)$$

or

$$k_c(\xi) = \sqrt{\frac{S}{6\bar{A}_s}}\,(1 - \mathrm{i}\xi) = \frac{\mathrm{e}^{-\mathrm{i}\frac{\pi}{4}}(\xi + \mathrm{i})^{\frac{7}{4}}(\xi - \mathrm{i})^{\frac{5}{4}}}{\sqrt{6f(\xi)}}$$

$$\left(\mathrm{Re}\{k_c\} > 0\right),$$

$$(7.58)$$

equation (7.51) can be changed to

$$\varepsilon^{3-3\alpha}\Omega_3\frac{\mathrm{d}^3\hat{W}}{\mathrm{d}\xi_*^3} + \mathrm{i}\varepsilon^{2-2\alpha}\Omega_2\frac{\mathrm{d}^2\hat{W}}{\mathrm{d}\xi_*^2} + \mathrm{i}S^2\left\{\sigma_0 - \Sigma_c(\xi)\right\}\hat{W}$$

$$= \text{(higher-order terms)}$$

$$(7.59)$$

where

$$\Omega_3 = -\frac{2\bar{A}_s(\xi)}{S};$$

$$\Omega_2 = -\frac{6k_c\bar{A}_s(\xi)}{S} = -\mathrm{e}^{-\mathrm{i}\frac{\pi}{4}}\frac{6^{\frac{1}{2}}f^{\frac{1}{2}}}{S^2}\left(\frac{\xi + \mathrm{i}}{\xi - \mathrm{i}}\right)^{\frac{1}{4}}$$

$$(7.60)$$

$$\Sigma_c(\xi) = \Sigma(k_c,\xi) = \mathrm{e}^{-\mathrm{i}\frac{3\pi}{4}}\sqrt{\frac{2}{27}}\frac{(\xi + \mathrm{i})^{\frac{7}{4}}(\xi - \mathrm{i})^{\frac{1}{4}}}{\sqrt{f(\xi)}}.$$

Clearly, at $\xi = \xi_c$,

$$\sigma_0 - \Sigma_c(\xi_c) = 0. \tag{7.61}$$

Hence, the singular point ξ_c is a turning point of this equation. The variable ξ in the coefficient functions of (7.59) also needs to be changed to the inner variable ξ_* by using $\xi = \xi_c + \varepsilon^\alpha \xi_*$. Next, we shall make Taylor expansions around ξ_c for the coefficient functions and balance the leading terms on the left-hand side of (7.59). Note that

$$\frac{S^2(\xi)}{\Omega_2(\xi)} \Sigma_c'(\xi_c)(\xi - \xi_c) = \frac{(\xi - \xi_c)}{6} \times \frac{(\xi - i)^{\frac{9}{4}}}{(\xi_c - i)^{\frac{3}{4}}} \times \frac{(\xi + i)^{\frac{7}{4}}(\xi_c + i)^{\frac{3}{4}}}{f^{\frac{1}{2}}(\xi) f^{\frac{1}{2}}(\xi_c)}$$

$$\times \left(1 + i\alpha_4 \frac{16}{3} \frac{\xi_c(\xi_c^2 - 1)}{f(\xi_c)}\right). \tag{7.62}$$

The function $f(\xi)$ defined by (7.12) has four imaginary zeros $(\pm ia_1, \pm ia_2)$, where

$$\begin{cases} a_1^2 = 1 + 3\alpha_4 - 2\sqrt{2}\alpha_4^{\frac{1}{2}}(1 + \alpha_4)^{\frac{1}{2}} \\ a_2^2 = 1 + 3\alpha_4 + 2\sqrt{2}\alpha_4^{\frac{1}{2}}(1 + \alpha_4)^{\frac{1}{2}}. \end{cases} \tag{7.63}$$

As $\alpha_4 \ll 1$,

$$a_1 \approx 1 - \sqrt{2\alpha_4} \quad a_2 \approx 1 + \sqrt{2\alpha_4}. \tag{7.64}$$

Hence, one can rewrite

$$f(\xi) = (1 - \alpha_4)(\xi^2 + a_1^2)(\xi^2 + a_2^2). \tag{7.65}$$

It is seen that, besides ξ_c, the inner equation has two more turning points: $\xi = \pm i$, and four other singular points: $\xi = (\pm ia_1, \pm ia_2)$. The relative positions of these singular points compared to ξ_c are related to the values of the parameters σ_0 and α_4. We may set the branch cut along the Stokes line (L_1) and assume the turning point ξ_c has $\text{Re}\{\xi_c\} > 0$ and $\text{Im}\{\xi_c\} < 0$. The singular points $\xi = (i, ia_1, ia_2)$ are always away from ξ_c. Consequently, the influence of these singular points on the inner solution is negligible. However, as $\sigma_0 \to 0$, $\xi_c \to -i$, while as $\alpha_4 \to 0$, $(-ia_1, -ia_2) \to -i$. The singular points $\xi = -i, -ia_1$ and $-ia_2$ may enter the inner region of ξ_c, and consequently influence the behavior of the inner solution under some circumstances.

We rewrite the inner equation (7.59) in the form

$$\varepsilon^{3-3\alpha} \underset{(1)}{\frac{i}{3k_c} \frac{d^3\hat{W}}{d\xi_*^3}} - \varepsilon^{2-2\alpha} \underset{(2)}{\frac{d^2\hat{W}}{d\xi_*^2}} + \underset{(3)}{A^2(\xi)\hat{W}} = \text{(higher-order terms)},$$

$$\tag{7.66}$$

where

$$A^2(\xi) = A_0^2 \frac{(\xi_c - i)^{\frac{3}{2}}}{(\xi_c - ia_1)(\xi_c - ia_2)} \times \frac{(\xi + i)^{\frac{7}{4}}(\xi_c + i)^{\frac{3}{4}}}{(\xi + ia_1)^{\frac{1}{2}}(\xi + ia_2)^{\frac{1}{2}}}$$

$$\times \frac{(\xi - \xi_c)}{(\xi_c + ia_1)^{\frac{1}{2}}(\xi_c + ia_2)^{\frac{1}{2}}} \times \left(1 + i\alpha_4 \frac{16}{3} \frac{\xi_c(\xi_c^2 - 1)}{f(\xi_c)}\right) \tag{7.67}$$

and

$$A_0^2 = \frac{1}{6(1 - \alpha_4)}. \tag{7.68}$$

In the leading order approximation, the three terms on the left-hand side of (7.66) must be balanced in some way. Four types of balances, in general, are possible:

(A). $(1) \approx (2) \approx (3)$;   (B). $(1) \ll (2) \approx (3)$;

(C). $(1) \approx (3) \gg (2)$;   (D). $(1) \approx (2) \gg (3)$.

With the balance (D), (7.66) will be reduced to a first-order equation and so the inner solutions cannot be matched to the outer solutions (7.38). Thus, only the balances (A)–(C) are possible.

In what follows we shall discuss several different cases.

7.3.1 Case I: $\sigma_0 = O(1)$

In this case, all the singular points $\xi = \pm i$ and $\xi = \pm ia_i$ $(i = 1, 2)$ are away from ξ_c, namely, $|\xi_c + i| \gg O(\varepsilon^\alpha)$ and $|\xi_c + ia_i| \gg O(\varepsilon^\alpha)$ $(i = 1, 2)$.

As for the isotropic case, we find that, in the leading order approximation, the left-hand side of (7.66) has the balance: $(2) \approx (3) \gg (1)$, and so we obtain $\alpha = \frac{2}{3}$. Condition (7.52) is satisfied. Moreover, in the leading order approximation, the inner equation is reduced to the Airy equation:

$$\frac{d^2 \hat{W}_0}{d\hat{\xi}_*^2} + \hat{\xi}_* \hat{W}_0 = 0, \tag{7.69}$$

where

$$\hat{\xi}_* = \frac{B}{\varepsilon^{\frac{2}{3}}}(\xi - \xi_c), \tag{7.70}$$

and

$$B = \hat{A}_1^{\frac{2}{3}}$$

$$\hat{A}_1 = iA_0\sqrt{\frac{1}{6}}\left(\frac{\xi_c + i}{\xi_c - i}\right)^{\frac{1}{4}} F^{\frac{1}{2}}(\alpha_4, \xi_c) \tag{7.71}$$

$$F(\alpha_4, \xi_c) = \frac{S^4}{f(\xi_c)}\left(1 + i\alpha_4 \frac{16}{3}\frac{\xi_c(\xi_c^2 - 1)}{f^2(\xi_c)}\right).$$

Thus,

$$\frac{\pi}{2} < \arg(\hat{A}_1) \leq \frac{3\pi}{4}, \quad \text{and} \quad \frac{\pi}{3} < \arg(B) \leq \frac{\pi}{2}. \tag{7.72}$$

The structure of the Stokes lines for this case is similar to that shown in Fig. 6.2.

7.3.2 Case II: $|\sigma_0| \ll 1$

In this case, the turning point ξ_c will be very close to $-i$ and the form of the inner equation and the behavior of the inner solutions are closely related to the asymptotic behavior of σ_0, α_4, and ξ_c as $\varepsilon \to 0$.

Let

$$\sigma_0 = \hat{\sigma}_0 \varepsilon^{\nu_0}, \quad \alpha_4 = \hat{\alpha}_4 \varepsilon^{\mu_0}, \quad \text{and} \quad \xi_c = -i + \hat{\delta}\varepsilon^\beta \quad (\text{as } \varepsilon \to 0), \tag{7.73}$$

where the indices $\nu_0, \mu_0, \beta > 0$. From (7.40) we derive

$$\hat{\sigma}_0 \varepsilon^{\nu_0} = -\frac{e^{-i\frac{3\pi}{8}}}{2^{\frac{1}{4}}3^{\frac{3}{2}}} \frac{\hat{\delta}^{\frac{7}{4}}\varepsilon^{\frac{7\beta}{4}}}{\left(\hat{\delta}^2\varepsilon^{2\beta} + 2\hat{\alpha}_4\varepsilon^{\mu_0}\right)^{\frac{1}{2}}} + (\text{higher-order terms}). \tag{7.74}$$

This leads to the following cases:

(1) When $\frac{\mu_0}{2} > \beta$, one has

$$\nu_0 = \frac{3}{4}\beta \quad \text{and} \quad \hat{\sigma}_0 = -\frac{\hat{\delta}^{\frac{3}{4}}e^{-i\frac{3\pi}{8}}}{2^{\frac{1}{4}}3^{\frac{3}{2}}}. \tag{7.75}$$

This case includes the case of isotropic surface tension.

(2) When $\frac{\mu_0}{2} < \beta$, one has

$$\nu_0 = \frac{7}{4}\beta - \frac{\mu_0}{2} \quad \text{and} \quad \hat{\sigma}_0 = -\frac{\hat{\delta}^{\frac{7}{4}}e^{-i\frac{3\pi}{8}}}{2^{\frac{3}{4}}3^{\frac{3}{2}}\hat{\alpha}_4^{\frac{1}{2}}}. \tag{7.76}$$

(3) Finally, when $\frac{\mu_0}{2} = \beta$, one has

$$\nu_0 = \frac{7}{4}\beta - \frac{\mu_0}{2} \quad \text{and} \quad \hat{\sigma} = -\frac{\hat{\delta}^{\frac{7}{4}}e^{-i\frac{3\pi}{8}}}{2^{\frac{1}{4}}3^{\frac{3}{2}}(2\hat{\alpha}_4 + \hat{\delta}^2)^{\frac{1}{2}}}. \tag{7.77}$$

It is found that

$$\begin{cases} \xi_c + i = \hat{\delta}\varepsilon^\beta \\[2mm] \xi + i = \varepsilon^\alpha \xi_* + \hat{\delta}\varepsilon^\beta. \end{cases} \tag{7.78}$$

Furthermore, from (7.73),

$$\begin{cases} \xi_c + ia_i = \hat{\delta}\varepsilon^\beta + i\hat{a}_i\varepsilon^{\frac{\mu_0}{2}} \\[2mm] \xi + ia_i = \varepsilon^\alpha \xi_* + \hat{\delta}\varepsilon^\beta + i\hat{a}_i\varepsilon^{\frac{\mu_0}{2}} \\[2mm] \hat{a}_1 = -\sqrt{2\hat{\alpha}_4}, \quad \hat{a}_2 = \sqrt{2\hat{\alpha}_4}. \end{cases} \tag{7.79}$$

To examine case II, several situations need to be separately discussed.

(a) Only the singular point $\xi = -i$ is in the inner region of ξ_c. This implies that the turning point ξ_c will be near the singular points $\xi = -i$, but it is still away from other singular points, such as $\xi = -ia_i$ $(i = 1, 2)$. Namely, we have $|\xi_c + i| \le O(\varepsilon^\alpha)$ and $|\xi_c + ia_i| \gg O(\varepsilon^\alpha)$ $(i = 1, 2)$, so that $\beta > \frac{\mu_0}{2}$.

In this case, the orders of magnitudes of each term on the left-hand side of (7.66) are

$$(1) = O(\varepsilon^{3 - \frac{19\alpha}{4}}), \quad (2) = O(\varepsilon^{2 - 2\alpha}) \quad \text{and} \quad (3) = O(\varepsilon^{\frac{11\alpha}{4} + \frac{3\beta}{4}}).$$

It turns out that if $\beta = \frac{4}{11} = \alpha$, the balance of (7.66) will be $(1) \approx (2) \approx (3)$. Consequently, the leading order approximation of the inner equation is

$$\frac{K_2}{(\xi_* + \hat{\delta})^{\frac{7}{4}}} \frac{\mathrm{d}^3 \hat{W}_0}{\mathrm{d}\xi_*^3} + \frac{\mathrm{d}^2 \hat{W}_0}{\mathrm{d}\xi_*^2} + A_2^2 \hat{\delta}^{\frac{3}{4}} \xi_* (\xi_* + \hat{\delta})^{\frac{7}{4}} \hat{W}_0 = 0, \tag{7.80}$$

where $K_2 = O(1), A_2 = O(1)$ are some constants.

On the other hand, if $\beta > \frac{4}{11} > \alpha$, the balance of (7.66) will be $(1) \ll (2) \approx (3)$ and the leading order approximation of the inner equation becomes

$$\frac{\mathrm{d}^2 \hat{W}_0}{\mathrm{d}\xi_*^2} + A_2^2 \hat{\delta}^{\frac{3}{4}} \xi_*^{\frac{11}{4}} \hat{W}_0 = 0. \tag{7.81}$$

For the above two cases, as $\xi_* \to \infty$, the inner equation in the far-field of the inner region has the same form:

$$\frac{\mathrm{d}\hat{W}_0}{\mathrm{d}\hat{\xi}_*^2} + \hat{\xi}_*^{\frac{11}{4}} \hat{W}_0 = 0, \tag{7.82}$$

where

$$\hat{\xi}_* = \frac{B}{\varepsilon^{\frac{4}{11}}} (\xi - \xi_c)$$

$$B = \hat{A}_2^{\frac{8}{19}} \hat{\delta}^{\frac{3}{19}} \tag{7.83}$$

$$\hat{A}_2 = i \frac{A_0 (\xi_c - i)^{\frac{3}{4}}}{(\xi_c^2 + a_1^2)^{\frac{1}{2}} (\xi_c^2 + a_2^2)^{\frac{1}{2}}} \left(1 + i\alpha_4 \frac{16}{3} \frac{\xi_c(\xi_c^2 - 1)}{f(\xi_c)} \right)^{\frac{1}{2}}.$$

Moreover, for both cases, since $\beta \ge \frac{4}{11} \ge \alpha > \frac{\mu_0}{2}$, and from (7.76) it follows that $\nu_0 = \frac{7\beta}{4} - \frac{\mu_0}{2}$. Thus, we can write

$$\nu_0 > \frac{7}{11} - \frac{\mu_0}{2} > \frac{11}{3}, \quad \text{and} \quad \mu_0 < 2\alpha < \frac{8}{11}.$$

In other words

$$\sigma_0 \ll O(\varepsilon^{\frac{11}{3}}) \quad \text{and} \quad \alpha_4 \gg O(\varepsilon^{\frac{8}{11}}). \tag{7.84}$$

From the above results we derive

$$\alpha_4^{\frac{7}{8}} \gg O(\varepsilon^{\frac{7}{11}}) \gg O(\varepsilon)$$

and

$$\sigma_0^{\frac{11}{7}} \alpha_4^{\frac{2}{7}} \gg O(\varepsilon^{\frac{7}{11}}) \gg O(\varepsilon) .$$

However, it will be seen that this relationship for the orders of magnitude of σ_0, ε, and α_4 is inconsistent with the quantization condition that we shall obtain later. Hence, situation (a) will actually never occur.

(b) The singular points $\xi = -i, -ia_1, -ia_2$ are all in the inner region of ξ_c. This implies that $|\xi_c + i| = O(\varepsilon^\alpha)$ and $|\xi_c + ia_i| = O(\varepsilon^\alpha)$ $(i = 1, 2)$, or say, $\frac{\mu_0}{2} = \alpha = \beta$.

In this case, the orders of magnitude of each term on the left-hand side of (7.66) are

$$(1) = O(\varepsilon^{3-\frac{15\alpha}{4}}), \quad (2) = O(\varepsilon^{2-2\alpha}) \quad \text{and} \quad (3) = O(\varepsilon^{\frac{7\alpha}{4}-\frac{\beta}{4}}).$$

Moreover, the balance on the left-hand side of (7.66) will be

$$(1) \approx (2) \approx (3).$$

We thus have

$$\alpha = \frac{4}{7}, \quad \nu_0 = \frac{3}{7} \quad \text{and} \quad \mu_0 = \frac{8}{7} . \tag{7.85}$$

In other words,

$$\sigma_0 = O(\varepsilon^{\frac{3}{7}}), \quad \text{and} \quad \alpha_4 = O(\varepsilon^{\frac{8}{7}}) . \tag{7.86}$$

The leading order approximation of the inner equation is

$$\frac{K_3(\xi_* + \hat{\delta} + i\hat{a}_1)^{\frac{1}{2}}(\xi_* + \hat{\delta} + i\hat{a}_2)^{\frac{1}{2}}}{(\xi_* + \hat{\delta})^{\frac{7}{4}}} \frac{\mathrm{d}^3 \hat{W}_0}{\mathrm{d}\xi_*^3} + \frac{\mathrm{d}^2 \hat{W}_0}{\mathrm{d}\xi_*^2}$$

$$+ \frac{A_3^2 \hat{\delta}^{\frac{3}{4}}}{(\hat{\delta} + i\hat{a}_1)^{\frac{1}{2}}(\hat{\delta} + i\hat{a}_2)^{\frac{1}{2}}} \frac{\xi_*(\xi_* + \hat{\delta})^{\frac{7}{4}}}{(\xi_* + \hat{\delta} + i\hat{a}_1)^{\frac{1}{2}}(\xi_* + \hat{\delta} + i\hat{a}_2)^{\frac{1}{2}}} \hat{W}_0 = 0 \tag{7.87}$$

where $K_3 = O(1)$, and $A_3 = O(1)$ are some constants. In the far-field of the inner region, as $\xi_* \to \infty$, this inner equation reduces to

$$\frac{\mathrm{d}\hat{W}_0}{\mathrm{d}\hat{\xi}_*^2} + \hat{\xi}_*^{\frac{7}{4}} \hat{W}_0 = 0 , \tag{7.88}$$

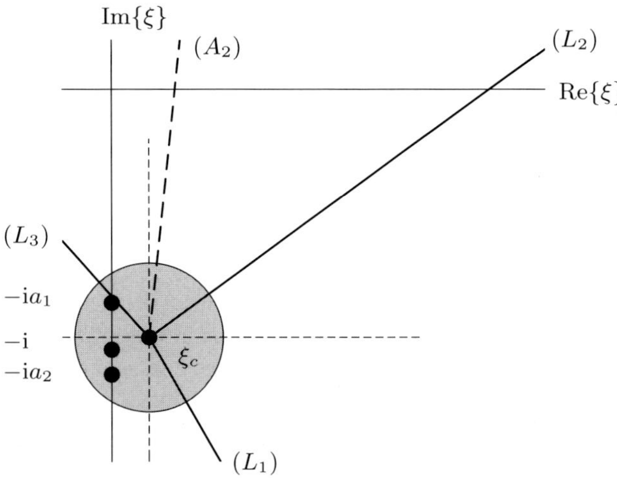

Fig. 7.1. Sketch of the Stokes lines (L_1), (L_2), (L_3) and the anti-Stokes line (A_2) emanating from the turning point ξ_c, in the case $|\sigma| \ll 1$

where

$$\hat{\xi}_* = \frac{B}{\varepsilon^{\frac{4}{7}}}(\xi - \xi_c)$$

$$B = \frac{A_3^{\frac{8}{15}}\hat{\delta}^{\frac{1}{5}}}{(\hat{\delta} + i\hat{a}_1)^{\frac{2}{15}}(\hat{\delta} + i\hat{a}_2)^{\frac{2}{15}}} \tag{7.89}$$

$$A_3 = i\frac{A_0(\xi_c - i)^{\frac{3}{4}}}{(\xi_c - ia_1)^{\frac{1}{2}}(\xi_c - ia_2)^{\frac{1}{2}}}\left\{1 + i\alpha_4 \frac{16}{3}\frac{\xi_c(\xi_c^2 - 1)}{f(\xi_c)}\right\}^{\frac{1}{2}}.$$

One can estimate that

$$\arg(\hat{A}_3) \approx \frac{5\pi}{8} \quad \text{and} \quad \arg(B) \approx \frac{\pi}{3}. \tag{7.90}$$

From the above results we derive

$$\alpha_4^{\frac{7}{8}} = O(\varepsilon) \quad \text{and} \quad \sigma_0^{\frac{11}{7}}\alpha_4^{\frac{2}{7}} = O(\varepsilon). \tag{7.91}$$

It will be seen that this case is consistent with the quantization condition obtained later.

The structure of the Stokes lines for case IIb is sketched in Fig. 7.1. The open angle between (L_1) and (L_2) is $\theta = \frac{8}{15}\pi$.

7.3.3 A Brief Summary

From the results derived above, one sees that for case I and II, the inner equation in the far field of the inner region as $\xi_* \to \infty$, can be approximately written in the following unified form:

$$\frac{d^2 \hat{W}_0}{d\hat{\xi}_*^2} + \hat{\xi}_*^{p_0} \hat{W}_0 = 0 \, , \tag{7.92}$$

where

$$\hat{\xi}_* = \frac{B}{\varepsilon^\alpha}(\xi - \xi_c) \, . \tag{7.93}$$

For different cases, the constants B and p_0 and the exponent number α are different. Moreover, for these two cases, the anti-Stokes line (A_2) which is tangent to the direction $\arg(\zeta) = \frac{3\pi}{2}$ at the turning point ξ_c always intersects the real axis of ξ at a point $\xi'_c > 0$, and divides the real axis into two parts. One part is in the sector (S_1) while the other is in (S_2). The open angle at the turning point ξ_c between two neighboring Stokes lines is

$$\theta = \frac{2}{p_0 + 2}\pi \, . \tag{7.94}$$

The general solution of (7.92) is

$$\hat{W}_0 = D_1 \hat{\xi}_*^{\frac{1}{2}} H_\nu^{(1)}(\zeta) + D_2 \hat{\xi}_*^{\frac{1}{2}} H_\nu^{(2)}(\zeta) \quad (\zeta = 2\nu \hat{\xi}_*^{\frac{1}{2\nu}}), \tag{7.95}$$

where $\nu = \frac{1}{p_0+2}$ and $H_\nu^{(1)}(\zeta)$ and $H_\nu^{(2)}(\zeta)$ are the νth-order Hankel functions of first and second kind, respectively. For the future reference, we summarize the results below:

- Case I : $\sigma_0 = O(1)$, and

$$\alpha = \frac{2}{3} \, , \quad p_0 = 1 \, , \quad \nu = \frac{1}{3} \, .$$

- Case IIa : $|\sigma_0| \ll O(\varepsilon^{\frac{3}{11}})$, $\alpha_4 \gg O(\varepsilon^{\frac{8}{11}})$, and

$$\frac{\mu_0}{2} < \alpha \le \frac{4}{11} \le \beta \, , \quad p_0 = \frac{11}{4} \, , \quad \nu = \frac{4}{19} \, .$$

- Case IIb: $|\sigma_0| = O(\varepsilon^{\frac{3}{7}})$, $\alpha_4 = O(\varepsilon^{\frac{8}{7}})$, and

$$\alpha = \frac{4}{7} \, , \quad p_0 = \frac{7}{4} \, , \quad \nu = \frac{4}{15} \, .$$

7.4 Matching Conditions

We now turn to match the inner solution (7.95) with the outer solution (7.38) in the intermediate regions in each sector. Again, we apply the connection formula for the Hankel functions: When $\pi < \arg(\zeta_1) < 2\pi$,

$$H_\nu^{(2)}(\zeta_1) = H_\nu^{(2)}(\zeta e^{i\pi}) = 2\cos(\nu\pi)H_\nu^{(2)}(\zeta) + e^{i\nu\pi}H_\nu^{(1)}(\zeta) \qquad (7.96)$$

$$(0 \le \arg(\zeta) \le \pi),$$

and the asymptotic expansions, as $|\zeta| \to \infty$,

$$H_\nu^{(1)}(\zeta) \sim \sqrt{\frac{2}{\pi\zeta}}\, e^{i\zeta - i(\nu/2 + 1/4)\pi} \quad (-\pi < \arg(\zeta) \le 2\pi),$$

$$H_\nu^{(2)}(\zeta) \sim \sqrt{\frac{2}{\pi\zeta}}\, e^{-i\zeta + i(\nu/2 + 1/4)\pi} \quad (-2\pi < \arg(\zeta) \le \pi). \qquad (7.97)$$

First match the inner solution (7.95) with the outer solution (7.38) as $\hat{\xi}_* \to \infty$ in the sector (S_2). Note that in this sector, the outer solution satisfying the root condition (7.27) has $D_1' = 0$. Thus, to match with the outer solution, one must set $D_1 = 0$ and $D_2 = D = D_3' \times$ constant $\neq 0$ in (7.95).

Furthermore, one can match the inner solution with the outer solution in the sector (S_1), as conducted in Chap. 6, which leads to the following connection condition:

$$\frac{D_1}{D_3} = i2\cos(\nu\pi). \qquad (7.98)$$

So far we have not applied the tip condition. When the tip condition (7.28) or (7.29) is applied, the parameter σ_0 will be determined as a function of α_4 and ε. We shall derive these results in the next section.

7.5 The Spectra of Eigenvalues and Instability Mechanisms

We find that the system allows two different types of eigenvalue spectra.

1. The complex spectrum: $\sigma_0 = \sigma_R - i\omega$ $(\omega > 0)$.
2. The real spectrum (within the accuracy of our asymptotic solution): $\sigma_0 = \sigma_R$. The exact eigenvalues σ have not been proved to be real numbers yet. As such σ might have a higher-order small imaginary part.

Consequently, the system is subject to two different types of instability mechanism:

1. The global trapped-wave (GTW) instability induced by perturbations with a high frequency ($|\omega| = O(1)$).
2. The low-frequency (LF) instability induced by perturbations with a low frequency $\left(\omega \leq O(\varepsilon)\right)$. Note that to date, the validity of the exchange principle of stability is not certain, so we use the term 'low-frequency' to describe the second type of instability mechanism.

7.5.1 The Global Trapped-Wave Instability

Consider $\sigma_0 = \sigma_R - i\omega$ and $\omega > 0$. With the complex eigenvalue σ_0, the physical solution in the outer region is

$$\mathrm{Re}\left\{\tilde{h}_0(\xi, t)\right\} = \mathrm{Re}\left\{H(\xi)\mathrm{e}^{\frac{\sigma_0 t}{\varepsilon \eta_0^2}}\right\},\tag{7.99}$$

where

$$H(\xi) = D_1 H_1 + D_3 H_3\,.\tag{7.100}$$

We define

$$\begin{cases} d_1 = D_1 \mathrm{e}^{-i\chi_1} \\ d_3 = D_3 \mathrm{e}^{-i\chi_3}\,, \end{cases}\tag{7.101}$$

where

$$\begin{aligned} \chi_1 &= \frac{1}{\varepsilon}\int_0^{\xi_c} \hat{k}_0^{(1)}\mathrm{d}\xi \\ \chi_3 &= \frac{1}{\varepsilon}\int_0^{\xi_c} \hat{k}_0^{(3)}\mathrm{d}\xi\,. \end{aligned}\tag{7.102}$$

Thus, it follows from the connection formula (7.98) that

$$\frac{d_1}{d_3} = i2\cos(\nu\pi)\mathrm{e}^{-i\chi},\tag{7.103}$$

where

$$\chi = \frac{1}{\varepsilon}\int_0^{\xi_c} \left(\hat{k}_0^{(1)} - \hat{k}_0^{(3)}\right)\mathrm{d}\zeta\,.\tag{7.104}$$

To satisfy the smooth tip conditions, the parameter σ_0 must be a proper function of ε, such that the tip $\xi = 0$ is located on the Stokes line (L_3) and, as $\varepsilon \to 0$, the following conditions hold:

(i) For the symmetrical S-modes,

$$d_1 \hat{k}_0^{(1)}(0) + d_3 \hat{k}_0^{(3)}(0) = 0 \, , \qquad (7.105)$$

which gives

$$\mathrm{i}2\cos(\nu\pi)\mathrm{e}^{-\mathrm{i}\chi} = -\frac{\hat{k}_0^{(3)}(0)}{\hat{k}_0^{(1)}(0)} \, . \qquad (7.106)$$

(ii) For the anti-symmetrical A-modes,

$$d_1 + d_3 = 0 \, , \qquad (7.107)$$

which gives

$$\mathrm{i}2\cos(\nu\pi)\mathrm{e}^{-\mathrm{i}\chi} = -1 \, . \qquad (7.108)$$

Thus, we obtain the following quantization conditions:

$$\chi = \frac{1}{\varepsilon} \int_0^{\xi_c} \left(\hat{k}_0^{(1)} - \hat{k}_0^{(3)} \right) \mathrm{d}\xi = \left(2n + 1 + \frac{1}{2} + \theta_0\right)\pi$$

$$- \mathrm{i}\left\{ \ln \alpha_0 + \ln\left[2\cos(\nu\pi)\right] \right\}, \qquad (n = 0, \pm 1, \pm 2, \cdots) \, , \qquad (7.109)$$

where

$$\begin{cases} \alpha_0\, \mathrm{e}^{\mathrm{i}\theta_0\pi} = \dfrac{\hat{k}_0^{(1)}(0)}{\hat{k}_0^{(3)}(0)} & \text{for the S-modes;} \\[4mm] \alpha_0 = 1, \quad \theta_0 = 0 & \text{for the A-modes.} \end{cases} \qquad (7.110)$$

Let us first consider the complex eigenvalues: $|\sigma_0| = O(1)$. This branch corresponds to case I discussed in the last section; hence we must set $\nu = \frac{1}{3}$. This branch contains two discrete sets of complex eigenvalues σ_{0n} ($n = 0, \pm 1, \pm 2, \cdots$) for S-modes and A-modes, respectively, given by the quantization conditions (7.109) as functions of ε and α_4.

In Fig. 7.2, we show the leading order approximation of eigenvalues σ_0 of the A-modes ($n = 0, 1, 2,$) versus ε, for the two cases: $\alpha_4 = 0$ and 0.1. The smaller symbols describe the case $\alpha_4 = 0$, while the larger symbols describe the case $\alpha_4 = 0.1$. As we have seen for the case of a pure melt in Chap. 6, these eigenvalues for $n = 0, 1, 2, \cdots$ are all located on the same curve in the complex σ-plane, and they all tend to the point $\sigma_0 = \sigma_{\max}$ as $\varepsilon \to 0$. As $\alpha_4 = 0$, $\hat{\alpha} = 1$, $\sigma_{\max} = 0.2722$, while as $\alpha_4 = 0.1$, $\hat{\alpha} = 1$, $\sigma_{\max} = 0.2869$.

The system, for each n, allows a unique, neutrally stable A-mode, and a unique, neutrally stable S-mode; these will be denoted by $\varepsilon_{*n}^{(\mathrm{a})}$ and $\varepsilon_{*n}^{(\mathrm{s})}$, respectively. It can be shown that

$$\varepsilon_{*0}^{(\mathrm{a})} > \varepsilon_{*0}^{(\mathrm{s})} > \varepsilon_{*1}^{(\mathrm{a})} > \varepsilon_{*1}^{(\mathrm{s})} > \varepsilon_{*2}^{(\mathrm{a})} > \varepsilon_{*2}^{(\mathrm{s})} > \cdots .$$

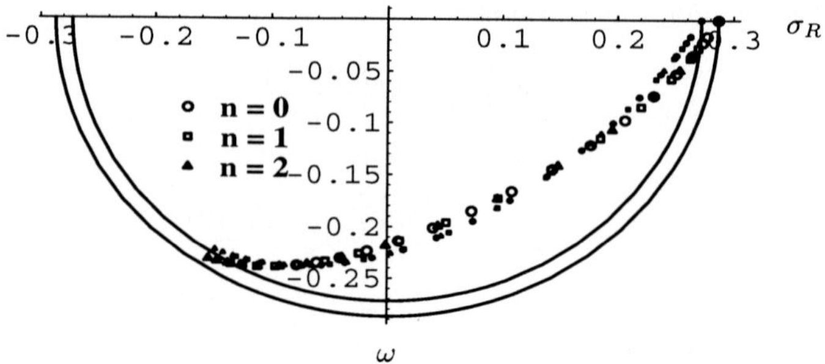

Fig. 7.2. The variations with ε of $\sigma_0 = \sigma_R - i\omega$ of GTW A-modes ($n = 0, 1, 2$) in the complex σ_0-plane. The smaller symbols describe the case $\alpha_4 = 0$, while the larger symbols correspond to $\alpha_4 = 0.1$

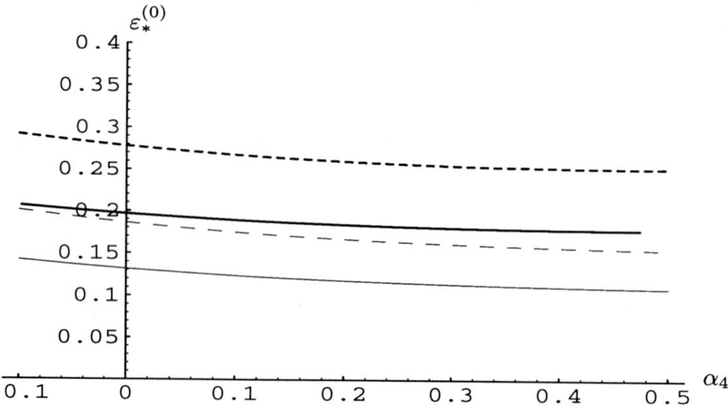

Fig. 7.3. The variations of $\varepsilon_*^{(0)}$ of the GTW A-modes and S-modes ($n = 0$) with α_4. The thin lines are used for the S-modes, the thick lines for the A-modes; the solid lines are for the case $\hat{\alpha} = 1$, and the dashed lines for $\hat{\alpha} = 0$

Hence, the neutral A-mode ($n = 0$) is more stable then the neutral S-mode ($n = 0$).

In Fig. 7.3 we show the critical numbers ε_* corresponding to neutral GTW modes versus α_4, while in Fig. 7.4 we show the eigenvalue ω_* of the neutral GTW modes versus α_4. In these figures, the thin lines are used for the S-modes, the thick lines for the A-modes; the solid lines are for the case $\hat{\alpha} = 1$, and the dashed lines for the case $\hat{\alpha} = 0$.

For the first-order approximation solutions, one can follow the same procedure as in Chap. 6. Strictly speaking, the anisotropy parameter α_4 will also affect the first-order approximation solution through the coefficient functions $\bar{A}_s$, $\tilde{A}_{s0}$, and $\tilde{A}_{s1}$ in the Gibbs–Thomson condition (7.17). However, for the small anisotropy case ($\alpha_4 \ll 1$), one may neglect the small quantities of

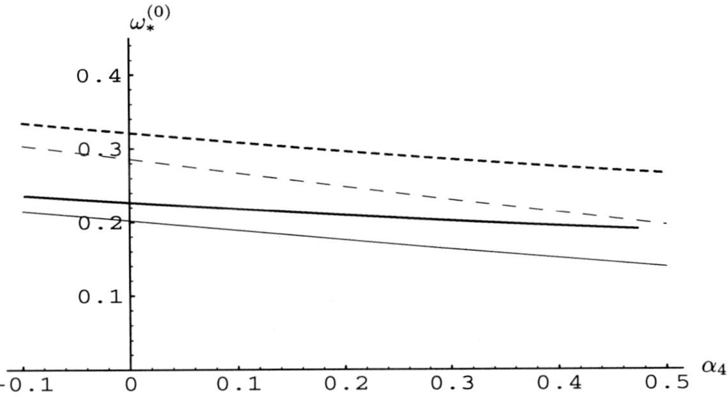

Fig. 7.4. The variations of $\omega_*^{(0)}$ of the corresponding GTW A-modes and S-modes $(n = 0)$ with α_4

$O(\varepsilon\alpha_4)$ to obtain

$$\sigma_1 = \frac{1}{1 + \xi_c^2}\left\{ \frac{\eta_0^2}{2}\left(\frac{2}{3} + \frac{i\xi_c}{3}\right)^2 - (1 + \eta_0^2) + \frac{\eta_0^2}{4\Lambda_1} \right\}. \tag{7.111}$$

Assuming $\sigma = \sigma_0 + \varepsilon\sigma_1$, we can compute the first-order approximation for the critical number ε_* and the corresponding eigenvalues of the neutral A-modes and S-modes $(n = 0)$, which depend on the Peclet number η_0^2. It is noticed that, in the small undercooling regime, the critical number ε_* is insensitive to a change of undercooling, while in the large undercooling regime, as $\mathrm{Pe}_0 > 1.0$, the critical number ε_* becomes a very sensitive function of the undercooling temperature. It decreases rapidly as the Peclet number Pe_0 increases. In Figs. 7.5 and 7.6, we show the numerical results for the cases: $\eta_0^2 = 0.001, 0.1, 1.0$ and $\hat{\alpha} = 0, 1$, respectively. Here, we use $\varepsilon_*^{(1)}$ to denote the first-order approximation of the critical number ε_* for the neutral A-mode $(n = 0)$ and $\omega_{(*)}^{(1)}$ to denote its corresponding frequency. The numerical results for the cases $\eta_0^2 = 0.001$ and $\hat{\alpha} = 0, 1$ respectively are also listed in Tables 7.1 and 7.2.

Now we consider the possibility of complex eigenvalues with $|\sigma_0| \ll 1$, which corresponds to case II. We shall show that the system does not allow such a branch of complex eigenvalues. Let us first simplify the quantization condition (7.109) with the assumption $|\sigma_0| \ll 1$. From the local dispersion formula (7.34) one can show that, as $\sigma_0 \to 0$,

$$\left(\hat{k}_0^{(1)} - \hat{k}_0^{(3)}\right) = \frac{1}{\sqrt{2}}\frac{(1 - i\xi)^{\frac{7}{4}}(1 + i\xi)^{\frac{5}{4}}}{f^{\frac{1}{2}}(\xi)} - \frac{3}{2}(1 + i\xi)\sigma_0 + O(\sigma_0^2). \tag{7.112}$$

Table 7.1. The critical numbers, $\varepsilon_*^{(1)}$, and the corresponding frequencies, $\omega_*^{(1)}$, of 2D neutral GTW A-modes ($n = 0$) for the case $\eta_0^2 = 0.001$, $\hat{\alpha} = 0$

α_4	$\varepsilon_*^{(1)}$	$\omega_*^{(1)}$	V_{p}
-0.05	0.1832	0.3320	1.031
-0.01	0.1808	0.3263	1.029
0.0	0.1794	0.3226	1.028
0.01	0.1787	0.3208	1.028
0.04	0.1769	0.3155	1.026
0.07	0.1753	0.3105	1.025
0.10	0.1740	0.3057	1.023
0.13	0.1727	0.3011	1.022
0.16	0.1718	0.2968	1.021
0.19	0.1708	0.2926	1.020
0.22	0.1701	0.2887	1.019
0.25	0.1695	0.2849	1.018

Table 7.2. The critical numbers, $\varepsilon_*^{(1)}$, and the corresponding frequencies, $\omega_*^{(1)}$, of 2D neutral GTW A-modes ($n = 0$) for the case $\eta_0^2 = 0.001$, $\hat{\alpha} = 1$

α_4	$\varepsilon_*^{(1)}$	$\omega_*^{(1)}$	V_{p}
-0.05	0.1296	0.2348	1.031
-0.01	0.1279	0.2307	1.029
0.0	0.1268	0.2281	1.028
0.01	0.1264	0.2268	1.028
0.04	0.1251	0.2231	1.026
0.07	0.1240	0.2196	1.025
0.10	0.1230	0.2162	1.023
0.13	0.1222	0.2129	1.022
0.16	0.1215	0.2099	1.021
0.19	0.1208	0.2069	1.020
0.22	0.1203	0.2041	1.019
0.25	0.1198	0.2014	1.018

We find that

$$
\chi = \frac{1}{\varepsilon} \int_0^{\xi_c} \left(\hat{k}_0^{(1)} - \hat{k}_0^{(3)} \right) \mathrm{d}\xi = \left(\int_0^{-\mathrm{i}(a_1-0)} \right) + \left(\int_{(C_r)} \right) + \left(\int_{-\mathrm{i}(a_1+0)}^{-\mathrm{i}} \right) + \left(\int_{-\mathrm{i}}^{\xi_c} \right)
$$

$$
= \frac{1}{\varepsilon} \left[\frac{\mathrm{i}}{\sqrt{2}} \mathcal{R} - \frac{1}{\sqrt{2}} \gamma_0 + \mathrm{i}\frac{9}{4}\sigma_0 + 3(\sqrt{3} - 1)\sigma_0(\xi_c + \mathrm{i}) \right]
$$

$$
+ \text{(higher-order terms)}, \tag{7.113}
$$

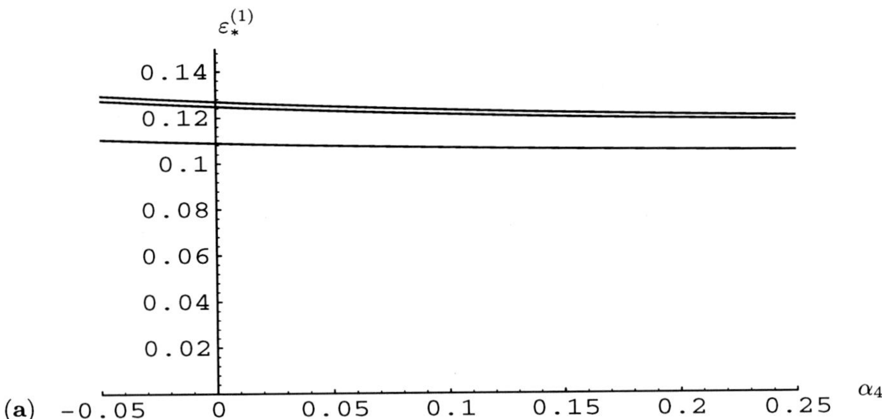

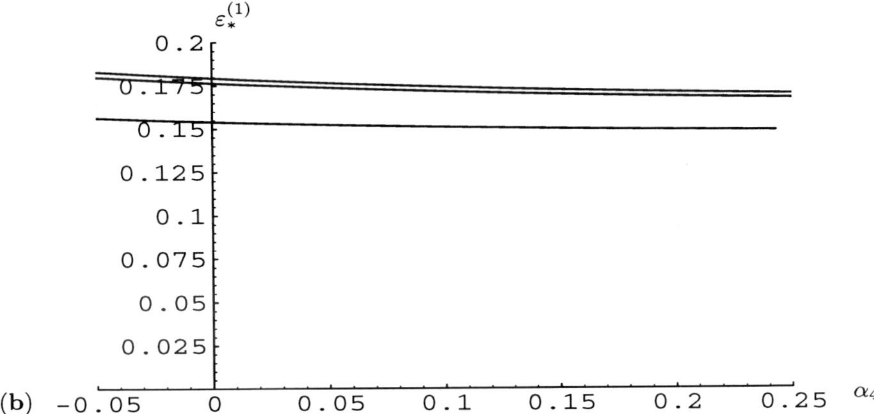

Fig. 7.5. The variations of $\varepsilon_*^{(1)}$ of the neutral GTW A-modes ($n = 0$) in the first-order approximation with α_4 for various Peclet numbers $\eta_0^2 = 0.001, 0.1, 1.0$, from top to bottom: **(a)** in the case $\hat{\alpha} = 1$, **(b)** in the case $\hat{\alpha} = 0$

where the integral path (C_r) is a semicircle, connecting the point $\xi = -\mathrm{i}(a_1 - 0)$ and the point $\xi = -\mathrm{i}(a_1 + 0)$ clockwisely, and

$$\gamma_0 = C_0 \alpha_4^{\frac{7}{8}}, \quad C_0 \approx 1.80205, \quad \text{and} \quad \mathcal{R} \approx 0.615622. \tag{7.114}$$

By substituting (7.113) into the quantization condition (7.109), one obtains the eigenvalues $\sigma_0 = O(1)$. This contradicts the original assumption that $\sigma_0 \ll 1$. Therefore, we conclude that the system only allows one branch of the complex spectrum of eigenvalues with $|\sigma_0| = O(1)$.

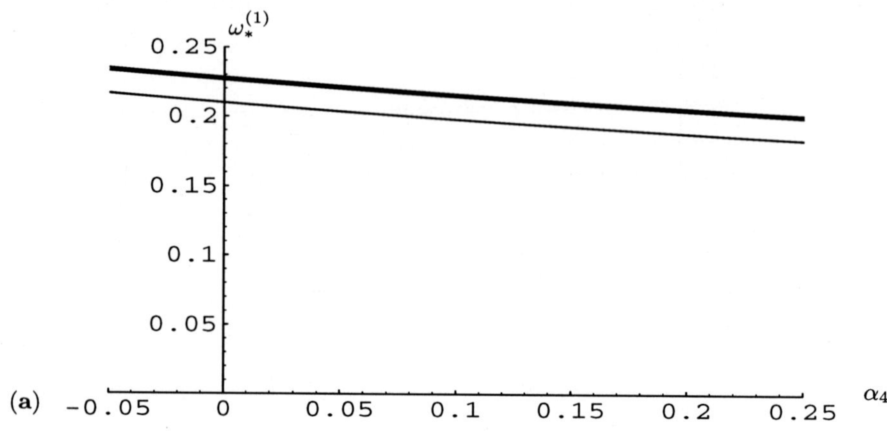

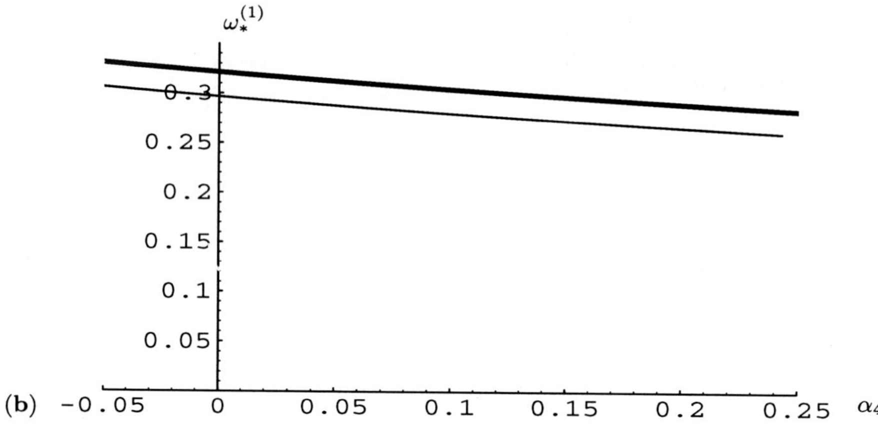

Fig. 7.6. The variations of $\omega_*^{(1)}$ of the neutral GTW A-modes ($n = 0$) in the first-order approximation with α_4 for various Peclet numbers $\eta_0^2 = 0.001, 0.1, 1.0$, from top to bottom: **(a)** in the case $\hat{\alpha} = 1$, **(b)** in the case $\hat{\alpha} = 0$

7.5.2 The Low-Frequency Instability

We now turn to derive the spectrum of real eigenvalues σ_0. In this case, the physical solution in the outer region becomes

$$\mathrm{Re}\{\tilde{h}(\xi, t)\} = \mathrm{Re}\{H(\xi)\} \exp\left\{\frac{\sigma_0 t}{\varepsilon \eta_0^2}\right\}. \tag{7.115}$$

The tip condition for symmetric modes is now

$$\mathrm{Re}\{H(\xi)\}'(0) = \mathrm{Re}\{H'(0)\} = \mathrm{Re}\left\{\frac{i d_1}{\varepsilon} \hat{k}_0^{(1)}(0) + \frac{i d_3}{\varepsilon} \hat{k}_0^{(3)}(0)\right\} = 0. \tag{7.116}$$

Note that the eigenmodes are determined to within an arbitrary multiplicative constant. Hence, without loss of generality, we can assume that d_1 is a positive real number and write

$$d_1 > 0, \quad d_3 = |d_3|e^{i\chi_0\pi}. \tag{7.117}$$

Note also that for a real eigenvalue σ_0, $\hat{k}_0^{(1)}(0)$ and $\hat{k}_0^{(3)}(0)$ are both real. Therefore, the tip condition (7.116) leads to $d_{3I} = 0$, in other words,

$$\frac{d_1}{d_3} = \left|\frac{d_1}{d_3}\right|e^{-i\chi_0\pi} = i2\cos(\nu\pi)e^{-i\chi} \quad (\chi_0 = 0, 1). \tag{7.118}$$

It follows that

$$\mathrm{Re}\left[\frac{1}{\varepsilon}\int_0^{\xi_c}\left(\hat{k}_0^{(1)} - \hat{k}_0^{(3)}\right)d\xi\right] = \left(2n + \frac{1}{2} + \chi_0\right)\pi \tag{7.119}$$

$$(n = 0, \pm1, \pm2, \cdots),$$

$$\left|\frac{d_1}{d_3}\right| = 2\cos(\nu\pi)e^{\mathrm{Im}\{\chi\}}. \tag{7.120}$$

For the antisymmetric modes, the tip condition is

$$\mathrm{Re}\{H(0)\} = \mathrm{Re}\{d_1 + d_3\} = 0 \tag{7.121}$$

which leads to

$$d_1 + |d_3|\cos(\chi_0\pi) = 0$$

or

$$\left|\frac{d_1}{d_3}\right| = -\cos(\chi_0\pi) \leq 1. \tag{7.122}$$

From the above, it follows that

$$\frac{d_1}{d_3} = \left|\frac{d_1}{d_3}\right|e^{-i\chi_0\pi} = -\cos(\chi_0\pi)e^{-i\chi_0\pi} = i2\cos(\nu\pi)e^{-i\chi}. \tag{7.123}$$

It follows that

$$\mathrm{Re}\left[\frac{1}{\varepsilon}\int_0^{\xi_c}\left(\hat{k}_0^{(1)} - \hat{k}_0^{(3)}\right)d\xi\right] = \left(2n + \frac{1}{2} + \chi_0\right)\pi \tag{7.124}$$

$$(n = 0, \pm1, \pm2, \cdots)$$

and

$$\left|\frac{d_1}{d_3}\right| = 2\cos(\nu\pi)e^{\mathrm{Im}\{\chi\}}. \tag{7.125}$$

The above two quantization conditions can be written in the same form as follows:

$$\mathrm{Re}\left[\frac{1}{\varepsilon}\int_0^{\xi_c}\left(\hat{k}_0^{(1)}-\hat{k}_0^{(3)}\right)\mathrm{d}\xi\right]=\left(2n+\frac{1}{2}+\chi_0\right)\pi \qquad (7.126)$$

$$(n=0,\pm 1,\pm 2,\cdots)\,;$$

$$\left|\frac{d_1}{d_3}\right|=2\cos(\nu\pi)\mathrm{e}^{\mathrm{Im}\{\chi\}} \qquad (7.127)$$

where

$$\chi_0=\begin{cases} 0\ \text{or}\ 1, & \text{(for S-modes);} \\[2mm] 1+\dfrac{\cos^{-1}\left|\frac{d_1}{d_3}\right|}{\pi}, & \text{(for A-modes).} \end{cases} \qquad (7.128)$$

The quantization condition (7.126) determines the eigenvalues $\sigma_0^{(n)}$, while formula (7.127) determines the corresponding eigenfunctions. We did not find the real eigenvalues with $|\sigma_0|=O(1)$. However, we found that the system does allow a branch of the real spectrum with $|\sigma_0|\ll 1$, which corresponds to case IIb. We are much concerned with neutral modes in the study of stability, hence $|\sigma|\ll 1$ is the case of most interest. To derive this branch of eigenvalues, we first simplify the quantization condition (7.126) as we did for the case of a complex spectrum. In terms of (7.113), we obtain

$$\mathrm{Re}\{\chi\}=\frac{1}{\varepsilon}\left[-\frac{\gamma_0}{\sqrt{2}}+C_1\sigma_0^{\frac{11}{7}}\alpha_4^{\frac{2}{7}}\right], \qquad (7.129)$$

$$\mathrm{Im}\{\chi\}=\frac{1}{\varepsilon}\left[\frac{\mathcal{R}}{\sqrt{2}}+\frac{9}{4}\sigma_0+C_2\sigma_0^{\frac{11}{7}}\alpha_4^{\frac{2}{7}}\right] \qquad (7.130)$$

where

$$\begin{cases} C_1=3(\sqrt{3}-1)2^{\frac{3}{7}}3^{\frac{6}{7}}\cos(\frac{5\pi}{14})\approx 3.2886 \\[2mm] C_2=3(\sqrt{3}-1)2^{\frac{3}{7}}3^{\frac{6}{7}}\sin(\frac{5\pi}{14})\approx 6.2883\,. \end{cases} \qquad (7.131)$$

By using these results, we derive the following simplified quantization conditions:

$$C_1\sigma_0^{\frac{11}{7}}\alpha_4^{\frac{2}{7}}=\frac{C_0}{\sqrt{2}}\alpha_4^{\frac{7}{8}}+\varepsilon\left(2n+\frac{1}{2}+\chi_0\right)\pi\,, \qquad (7.132)$$

$$\left|\frac{d_1}{d_3}\right|=2\cos(\nu\pi)\mathrm{e}^{\mathrm{Im}\{\chi\}}\,, \qquad (7.133)$$

where

$$\chi_0=\begin{cases} 0\ \text{or}\ 1 & \text{(for the S-modes);} \\[2mm] \cos(\chi_0\pi)=-2\cos(\nu\pi)\mathrm{e}^{\mathrm{Im}\{\chi\}} & \text{(for the A-modes).} \end{cases} \qquad (7.134)$$

The quantization condition (7.132) shows that as $\varepsilon\to 0$, the asymptotic relations $\alpha_4^{\frac{7}{8}}=O(\varepsilon)$ and $\sigma_0^{\frac{11}{7}}\alpha_4^{\frac{2}{7}}=O(\varepsilon)$ must hold. Obviously, only case

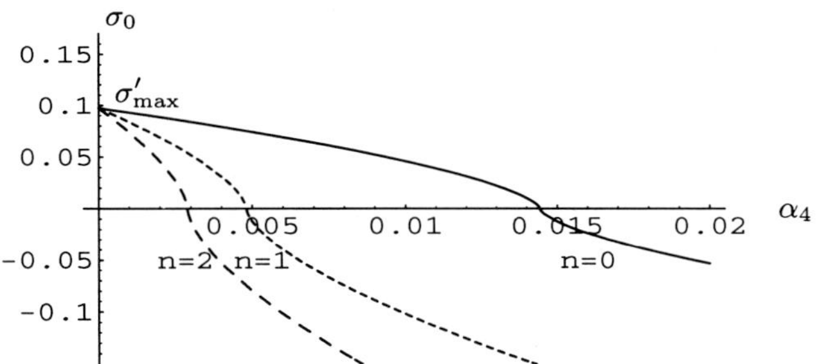

Fig. 7.7. The spectrum of eigenvalues of LF instability mechanism for the case $\alpha = 1, \alpha_4 = 0.01$

IIb, discussed in the last section, is consistent with this requirement. Hence, we must have $\nu = \frac{4}{15}$. Moreover, it is seen from (7.113) that when $\sigma_0 \geq 0$, $\mathrm{Im}\{\chi\} > 0$. Consequently, from (7.133) we have $\left|\frac{d_1}{d_3}\right| > 1$. Thus, no solution χ_0 can be found to satisfy (7.134) for the A-modes. From this we conclude that the system does not allow any growing LF A-mode; only growing S-modes are allowed. Moreover, the quantization condition for these S-modes can be written as follows:

$$C_1 \sigma_0^{\frac{11}{7}} \alpha_4^{\frac{2}{7}} = \frac{C_0}{\sqrt{2}} \alpha_4^{\frac{7}{8}} - \varepsilon \left(n + \frac{1}{2}\right)\pi , \tag{7.135}$$

$$\left|\frac{d_1}{d_3}\right| = 2\cos\left(\frac{4\pi}{15}\right) \mathrm{e}^{\mathrm{Im}\{\chi\}} , \tag{7.136}$$

where

$$\mathrm{sign}\left\{\frac{d_1}{d_3}\right\} = \begin{cases} 1 \text{ when } n \text{ is an even integer} \\ -1 \text{ when } n \text{ is an odd integer.} \end{cases} \tag{7.137}$$

For any fixed ε and α_4, from the quantization condition (7.135), one can solve for a discrete set of the eigenvalues $\{\sigma_n\}$ $(n = 0, 1, 2, 3, \cdots)$ as shown in Fig. 7.7. It is seen that there is a discrete set of neutral stable modes $(\sigma_n = 0)$ corresponding to $\varepsilon'_0 > \varepsilon'_1 > \varepsilon'_2 > \cdots > \varepsilon'_n > \cdots$. These neutral modes coincide with the steady needle crystal growth solutions predicted by the MSC theory.

For the neutral mode $n = 0$, we obtain

$$\varepsilon = \varepsilon_a = \varepsilon'_0 = \mathcal{K}_0 \alpha_4^{\frac{7}{8}} , \tag{7.138}$$

where $\mathcal{K}_0 = 0.81120$. In the MSC theory for the steady needle solutions with the largest tip velocity $(n = 0)$, this coefficient has a different value of 1.09.

We remark here that, in the above, we obtained only the leading order approximation of the eigenvalues, σ_0. For more accurate numerical values of these critical numbers, one needs to include higher-order approximations. For instance, one may solve ε'_n from the equation:

$$\sigma_0 + \varepsilon\sigma_1 = 0. \tag{7.139}$$

For the higher-order approximation, one can follow the approach described in Sect. 6.3. We shall not carry out such a derivation here.

There are $m + 1$ purely growing modes and infinitely many decaying modes, when $\varepsilon_{m+1} < \varepsilon < \varepsilon_m$. As $\varepsilon \to 0$, the first n eigenvalues σ_k $(k = 0, 1, 2, \cdots, n)$ all tend to the upper limit $\sigma'_{\max}$, which can be calculated as

$$\sigma'_{\max} = \alpha_4^{\frac{3}{8}} \left(\frac{C_0}{\sqrt{2}C_1} \right)^{\frac{7}{11}} \approx 0.5470\alpha_4^{\frac{3}{8}} . \tag{7.140}$$

When $\varepsilon > \varepsilon_a = \varepsilon'_0$, all modes are purely decaying. Thus, the system will be stable provided no other mechanism can cause the instability. As $\alpha_4 \to 0$, the low-frequency stability disappears as the upper limit $\sigma'_{\max} \to 0$.

The above low-frequency (LF) instability was first discovered by Kessler and Levine numerically in 1986 [7.1]. It was later confirmed by Bensimon et al. in an analytical way [7.2]. The quantization condition obtained by Bensimon et al. involves some errors but, with their quantization condition, the authors were still able to draw the same conclusion as was drawn here. More precisely, it was concluded that the steady needle solution predicted by MSC theory is neutrally stable (under the LF instability mechanism).

The physical implications of the LF and GTW instability can also be clarified. Assume that a small initial perturbation of the interface shape is given. Due to the low-frequency instability, this initial perturbation of the interface shape will be purely growing or decaying without propagating and oscillating. As a consequence, the perturbed interface will diverge from or be restored to its original shape without oscillation, depending on whether it is unstable or stable. On the other hand, under the GTW instability mechanism, the initial perturbation will propagate and oscillate as a traveling wave with either growing or decaying amplitude. As such the perturbed interface will depart from the original one with oscillation if it is unstable under the GTW mechanism. On the other hand, it will be restored to its original state with oscillation if it is stable under the GTW mechanism.

7.6 Low-Frequency Instability
for Axially Symmetric Dendrite Growth

The low-frequency instability can also be obtained for axially symmetric dendrite growth if one only considers the axial component of the anisotropy of surface tension and assumes that the azimuthal component of the anisotropy is zero.

For the leading order approximation, the local dispersion formula (7.34) is still valid for three-dimensional growth. Moreover, the entire derivation for the quantization conditions (7.109) and (7.126) is also valid for three-dimensional growth, except that in the axially symmetric case, one needs to apply a different tip matching condition (6.138), instead of (7.106)–(7.108) and (7.116)–(7.122).

Note that the 3D dendrite growth system with axially symmetric basic state allows non-axially symmetric perturbation modes. But here, we shall only discuss axially symmetric S-modes. We list the corresponding results below without derivation. When σ_0 is real, $\hat{k}_0^{(1)}(0)$ and $\hat{k}_0^{(3)}(0)$ will be also real. We have

$$\alpha_0 = \frac{\hat{k}_0^{(1)}(0)}{\hat{k}_0^{(3)}(0)} > 0, \quad \text{and} \quad \theta_0 = 0 \,. \tag{7.141}$$

One can derive

$$\text{Re}\left\{\frac{1}{\varepsilon}\int_0^{\xi_c}\left(\hat{k}_0^{(1)} - \hat{k}_0^{(3)}\right)\mathrm{d}\xi\right\} = \left(n + \frac{1}{2}\right)\pi \,, \tag{7.142}$$

or

$$C_1\sigma_0^{\frac{11}{7}}\alpha_4^{\frac{2}{7}} = \frac{C_0}{\sqrt{2}}\alpha_4^{\frac{7}{8}} - \varepsilon\left(n + \frac{1}{2}\right)\pi \tag{7.143}$$

$$(n = 0, \pm 1, \pm 2, \cdots) \,.$$

7.7 The Selection Conditions for Dendrite Growth

In the above, we have studied the stability properties of the nonclassic steady, or 'nearly' steady, needle solutions. The asymptotic results for the spectrum of complex eigenvalues have been verified by the numerical solutions in [7.5]. It is found that the dendrite growth system is controlled by entirely new instability mechanisms, as opposed to the well-known Mullins–Sekerka instability. These new instability mechanisms are the GTW and LF instabilities. Based on an understanding of these instability mechanisms, the selection criterion for dendrite growth at the later stage of evolution can naturally be derived.

In Fig. 7.8, we plot the neutral curve $\{\gamma_0\}$ of the GTW A-mode ($n = 0$), in the first-order approximation, for the small undercooling case with $\alpha = 1.0, \text{Pe}_0 = 0.001$. In the same figure, we also plot the neutral curve $\{\mathcal{C}_0\}$ of the LF S-mode ($n = 0$) in the zeroth-order approximation. These two neutral curves intersect each other at a critical number $\alpha_c = 0.1153$. Recall that the critical number ε_* for the GTW instability is a sensitive function of the undercooling temperature in the large undercooling regime. Hence, in the regime of large undercooling T_∞ (or Peclet number Pe_0), the critical number α_c is also a sensitive function of the undercooling temperature, T_∞. It decreases as Pe_0 increases. For instance, when the Peclet number $\text{Pe}_0 = 1.0$, we

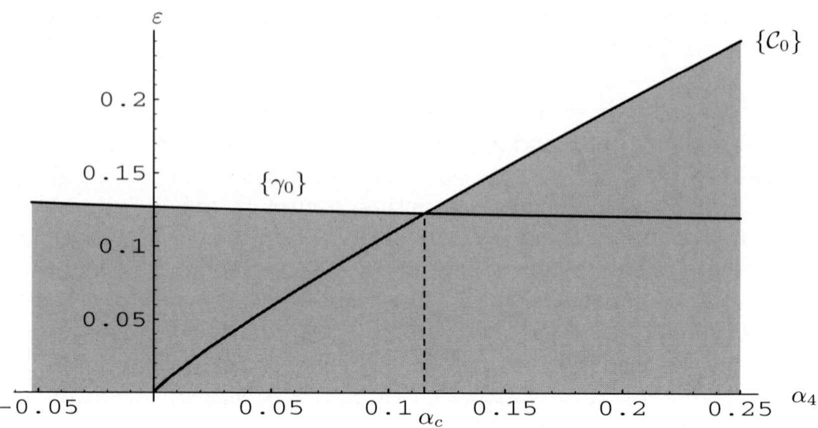

Fig. 7.8. The neutral curves $\{\gamma_0\}$, $\{\mathcal{C}_0\}$ and the stability diagram of dendrite growth in the (α_4, ε) plane for the case of small undercooling

have $\alpha_c = 0.085$. Of course, if the first-order correction to the low-frequency instability mechanism is included, the numerical value of the critical number α_c might be further changed.

The shaded region below the two curves in Fig. 7.8 is the unstable region, whereas the remaining region is stable.

One can examine the basic state of the system in the (α_4, ε) plane. For a given operation condition, α_4 is fixed whereas the parameter ε associated with the tip velocity of the dendrite under investigation may slowly vary. Hence, the representative point of the basic state in the (α_4, ε)-plane moves with time. In general, when $t \to \infty$, the system is expected to exhibit one of the following three behaviors:

1. It may approach a steady solution describing a smooth growing needle.
2. It may approach a time periodic solution describing an oscillatory growing dendrite.
3. It may have no limit solution. The solution evolves with short time scales and exhibits a chaotic pattern.

On the other hand, as we know, if the system has a steady limit solution when $t \to \infty$, this fixed point in the parameter plane must correspond to a classic steady needle solution, which must be one of the neutral LF modes $(n = 0, 1, \cdots)$ and must be on the neutral curve $\{\mathcal{C}_0\}$.

Therefore, we draw the following conclusions:

1. If the dendrite growth system exhibits a steady pattern as $t \to \infty$, the corresponding, steady limit solution must be on the neutral curve $\{\mathcal{C}_0\}$, and it occurs only when $\alpha_4 \geq \alpha_c$. In other words, for small anisotropy of surface tension, when $\alpha_4 < \alpha_c$, the steady needle solution is not observable

due to the existence of a number of growing oscillatory GTW modes. The selection criterion given by the MSC theory is apparently not applicable in this range.

2. If the dendrite exhibits a time periodic, oscillatory pattern as $t \to \infty$, within a linear theory, the limit solution must correspond to a GTW neutral mode on the neutral curve $\{\gamma_0\}$. This occurs only when $0 \leq \alpha_4 \leq \alpha_c$. The unsteady oscillatory pattern determined by the GTW neutral mode is self-sustaining. It can be stimulated by an imposed initial perturbation and does not require a continuously acting noise for its persistence. We emphasize that this statement has two implications. First, it implies that for the large surface tension anisotropy case ($\alpha_4 > \alpha_c$) no self-sustaining, oscillatory dendrite is possible. In this range, one may still see a time-dependent, oscillatory structure on a steady smooth dendrite interface induced by the decaying GTW modes. But this pattern can only be sustained by some external, continuously acting forces such as noises. Once these external forces cease, such an oscillatory structure disappears with time. Second, the criterion $\alpha_4 < \alpha_c$ is the necessary condition for the occurrence of a time-dependent, oscillatory dendrite growth. It has not been proven as being a sufficient condition for such a linear neutrally stable mode to actually be observable with any initial conditions.

3. If the dendrite growth system exhibits a chaotic pattern as $t \to \infty$, the system evidently remains in the unstable region.

The above conclusions appear to be in good agreement with experimental observations. So far most experiments on dendrite growth from a pure melt are three-dimensional. These experimental results show little correlation between the selected values $\varepsilon = \varepsilon_*$ and the anisotropy [7.7]. These results, in agreement with the theory, suggest that the anisotropy for these materials lies in the range of $0 \leq \alpha_4 \leq \alpha_c$ and the realistic dendrite growth is attracted to the neutral GTW mode, as $t \to \infty$.

During recent years, several groups of researchers have performed numerical simulations of two-dimensional dendrite growth for the initial value problem [7.9], [7.10]. The results of IFW theory are also in agreement with their numerical results.

Ihle and Müller-Krumbhaar (1994) have performed numerical simulations for the large undercooling case, $|T_\infty| = 0.5$, with various anisotropy [7.9]. Due to the numerical difficulty, they were not able to conduct the numerical simulation for the small undercooling case. They show that when $\alpha_4 = 0.15$, the numerical solutions are attracted to the LF neutral mode. At the later stage of evolution, the dendrite has a smooth interface with no side-branching. Then, when $\alpha_4 = 0.1$, a strong time-oscillatory instability occurs and the tip radius has up to 10 % fluctuation with time. When $\alpha_4 = 0.05$, the initially steady needle solution undergoes 'strongly irregular side-branching and large fluctuations in tip radius and velocity'. At the end of the computation for this case, the tip velocity is still noticeably changing and the numerical results on the tip radius R_{tip} and V_{tip} show the inconsistency with the scaling law (7.138). Although these numerical simulation results have been given a different implication, they may be properly interpreted as evidence for the existence of the GTW instability mechanism and suggest that the system is dominated by this GTW mechanism when α_4 is smaller than a critical number in the range $0.05 < \alpha_c \leq 0.1$. This critical number is in a good agreement with the theoretical prediction $\alpha_c = 0.085$ for the case of large undercooling ($\text{Pe}_0 = 1.0$). The numerical simulations [7.10] conducted by Brener, Müller-Krumbhaar, Saito, and Schiraishi (1994) in terms of the quasi-static approximation show the same scenario. Their results suggest $0.068 < \alpha_c \leq 0.125$.

The GTW neutral mode was not obtained in the numerical simulations of the initial value problem conducted by the above authors. This is quite understandable when the steady Ivantsov solution is used in the simulations as the initial condition. If the initial condition used is 'too far' from the GTW neutral mode, it might need too much time to reach the GTW neutral mode. This will be beyond the capacity of the computer. Moreover, with some initial conditions, the numerical solutions might never reach the GTW neutral mode. Indeed, a general dynamic system that has an isolated limit cycle solution may not be attracted to this limit cycle under some initial conditions.

References

7.1   D. A. Kessler and H. Levine, "Stability of Dendritic Crystals", Phys. Rev. Lett. **57**, pp. 3069–3072, (1986).

7.2   D. Bensimon, P. Pelce and B. I. Shraiman, "Dynamics of Curved Fronts and Pattern Selection", J. Physique **48**, pp. 2081–2087, (1987).

7.3   J. J. Xu, "Two-Dimensional Dendritic Growth with Anisotropy of Surface Tension", Physics (D), **51**, pp. 579–595, (1991).

7.4   J. J. Xu, "Interfacial Wave Theory of Two-Dimensional Dendritic Growth with Anisotropy of Surface Tension", Canad. J. Phys. **69**, No. 7, pp. 789–800, (1991).

7.5   J. J. Xu and Z. X. Pan, "Numerical Investigation of Global Instability Mechanism of Dendrite Growth", Report from Mathematics and Statistics, McGill University, ISSN 0824-4944, No. 93-05, (1993).

7.6   J. J. Xu, "Generalized Needle Solutions, Interfacial Instabilities, and Pattern Formation", Phys. Rev. E **53**, No. 5, pp. 5031–5062, (1996).

7.7   M. E. Glicksman and S. P. Marsh, "The Dendrite" in '*Handbook of Crystal Growth, Volume 1: Fundamentals, Part B: Transport and Stability*', Ed. by D.T.J. Hurle, (Elsevier Science, North–Holland, Amsterdam 1993).

7.8   E. A. Brener and V. I. Melnikov, "Pattern Selection in Two Dimensional Dendritic Growth", Adv. Phys. **40**, pp. 53–97, (1991).

7.9   T. Ihle and H. Müller-Krumbhaar, "Fractal and Compact Growth Morphologies in Phase Transitions with Diffusion Transport", Phy. Rev. E **49**, pp. 2972–2991, (1994).

7.10   E. Brener, T. Ihle, H. Müller-Krumbhaar, Y. Saito, and K. Shiraishi, "Fluctuation Effects on Dendritic Growth Morphology", Physica A **204**, pp. 96–110, (1994).

8. Three-Dimensional Dendrite Growth from Binary Mixtures

Up until now, we have only considered dendrite growth from a pure melt. Such a dendrite is also called a thermal dendrite. The dendritic structure on the interface may describe the temperature profile in the melt. However, such a profile will not be traceable after the solidification is completed. A more practical dendrite growth system is the binary alloy system. The dendritic structure not only describes the interface shape but, at the same time, also describes the variation of impurity concentration. After the solidification is completed, the impurity distribution profile remains in the solid state and has a profound significance on the properties of the final materials. This chapter is devoted to this important subject.

We are going to consider axially symmetric growth of a single dendrite into an undercooled dilute binary mixture.

The system is more complicated now because it involves both the heat transport process and mass transport in the solute. As demonstrated in Chap. 1, in a binary mixture the concentration of the dilute solute undergoes a jump at the interface due to the different solubilities in the solid and liquid phases. This is measured by the segregation coefficient κ defined by $C_\mathrm{S} = \kappa C_\mathrm{L}$ which relates the concentrations C_S and C_L at the interface of the solid and liquid sides, respectively. If $\kappa < 1$, the solute is rejected by the solid state. Assume that the concentration of solute in the liquid state, away from the interface, is C_∞, the solute diffusivity in the liquid state is D, while the diffusion of solute in the solid state is negligible.

There are two different length scales in the present system: the thermal diffusion length ℓ_T and solute diffusion length ℓ_D. Normally $\ell_\mathrm{D} \ll \ell_\mathrm{T}$. Hence as done in Chap. 1, we shall utilize the solute diffusion length $\ell_\mathrm{D} = D/U$ as the length scale instead of the thermal diffusion length $\ell_\mathrm{T} = \kappa_\mathrm{T}/U$ and set $\lambda = \ell_\mathrm{D}/\ell_\mathrm{T}$. Furthermore, we use ℓ_D/U as the time scale and C_∞ as the scale for the concentration. As before, $\Delta H/(c_\mathrm{p}\rho)$ is used as the scale for the temperature. For simplicity, we assume that the surface tension is isotropic and that the system is free of convection in the liquid phase.

8.1 Mathematical Formulation of the Problem

The same moving paraboloidal coordinate system (ξ, η) as in Chap. 6 is adopted.

The unknown functions now consist of the temperature fields in the melt and solid state, $T(\xi, \eta, t)$, $T_{\mathrm{S}}(\xi, \eta, t)$, respectively, the concentration field in the liquid, $C(\xi, \eta, t)$, and the interface shape $\eta_{\mathrm{s}}(\xi, t)$. The unsteady growth process is subject to the heat conduction equation and mass diffusion equation, which can be non-dimensionalized as follows:

$$
\begin{aligned}
\left(\frac{\partial^2 T}{\partial \xi^2} + \frac{\partial^2 T}{\partial \eta^2} + \frac{1}{\xi}\frac{\partial T}{\partial \xi} + \frac{1}{\eta}\frac{\partial T}{\partial \eta}\right) &= \lambda \eta_0^2\left(\xi\frac{\partial T}{\partial \xi} - \eta\frac{\partial T}{\partial \eta}\right) \\
&\quad + \lambda \eta_0^4(\xi^2 + \eta^2)\frac{\partial T}{\partial t} \\
\left(\frac{\partial^2 C}{\partial \xi^2} + \frac{\partial^2 C}{\partial \eta^2} + \frac{1}{\xi}\frac{\partial C}{\partial \xi} + \frac{1}{\eta}\frac{\partial C}{\partial \eta}\right) &= \eta_0^2\left(\xi\frac{\partial C}{\partial \xi} - \eta\frac{\partial C}{\partial \eta}\right) \\
&\quad + \eta_0^4(\xi^2 + \eta^2)\frac{\partial C}{\partial t} \,.
\end{aligned}
\tag{8.1}
$$

The same boundary conditions as in Chap. 6 are applied to this system, except for the Gibbs–Thomson condition (6.28) or (5.84). This condition must be modified to include the effect of impurities at the interface on the melt temperature. Furthermore, as demonstrated in Chap. 1, one also needs to impose the mass balance condition for impurity in the mass transport process. Thus, the boundary condition are:

1. The up-stream condition: as $\eta \to \infty$,

$$
T \to T_\infty, \quad C \to 1\,.
\tag{8.2}
$$

2. The regularity condition in the solid state: as $\eta \to 0$,

$$
T_{\mathrm{S}} = O(1)\,.
\tag{8.3}
$$

3. The interface condition: at $\eta = \eta_{\mathrm{s}}(\xi, t)$:
   (i) the thermodynamic equilibrium condition

$$
T = T_{\mathrm{S}}\,,
\tag{8.4}
$$

   (ii) the modified Gibbs–Thomson condition

$$
T = -\frac{\Gamma}{\eta_0^2}\mathcal{K}\left\{\frac{\mathrm{d}}{\mathrm{d}\xi}, \frac{\mathrm{d}^2}{\mathrm{d}\xi^2}\right\}\eta_{\mathrm{s}} - MC\,,
\tag{8.5}
$$

   (iii) the heat balance condition

$$
\left(\frac{\partial}{\partial \eta} - \eta_{\mathrm{s}}'\frac{\partial}{\partial \xi}\right)(T - T_{\mathrm{S}}) + \lambda \eta_0^2(\xi\eta_{\mathrm{s}})' + \lambda \eta_0^4(\xi^2 + \eta^2)\frac{\partial \eta_{\mathrm{s}}}{\partial t} = 0\,,
\tag{8.6}
$$

(iv) the mass balance condition for impurities

$$\frac{\partial C}{\partial \eta} - \eta_s' \frac{\partial C}{\partial \xi} + (1 - \kappa)C\left\{\eta_0^2 (\xi\eta_s)' + \eta_0^4 (\xi^2 + \eta^2)\frac{\partial \eta_s}{\partial t}\right\} = 0 \,. \quad (8.7)$$

In addition to the above, there are the tip conditions and the root conditions as usual. Note that the above Gibbs–Thomson condition involves a new parameter, the so-called morphological parameter M, defined as

$$M = -\frac{mC_\infty}{\Delta H/(c_p\rho)}\,, \quad (8.8)$$

where $m < 0$ is the slope of the liquidus in the phase diagram of the binary system as demonstrated in Chap. 1. Furthermore, the surface tension parameter Γ is different from the one used for the pure melt system. Γ is now defined through the mass diffusion length as opposed to the thermal diffusion length. Namely,

$$\Gamma = \frac{\ell_c}{\ell_D}\,, \quad (8.9)$$

where ℓ_c is the capillary length proportional to the isotropic surface tension coefficient. Accordingly, the interfacial stability parameter $\varepsilon = \sqrt{\Gamma}/\eta_0^2$ is also different from the one used in Chap. 6 for the pure melt system.

It is seen that the present system contains six physical parameters: $\Gamma, \lambda, \kappa, M, T_\infty$, and η_0^2. It is obviously more complicated than that discussed in Chap. 6. Nevertheless, the ideas and approach developed so far can be readily applied to this system. Solutions of the present system have the same mathematical structure as those discussed before. The system (8.1)–(8.7) also allows an exact similarity solution, like the Ivantsov solution, for the case $\Gamma = \varepsilon = 0$. For the general case $\varepsilon \neq 0$, one can define the basic state, as before, by the steady nonclassic needle solutions or by the 'nearly' steady needle formation solutions addressed in the previous chapters.

8.2 Basic State Solution for the Case of Zero Surface Tension

The similarity solution of the system (8.1)–(8.7) is found when $\varepsilon = 0$:

$$T_* = T_*(\eta) = T_\infty + \frac{\lambda\eta_0^2}{2} e^{\frac{\lambda\eta_0^2}{2}} E_1\left(\frac{\lambda\eta_0^2\eta^2}{2}\right)$$

$$T_{S*} = T_*(1)$$

$$C_* = C_*(\eta) = 1 + A_0\left(\frac{\eta_0^2}{2}\right)E_1\left(\frac{\eta_0^2\eta^2}{2}\right)$$

$$\eta_* = 1\,,$$

$$(8.10)$$

where

$$A_0\left(\frac{\eta_0^2}{2}\right) = \frac{\frac{1}{2}(1-\kappa)\eta_0^2}{e^{-\frac{\eta_0^2}{2}} - (1-\kappa)\frac{\eta_0^2}{2}E_1\left(\frac{\eta_0^2}{2}\right)}$$

$$C_{S*} = \kappa C_*(1) \tag{8.11}$$

$$T_\infty + \frac{\lambda\eta_0^2}{2}e^{\frac{\lambda\eta_0^2}{2}}E_1\left(\frac{\lambda\eta_0^2}{2}\right) = -M\left\{1 + A_0\left(\frac{\eta_0^2}{2}\right)E_1\left(\frac{\eta_0^2}{2}\right)\right\}.$$

From the above, η_0^2 can be solved as a function of the undercooling temperature T_∞, the ratio of diffusivities λ, and the morphological parameter M. The tip radius of the paraboloidal interface ℓ_t is calculated to be

$$\ell_t = \eta_0^2 \ell_D . \tag{8.12}$$

So if we define the Peclet number of the binary mixture system as the ratio of the tip radius of the interface to the mass diffusion length, namely $\mathrm{Pe} = \frac{\ell_t}{\ell_D}$, then it follows that when $\varepsilon = 0$, we have $\mathrm{Pe}_0 = \eta_0^2$.

The jump in the concentration at the interface, i.e.,

$$\Delta C = (1-\kappa)C_*(1) = \frac{1}{\eta_0^2}\frac{\mathrm{d}C_*}{\mathrm{d}\eta}(1) \tag{8.13}$$

can be calculated once η_0^2 is known.

8.3 Linear Perturbed System for the Case of Nonzero Surface Tension

For the case $\varepsilon \neq 0$, we may consider infinitesimal perturbations around the basic state in the same way as formulated for previous systems. The linear perturbed system is derived as

$$\left(\frac{\partial^2}{\partial\xi^2} + \frac{\partial^2}{\partial\eta^2}\right)\tilde{T}$$
$$= \left\{\lambda\eta_0^4(\xi^2+\eta^2)\frac{\partial}{\partial t} + \lambda\eta_0^2\left(\xi\frac{\partial}{\partial\xi} - \eta\frac{\partial}{\partial\eta}\right) - \frac{1}{\xi}\frac{\partial}{\partial\xi} - \frac{1}{\eta}\frac{\partial^2}{\partial\eta}\right\}\tilde{T}$$

$$\left(\frac{\partial^2}{\partial\xi^2} + \frac{\partial^2}{\partial\eta^2}\right)\tilde{C}$$
$$= \left\{\eta_0^4(\xi^2+\eta^2)\frac{\partial}{\partial t} + \eta_0^2\left(\xi\frac{\partial}{\partial\xi} - \eta\frac{\partial}{\partial\eta}\right) - \frac{1}{\xi}\frac{\partial}{\partial\xi} - \frac{1}{\eta}\frac{\partial^2}{\partial\eta}\right\}\tilde{C}$$

(8.14)

with the boundary conditions:

1. As $\eta \to \infty$,

$$\tilde{T} \to 0, \quad \tilde{C} \to 0 .$$

(8.15)

2. As $\eta \to 0$,

$$\tilde{T}_S \quad \text{is regular} .$$

(8.16)

3. At the interface: $\eta = 1$:

(i) the thermodynamic equilibrium condition

$$\tilde{T} = \tilde{T}_S + \lambda \tilde{h} + O(\varepsilon^2) ,$$

(8.17)

(ii) the Gibbs–Thomson condition

$$\tilde{T}_S = \frac{\varepsilon^2}{S(\xi)} \left\{ \frac{\partial^2}{\partial \xi^2} + \left(\frac{1}{\xi} + \frac{\xi}{S^2(\xi)} \right) \frac{\partial}{\partial \xi} - \frac{1}{S^2(\xi)} \right\} \tilde{h}$$

$$- M(\tilde{C} - \Delta C \tilde{h}) + O(\varepsilon^2) ,$$

(8.18)

(iii) the heat balance condition

$$\frac{\partial}{\partial \eta} \left(\tilde{T} - \tilde{T}_S \right) + \lambda S^2(\xi) \frac{\partial \tilde{h}}{\partial t} + \lambda \xi \frac{\partial \tilde{h}}{\partial \xi} + \lambda \left(2 + \lambda \eta_0^2 \right) \tilde{h}$$

$$+ O(\varepsilon^2) = 0 ,$$

(8.19)

(iv) the mass balance condition for impurity

$$\frac{\partial \tilde{C}}{\partial \eta} + \Delta C \left\{ \eta_0^2 S^2(\xi) \frac{\partial \tilde{h}}{\partial t} + \xi \frac{\partial \tilde{h}}{\partial \xi} \right\} + \eta_0^2 (1 - \kappa) \tilde{C}$$

$$+ \tilde{h} \Delta C (2 + \eta_0^2) + O(\varepsilon^2) = 0 ,$$

(8.20)

where

$$S(\xi) = \sqrt{\xi^2 + 1} .$$

(8.21)

4. The root condition: at $\xi = \xi_R(\varepsilon) = O(\varepsilon^{\frac{1}{2}})$, one should have

$$\{\tilde{T}, \tilde{T}_S, \tilde{h}\} = 0 .$$

(8.22)

5. The tip smoothness condition: at $\xi = 0$,

$$\frac{\partial}{\partial \xi} \{\tilde{T}, \tilde{T}_S, \tilde{h}\} = 0 .$$

(8.23)

One can seek eigenfunctions and eigenvalues for the above linear system. As before, we use the MVE method to derive the outer solutions for this eigenvalue problem. Similar to the case of a pure melt, we define the fast, stretched variables (ξ_+, η_+, t_+) as follows:

$$
\begin{aligned}
\xi_+ &= \frac{1}{\varepsilon} \int_{\xi_c}^{\xi} k(\xi', \eta, \varepsilon)\mathrm{d}\xi' \\
\eta_+ &= \frac{1}{\varepsilon} \int_{1}^{\eta} g(\xi, \eta', \varepsilon)\mathrm{d}\eta' \\
t_+ &= \frac{t}{\eta_0^2 \varepsilon} \, .
\end{aligned}
\tag{8.24}
$$

Generally speaking, the above η_+ may only be suitable for the concentration field $\tilde{C}$. For the temperature fields $\tilde{T}$ and $\tilde{T}_{\mathrm{S}}$, one may need to adopt a different fast variable η_+. In particular, η_+ may be defined as

$$
\eta_+ = \frac{1}{\varepsilon} \int_{1}^{\eta} q(\xi, \eta', \varepsilon)\mathrm{d}\eta'
$$

and

$$
\eta_+ = \frac{1}{\varepsilon} \int_{1}^{\eta} q_{\mathrm{s}}(\xi, \eta', \varepsilon)\mathrm{d}\eta',
$$

respectively. However, as we have seen in the case of a pure melt, up until the first-order approximation, we can set $k = g = q = q_{\mathrm{s}}$. Since we are not interested in discussing asymptotic solutions higher than the first-order approximation, for simplicity, we shall set $k = g = q = q_{\mathrm{s}}$. Furthermore, we still assume

$$
\frac{\partial k}{\partial \eta} = m \frac{\partial k}{\partial \xi}
$$

and

$$
m = m_0 + m_1 \varepsilon + \cdots
$$

as $\varepsilon \to 0$. As before, at the interface $\eta = 1, \eta_+ = 0$, we use the notation

$$
\begin{aligned}
\xi_+ = \hat{\xi}_+ &= \frac{1}{\varepsilon} \int_{\xi_c}^{\xi} \hat{k}(\xi', \varepsilon)\mathrm{d}\xi' \\
\hat{k}(\xi, \varepsilon) &= k(\xi, 1, \varepsilon).
\end{aligned}
\tag{8.25}
$$

In terms of the above multiple variables, we make the following MVE for the perturbed states:

$$\tilde{T} = \left\{ \tilde{T}_0(\xi, \eta, \xi_+, \eta_+) + \varepsilon \tilde{T}_1(\xi, \eta, \xi_+, \eta_+) + \cdots \right\} e^{\sigma t_+}$$

$$\tilde{T}_S = \left\{ \tilde{T}_{S0}(\xi, \eta, \xi_+, \eta_+) + \varepsilon \tilde{T}_{S1}(\xi, \eta, \xi_+, \eta_+) + \cdots \right\} e^{\sigma t_+}$$

$$\tilde{C} = \left\{ \tilde{C}_0(\xi, \eta, \xi_+, \eta_+) + \varepsilon C_1(\xi, \eta, \xi_+, \eta_+) + \cdots \right\} e^{\sigma t_+}$$

$$\tilde{h} = \left\{ \tilde{h}_0(\xi, \hat{\xi}_+) + \varepsilon \tilde{h}_1(\xi, \hat{\xi}_+) + \cdots \right\} e^{\sigma t_+} \tag{8.26}$$

$$k = k_0 + \varepsilon k_1 + \varepsilon^2 k_2 + \cdots,$$

$$\hat{k} = \hat{k}_0 + \varepsilon \hat{k}_1 + \varepsilon^2 \hat{k}_2 + \cdots$$

$$\sigma = \sigma_0 + \varepsilon \sigma_1 + \varepsilon^2 \sigma_2 + \cdots.$$

Here,

$$\hat{k}_0(\xi) = k_0(\xi, 1), \quad \hat{k}_1(\xi) = k_1(\xi, 1), \quad \hat{k}_2(\xi) = k_2(\xi, 1), \cdots \tag{8.27}$$

and $\sigma = \sigma_R - i\omega$ ($\omega \geq 0$). In the first step, we set σ_0 as an arbitrary constant. With these multiple variables, the converted perturbed systems are listed below.

The governing equation for the temperature field:

$$\left(k^2 \frac{\partial^2}{\partial \xi_+^2} + k^2 \frac{\partial^2}{\partial \eta_+^2} \right) \tilde{T} = \varepsilon \lambda \eta_0^2 (\xi^2 + \eta^2) \frac{\partial \tilde{T}}{\partial t_+} - \varepsilon^2 \left(\frac{\partial^2}{\partial \xi^2} + \frac{\partial^2}{\partial \eta^2} \right) \tilde{T}$$

$$+ \varepsilon \lambda \eta_0^2 \xi \left(k \frac{\partial}{\partial \xi_+} + \varepsilon \frac{\partial}{\partial \xi} \right) \tilde{T} - \varepsilon \lambda \eta_0^2 \eta \left(k \frac{\partial}{\partial \eta_+} + \varepsilon \frac{\partial}{\partial \eta} \right) \tilde{T}$$

$$- \frac{\varepsilon}{\xi} \left(k \frac{\partial}{\partial \xi_+} + \varepsilon \frac{\partial}{\partial \xi} \right) \tilde{T} - \frac{\varepsilon}{\eta} \left(k \frac{\partial}{\partial \eta_+} + \varepsilon \frac{\partial}{\partial \eta} \right) \tilde{T}$$

$$- \varepsilon \left(2k \frac{\partial^2}{\partial \xi \partial \xi_+} + 2k \frac{\partial^2}{\partial \eta \partial \eta_+} + \frac{\partial k}{\partial \xi} \frac{\partial}{\partial \xi_+} + \frac{\partial k}{\partial \eta} \frac{\partial}{\partial \eta_+} \right) \tilde{T}. \tag{8.28}$$

The governing equation for the concentration field:

$$\left(k^2 \frac{\partial^2}{\partial \xi_+^2} + k^2 \frac{\partial^2}{\partial \eta_+^2} \right) \tilde{C} = \varepsilon \eta_0^2 (\xi^2 + \eta^2) \frac{\partial \tilde{C}}{\partial t_+} - \varepsilon^2 \left(\frac{\partial^2}{\partial \xi^2} + \frac{\partial^2}{\partial \eta^2} \right) \tilde{C}$$

$$+ \varepsilon \eta_0^2 \xi \left(k \frac{\partial}{\partial \xi_+} + \varepsilon \frac{\partial}{\partial \xi} \right) \tilde{C} - \varepsilon \eta_0^2 \eta \left(k \frac{\partial}{\partial \eta_+} + \varepsilon \frac{\partial}{\partial \eta} \right) \tilde{C}$$

$$- \frac{\varepsilon}{\xi} \left(k \frac{\partial}{\partial \xi_+} + \varepsilon \frac{\partial}{\partial \xi} \right) \tilde{C} - \frac{\varepsilon}{\eta} \left(k \frac{\partial}{\partial \eta_+} + \varepsilon \frac{\partial}{\partial \eta} \right) \tilde{C}$$

$$- \varepsilon \left(2k \frac{\partial^2}{\partial \xi \partial \xi_+} + 2k \frac{\partial^2}{\partial \eta \partial \eta_+} + \frac{\partial k}{\partial \xi} \frac{\partial}{\partial \xi_+} + \frac{\partial k}{\partial \eta} \frac{\partial}{\partial \eta_+} \right) \tilde{C}. \tag{8.29}$$

The converted forms of the up-stream condition, regularity condition in the solid state, smooth tip condition and far-field condition are trivial and the same as those shown in Chap. 6 for the pure melt system. Hence, we give here only the interface conditions: at $\eta_+ = 0$, $\eta = 1$,

(i)

$$\tilde{T} = \tilde{T}_{\mathrm{S}} + \lambda \tilde{h} + O(\varepsilon^2) \, , \tag{8.30}$$

(ii)

$$\tilde{T}_{\mathrm{S}} = \frac{1}{S(\xi)} \left\{ \left(k^2 \frac{\partial^2}{\partial \hat{\xi}_+^2} + 2\varepsilon k \frac{\partial^2}{\partial \hat{\xi}_+ \partial \xi} + \varepsilon \frac{\partial k}{\partial \xi} \frac{\partial}{\partial \hat{\xi}_+} + \varepsilon^2 \frac{\partial^2}{\partial \xi^2} \right) \right.$$

$$\left. + \varepsilon \left(\frac{1}{\xi} + \frac{\xi}{S^2(\xi)} \right) \left(k \frac{\partial}{\partial \hat{\xi}_+} + \varepsilon \frac{\partial}{\partial \xi} \right) - \frac{\varepsilon^2}{S^2(\xi)} \right\} \tilde{h}$$

$$- M(\tilde{C} - \Delta C \tilde{h}) + O(\varepsilon^2) \, , \tag{8.31}$$

(iii)

$$\left(k \frac{\partial}{\partial \eta_+} + \varepsilon \frac{\partial}{\partial \eta} \right) \left(\tilde{T} - \tilde{T}_{\mathrm{S}} \right) + \lambda S^2(\xi) \frac{\partial \tilde{h}}{\partial t_+} + \lambda \xi \left(k \frac{\partial \tilde{h}}{\partial \hat{\xi}_+} + \varepsilon \frac{\partial \tilde{h}}{\partial \xi} \right)$$

$$+ \varepsilon \lambda (2 + \lambda \eta_0^2) \tilde{h} + O(\varepsilon^2) = 0 \, , \tag{8.32}$$

(iv)

$$k \frac{\partial \tilde{C}}{\partial \eta_+} + \varepsilon \frac{\partial \tilde{C}}{\partial \eta} + \Delta C \left\{ S^2(\xi) \frac{\partial \tilde{h}}{\partial t_+} + \xi \left(k \frac{\partial \tilde{h}}{\partial \hat{\xi}_+} + \varepsilon \frac{\partial \tilde{h}}{\partial \xi} \right) \right\}$$

$$+ \varepsilon \eta_0^2 (1 - \kappa) \tilde{C} + \varepsilon \Delta C (2 + \eta_0^2) \tilde{h} + O(\varepsilon^2) = 0 \, . \tag{8.33}$$

8.4 The MVE Solutions in the Outer Region

8.4.1 The Zeroth-Order Approximation

In the zeroth-order, we obtain

$$\begin{aligned}
\left(\frac{\partial^2}{\partial \xi_+^2} + \frac{\partial^2}{\partial \eta_+^2} \right) \tilde{T}_0 &= 0 \, , \\
\left(\frac{\partial^2}{\partial \xi_+^2} + \frac{\partial^2}{\partial \eta_+^2} \right) \tilde{C}_0 &= 0 \, ,
\end{aligned} \tag{8.34}$$

and the boundary conditions:

1. As $\eta_+ \to \infty$,

$$\tilde{T}_0 \to 0 \quad \text{and} \quad \tilde{C}_0 \to 0 . \tag{8.35}$$

2. As $\eta_+ \to -\infty$,

$$\tilde{T}_{\mathrm{S0}} \to 0 . \tag{8.36}$$

3. At $\eta_+ = 0$, $\eta = 1$,

$$\tilde{T}_0 = \tilde{T}_{\mathrm{S0}} + \lambda \tilde{h}_0 , \tag{8.37}$$

$$\tilde{T}_{\mathrm{S0}} = \frac{\hat{k}_0^2}{S(\xi)} \frac{\partial^2 \tilde{h}_0}{\partial \hat{\xi}_+^2} - M(\tilde{C}_0 - \Delta C \tilde{h}_0) , \tag{8.38}$$

$$\hat{k}_0 \frac{\partial}{\partial \eta_+}\left(\tilde{T}_0 - \tilde{T}_{\mathrm{S0}} \right) + \lambda \sigma_0 S^2(\xi) \tilde{h}_0 + \lambda \xi \hat{k}_0 \frac{\partial \tilde{h}_0}{\partial \hat{\xi}_+} = 0 , \tag{8.39}$$

$$\hat{k}_0 \frac{\partial \tilde{C}_0}{\partial \eta_+} + \Delta C \left(\sigma_0 S^2(\xi) \tilde{h}_0 + \xi \hat{k}_0 \frac{\partial \tilde{h}_0}{\partial \hat{\xi}_+} \right) = 0 . \tag{8.40}$$

Hence, we have the normal mode solutions

$$\begin{aligned}
\tilde{T}_0 &= A_0(\xi, \eta) \exp(\mathrm{i}\xi_+ - \eta_+) \\
\tilde{T}_{\mathrm{S0}} &= A_{\mathrm{S0}}(\xi, \eta) \exp(\mathrm{i}\xi_+ + \eta_+) \\
\tilde{C}_0 &= B_0(\xi, \eta) \exp(\mathrm{i}\xi_+ - \eta_+) \\
\tilde{h}_0 &= \hat{D}_0 \exp(\mathrm{i}\hat{\xi}_+) .
\end{aligned} \tag{8.41}$$

We write

$$\hat{A}_0(\xi) = A_0(\xi, 1), \quad \hat{A}_{\mathrm{S0}}(\xi) = A_{\mathrm{S0}}(\xi, 1) \quad \text{and} \quad \hat{B}_0(\xi) = B_0(\xi, 1), \tag{8.42}$$

then from (8.30)–(8.33) we derive

$$\begin{aligned}
&\hat{A}_0 - \hat{A}_{\mathrm{S0}} - \lambda \hat{D}_0 = 0 \\
&\hat{A}_{\mathrm{S0}} + \left(\frac{\hat{k}_0^2}{S} - M\Delta C \right) \hat{D}_0 + M\hat{B}_0 = 0 \\
&-\hat{k}_0(\hat{A}_0 + \hat{A}_{\mathrm{S0}}) + \lambda(\sigma_0 S^2 + \mathrm{i}\xi \hat{k}_0)\hat{D}_0 = 0 \\
&-\hat{k}_0 \hat{B}_0 + \Delta C(\sigma_0 S^2 + \mathrm{i}\xi \hat{k}_0)\hat{D}_0 = 0 .
\end{aligned} \tag{8.43}$$

For a nontrivial solution, one must have

$$\Delta = \det \begin{pmatrix} 1 & -1 & -\lambda & 0 \\ 0 & 1 & (\hat{k}_0^2/S - M\Delta C) & M \\ -\hat{k}_0 & -\hat{k}_0 & \lambda(\sigma_0 S^2 + \mathrm{i}\xi \hat{k}_0) & 0 \\ 0 & 0 & \Delta C(\sigma_0 S^2 + \mathrm{i}\xi \hat{k}_0) & -\hat{k}_0 \end{pmatrix} = 0 . \tag{8.44}$$

From this we derive the local dispersion relation

$$\sigma_0 = \Sigma(\xi, \hat{k}_0) = \frac{\hat{k}_0}{\Lambda_0 S^2}\left(1 - \frac{\hat{k}_0^2}{\Lambda_1 S}\right) - \mathrm{i}\frac{\xi \hat{k}_0}{S^2} , \tag{8.45}$$

where

$$\begin{cases} \Lambda_0 = 1 \\ \Lambda_1 = \frac{\lambda}{2} + \Delta CM , \end{cases} \tag{8.46}$$

and

$$\begin{cases} \hat{A}_0 + \hat{A}_{S0} = \lambda\left(1 - \frac{2\hat{k}_0^2}{\Lambda_1 S}\right)\hat{D}_0 \\ \hat{A}_0 - \hat{A}_{S0} = \lambda\hat{D}_0 , \end{cases} \tag{8.47}$$

where $\hat{D}_0$ is an arbitrary constant. The above dispersion formula is a generalization of the dispersion relation for dendrite growth from a pure melt. In fact, the dispersion relation derived in Chap. 6 for dendrite growth from a pure melt coincides with the above formula for the special case

$$\Lambda_0 = 1, \quad \Lambda_1 = \frac{1}{2} . \tag{8.48}$$

For any given constant σ, one can find three roots of (8.45): $\hat{k}_0^{(1)}(\xi)$, $\hat{k}_0^{(2)}(\xi)$, and $\hat{k}_0^{(3)}(\xi)$. As before, only the short-wavelength branch, the H_1 wave, with the larger $\mathrm{Re}\{\hat{k}_0^{(1)}\} > 0$ and the long-wavelength branch, the H_3 wave, with the smaller $\mathrm{Re}\{\hat{k}_0^{(3)}\} > 0$ are physically meaningful. In order for the solutions to satisfy the boundary condition (8.15), the H_2 wave, with $\mathrm{Re}\{\hat{k}_0^{(2)}\} < 0$, must be ruled out. Thus, the general solution of the H wave is:

$$\tilde{h} = D_0^{(1)} \exp\left[\frac{\sigma t}{\varepsilon \eta_0^2} + \frac{\mathrm{i}}{\varepsilon}\int_{\xi_c}^{\xi}\left(\hat{k}_0^{(1)} + \varepsilon \hat{k}_1^{(1)} + \cdots\right)\mathrm{d}\xi_1\right]$$

$$+ D_0^{(3)} \exp\left[\frac{\sigma t}{\varepsilon \eta_0^2} + \frac{\mathrm{i}}{\varepsilon}\int_{\xi_c}^{\xi}\left(\hat{k}_0^{(3)} + \varepsilon \hat{k}_1^{(3)} + \cdots\right)\mathrm{d}\xi_1\right] + \cdots . \tag{8.49}$$

8.4.2 The First-Order Approximation

In the first-order approximation, we have the following system:

$$k_0^2\left(\frac{\partial^2}{\partial \xi_+^2} + \frac{\partial^2}{\partial \eta_+^2}\right)\tilde{T}_1 = a_0 e^{\mathrm{i}\xi_+ - \eta_+}$$

$$k_0^2\left(\frac{\partial^2}{\partial \xi_+^2} + \frac{\partial^2}{\partial \eta_+^2}\right)\tilde{T}_{S1} = a_{S0} e^{\mathrm{i}\xi_+ + \eta_+} \tag{8.50}$$

$$k_0^2\left(\frac{\partial^2}{\partial \xi_+^2} + \frac{\partial^2}{\partial \eta_+^2}\right)\tilde{C}_1 = b_0 e^{\mathrm{i}\xi_+ - \eta_+} ,$$

where

$$a_0 = 2k_0\left(\frac{\partial A_0}{\partial \eta} - \mathrm{i}\frac{\partial A_0}{\partial \xi}\right)$$
$$+ A_0\left[\lambda\sigma_0\eta_0^2(\xi^2 + \eta^2) + \lambda k_0\eta_0^2(\mathrm{i}\xi + \eta) + \frac{k_0}{\eta} - \mathrm{i}\frac{k_0}{\xi} + (m_0 - \mathrm{i})k_0'\right]$$

$$a_{\mathrm{S}0} = -2k_0\left(\frac{\partial A_{\mathrm{S}0}}{\partial \eta} + \mathrm{i}\frac{\partial A_{\mathrm{S}0}}{\partial \xi}\right)$$
$$+ A_{\mathrm{S}0}\left[\lambda\sigma_0\eta_0^2(\xi^2 + \eta^2) + \lambda k_0\eta_0^2(\mathrm{i}\xi - \eta) - \frac{k_0}{\eta} - \mathrm{i}\frac{k_0}{\xi} - (m_0 + \mathrm{i})k_0'\right]$$

$$b_0 = 2k_0\left(\frac{\partial B_0}{\partial \eta} - \mathrm{i}\frac{\partial B_0}{\partial \xi}\right)$$
$$+ B_0\left[\sigma_0\eta_0^2(\xi^2 + \eta^2) + k_0\eta_0^2(\mathrm{i}\xi + \eta) + \frac{k_0}{\eta} - \mathrm{i}\frac{k_0}{\xi} + (m_0 - \mathrm{i})k_0'\right]. \tag{8.51}$$

To ensure the uniform validity of the expansions as $\xi \to \infty$, one must eliminate the secular terms on the right-hand side of (8.50). In other words, we set

$$a_0 = a_{\mathrm{S}0} = b_0 = 0. \tag{8.52}$$

From these equations we derive

$$\hat{D}_0 Q_c = \frac{\partial B_0}{\partial \eta}\bigg|_{\eta=1} = \mathrm{i}\frac{\partial \hat{B}_0}{\partial \xi}$$
$$- \hat{B}_0\left[\frac{\sigma_0\eta_0^2}{2k_0}S^2(\xi) + \frac{\eta_0^2}{2}(1 + \mathrm{i}\xi) + \frac{1}{2} - \frac{\mathrm{i}}{2\xi} + \frac{(m_0 - \mathrm{i})\hat{k}_0'}{2\hat{k}_0}\right] \tag{8.53}$$

and

$$\hat{D}_0 Q_0 = \frac{\partial}{\partial \eta}\left(A_0 - A_{\mathrm{S}0}\right)\bigg|_{\eta=1} = \mathrm{i}\frac{\partial}{\partial \xi}\left(\hat{A}_0 + \hat{A}_{\mathrm{S}0}\right)$$
$$- \left(\hat{A}_0 + \hat{A}_{\mathrm{S}0}\right)\left[\frac{\lambda\sigma_0\eta_0^2}{2\hat{k}_0}S^2(\xi) + \frac{\mathrm{i}}{2}\left(\lambda\xi\eta_0^2 - \frac{1}{\xi} - \frac{\mathrm{d}\ln\hat{k}_0}{\mathrm{d}\xi}\right)\right]$$
$$- \left(\hat{A}_0 - \hat{A}_{\mathrm{S}0}\right)\left(\frac{1 + \lambda\eta_0^2}{2} + \frac{m_0\hat{k}_0'}{2\hat{k}_0}\right). \tag{8.54}$$

Noting that

$$\begin{cases} \hat{B}_0 = \Delta C\left(1 - \frac{\hat{k}_0^2}{\Lambda_1 S}\right)\hat{D}_0 \\[2mm] \hat{A}_0 + \hat{A}_{\mathrm{S}0} = \lambda\left(1 - \frac{\hat{k}_0^2}{\Lambda_1 S}\right)\hat{D}_0 \\[2mm] \hat{A}_0 - \hat{A}_{\mathrm{S}0} = \lambda\hat{D}_0, \end{cases} \tag{8.55}$$

we obtain

$$\begin{aligned}
Q_c &= -\mathrm{i}\frac{\Delta C}{\Lambda_1}\frac{d}{d\xi}\left(\frac{\hat{k}_0^2}{S}\right) - \Delta C\left(1 - \frac{\hat{k}_0^2}{\Lambda_1 S}\right) \\
&\quad \times \left[\frac{\sigma_0\eta_0^2}{2\hat{k}_0}S^2(\xi) + \frac{\eta_0^2}{2}(1+\mathrm{i}\xi) + \frac{1}{2} - \frac{\mathrm{i}}{2\xi} + \frac{(m_0-\mathrm{i})\hat{k}_0'}{2\hat{k}_0}\right] \\
Q_0 &= -\mathrm{i}\frac{\lambda}{\Lambda_1}\frac{d}{d\xi}\left(\frac{\hat{k}_0^2}{S}\right) - \lambda\left(1 - \frac{\hat{k}_0^2}{\Lambda_1 S}\right) \\
&\quad \times \left[\frac{\lambda\sigma_0\eta_0^2}{2\hat{k}_0}S^2(\xi) + \frac{\mathrm{i}\lambda\eta_0^2}{2}\xi - \frac{\mathrm{i}}{2\xi}\right] - \frac{\lambda}{2}(1+\lambda\eta_0^2) \\
&\quad - \frac{\lambda\hat{k}_0'}{2\hat{k}_0}\left(m_0 - \mathrm{i} + \frac{\mathrm{i}\hat{k}_0^2}{\Lambda_1 S}\right).
\end{aligned} \tag{8.56}$$

These formulas will be needed later. In terms of the conditions (8.52), one obtains the first-order approximate solution:

$$\begin{aligned}
\tilde{T}_1 &= A_1(\xi,\eta)\ \exp\{\mathrm{i}\xi_+ - \eta_+\} \\
\tilde{T}_{S1} &= A_{S1}(\xi,\eta)\ \exp\{\mathrm{i}\xi_+ + \eta_+\} \\
\tilde{C}_1 &= B_1(\xi,\eta)\ \exp\{\mathrm{i}\xi_+ - \eta_+\} \\
\tilde{h}_1 &= \hat{D}_1\ \exp\left\{\mathrm{i}\hat{\xi}_+\right\}.
\end{aligned} \tag{8.57}$$

Setting

$$\begin{cases}
\hat{A}_1(\xi) = A_1(\xi,1) \\[4pt]
\hat{A}_{S1}(\xi) = A_{S1}(\xi,1) \\[4pt]
\hat{B}_1(\xi) = B_1(\xi,1),
\end{cases} \tag{8.58}$$

from the interface boundary conditions, we find that

$$\begin{cases}
\hat{A}_1 - \hat{A}_{S1} - \lambda\hat{D}_1 = 0 \\[4pt]
\hat{A}_{S1} + \left(\frac{\hat{k}_0^2}{S} - M\Delta C\right)\hat{D}_1 + M\hat{B}_1 = I_2\hat{D}_0 \\[4pt]
-\hat{k}_0(\hat{A}_1 + \hat{A}_{S1}) + \lambda(\sigma_0 S^2 + \mathrm{i}\xi\hat{k}_0)\hat{D}_1 = I_3\hat{D}_0 \\[4pt]
\hat{k}_0\hat{B}_1 - \Delta C(\sigma_0 S^2 + \mathrm{i}\xi\hat{k}_0)\hat{D}_1 = I_4\hat{D}_0,
\end{cases} \tag{8.59}$$

where

$$\begin{cases}
I_2 = \frac{1}{S}\left\{\mathrm{i}\hat{k}_0' + \mathrm{i}\hat{k}_0\left(\frac{1}{\xi} + \frac{\xi}{S^2}\right) - 2\hat{k}_0\hat{k}_1\right\} \\[6pt]
I_3 = \frac{\lambda\hat{k}_1}{\hat{k}_0}(\sigma_0 S^2 + \mathrm{i}\xi\hat{k}_0) - \lambda(\sigma_1 S^2 + \mathrm{i}\xi\hat{k}_1) - Q_0 - \lambda(2 + \lambda\eta_0^2) \\[6pt]
I_4 = Q_c + \Delta C(\sigma_1 S^2 + \mathrm{i}\xi\hat{k}_1) + \Delta C(2 + \eta_0^2) \\[6pt]
\quad + \Delta C\left(1 - \frac{\hat{k}_0^2}{\Lambda_1 S}\right)\left\{\eta_0^2(1 - \kappa) - \hat{k}_1\right\}.
\end{cases} \tag{8.60}$$

The determinant of the coefficient matrix of the above inhomogeneous system (8.59) is zero. For a nontrivial solution for $\{\hat{A}_1, \hat{A}_{S1}, \hat{B}_1, \hat{D}_1\}$, one must have

$$\det \begin{pmatrix} 1 & -1 & 0 & 0 \\ 0 & 1 & I_2 & M \\ -\hat{k}_0 & -\hat{k}_0 & I_3 & 0 \\ 0 & 0 & I_4 & \hat{k}_0 \end{pmatrix} = 0, \tag{8.61}$$

which leads to the solvability condition:

$$I_3 + 2\hat{k}_0 I_2 - 2M I_4 = 0. \tag{8.62}$$

From this solvability condition we obtain

$$\hat{k}_1 = \frac{\mathrm{i}}{2\xi} + \frac{R_1(\xi)}{F(\xi)} + \frac{R_2(\xi)}{F(\xi)} \frac{\hat{k}_0'}{\hat{k}_0}, \tag{8.63}$$

where we used the notation

$$\begin{cases} F(\xi) = 1 - \mathrm{i}\xi - \dfrac{3\hat{k}_0^2}{\Lambda_1 S} \\[2mm] R_1(\xi) = S^2 \sigma_1 + \left(1 + \eta_0^2\right) - \dfrac{\eta_0^2}{2}\left(1 + \dfrac{\lambda^2 - \lambda}{2\Lambda_1}\right)\left(1 - \dfrac{\hat{k}_0^2}{\Lambda_1 S}\right)^2 \\[3mm] \qquad\quad - \dfrac{\eta_0^2 \lambda}{2\Lambda_1}\left(1 - \dfrac{\lambda}{2}\right) - \dfrac{M\Delta C}{2\Lambda_1}\left\{1 - \eta_0^2(1 - 2\kappa)\right\}\left(1 - \dfrac{\hat{k}_0^2}{\Lambda_1 S}\right) \\[3mm] R_2(\xi) = -\dfrac{m_0 - \mathrm{i}}{2} + \dfrac{\hat{k}_0^2}{2\Lambda_1 S}\left(\dfrac{M\Delta C}{\Lambda_1}m_0 - 7\mathrm{i}\right). \end{cases} \tag{8.64}$$

Therefore, similar to the case of the pure melt, the solution for $\hat{k}_1$ has a singularity at the zero point ξ_c of $F(\xi)$. Noting that

$$F(\xi) = (1 - \mathrm{i}\xi) - \frac{3\hat{k}_0^2}{\Lambda_1 S} = S^2\left(\frac{\partial \Sigma}{\partial \hat{k}_0}\right), \tag{8.65}$$

this ξ_c is also the root of the equation

$$\frac{\partial \Sigma(\xi, \hat{k}_0)}{\partial \hat{k}_0} = 0 \tag{8.66}$$

or

$$\sigma_0 = \sqrt{\frac{4\Lambda_1}{27}}\left(\frac{1 - \mathrm{i}\xi}{S}\right)^{\frac{3}{2}} = \mathrm{e}^{-\mathrm{i}\frac{3\pi}{4}}\sqrt{\frac{4\Lambda_1}{27}}\left(\frac{\xi + \mathrm{i}}{\xi - \mathrm{i}}\right)^{3/4}. \tag{8.67}$$

Note also that one can derive

$$\frac{\hat{k}_0'}{\hat{k}_0} = \frac{R_0(\xi)}{F(\xi)}$$

$$F'(\xi) = -\mathrm{i} + \frac{3\hat{k}_0^2 \xi}{\Lambda_1 S^3} - \frac{6\hat{k}_0^2}{\Lambda_1 S}\frac{R_0(\xi)}{F(\xi)}, \tag{8.68}$$

where

$$R_0(\xi) = \mathrm{i} + \frac{2\xi(1 - \mathrm{i}\xi)}{S^2} - 3\frac{\xi \hat{k}_0^2}{\Lambda_1 S^3} \,. \tag{8.69}$$

Thus, one deduces that as $\xi \to \xi_c$, $R_0(\xi) \sim \frac{1}{\xi_c - \mathrm{i}}$ and

$$F(\xi) \sim A(\xi - \xi_c)^{\frac{1}{2}},$$

where

$$A = \mathrm{i}\hat{k}_0 \sqrt{\frac{12 R_0}{\Lambda_1 S}}\Bigg|_{\xi = \xi_c} = \mathrm{i}\hat{k}_0 \frac{2 \cdot 3^{\frac{1}{2}}}{\Lambda_1} \frac{1}{(\xi_c - \mathrm{i})^{\frac{3}{4}}(\xi_c + \mathrm{i})^{\frac{1}{4}}} \,. \tag{8.70}$$

Based on these results, from (8.62) we find that, as $\xi \to \xi_c$,

$$k_1 \sim \frac{R_1(\xi_c)}{A(\xi - \xi_c)^{\frac{1}{2}}} + \frac{b_1}{(\xi - \xi_c)} \,, \tag{8.71}$$

where

$$b_1 = -\frac{\Lambda_1 S R_2}{12 \hat{k}_0^2}\Bigg|_{\xi = \xi_c} \,. \tag{8.72}$$

8.5 The Inner Solutions near the Singular Point ξ_c

As in Chap. 6, in the vicinity of ξ_c ($|\xi - \xi_c| \ll 1$; $|\eta - 1| \ll 1$), we introduce the inner variables:

$$\xi_* = \frac{\xi - \xi_c}{\varepsilon^\alpha} \,,$$
$$\eta_* = \frac{\eta - 1}{\varepsilon^\alpha} \,, \tag{8.73}$$

and set

$$\tilde{T}(\xi, \eta, t) = \varepsilon^\alpha \hat{T}(\xi_*, \eta_*, t)$$
$$\tilde{T}_{\mathrm{S}}(\xi, \eta, t) = \varepsilon^\alpha \hat{T}_{\mathrm{S}}(\xi_*, \eta_*, t)$$
$$\tilde{C}(\xi, \eta, t) = \varepsilon^\alpha \hat{C}(\xi_*, \eta_*, t) \tag{8.74}$$
$$\tilde{h}(\xi, t) = \varepsilon^\alpha \hat{h}(\xi_*, t) \,.$$

Then, at the interface we have

$$\eta_{*\mathrm{s}} = \frac{\hat{h}(\xi_*, t)}{\eta_0^2} \,. \tag{8.75}$$

In terms of these inner variables, the perturbed system (8.14)–(8.20) is expressed in the form:

$$\frac{\partial^2 \hat{T}}{\partial \xi_*^2} + \frac{\partial^2 \hat{T}}{\partial \eta_*^2} = \left\{ \varepsilon^{2\alpha-1} \lambda \sigma \eta_0^2 (\xi^2 + \eta^2) - \varepsilon^\alpha \lambda \eta_0^2 \left(\xi \frac{\partial}{\partial \xi_*} - \eta \frac{\partial}{\partial \eta_*} \right) \right.$$
$$\left. - \varepsilon^\alpha \left(\frac{1}{\xi} \frac{\partial}{\partial \xi_*} + \frac{1}{\eta} \frac{\partial}{\partial \eta_*} \right) \right\} \hat{T} , \tag{8.76}$$

$$\frac{\partial^2 \hat{C}}{\partial \xi_*^2} + \frac{\partial^2 \hat{C}}{\partial \eta_*^2} = \left\{ \varepsilon^{2\alpha-1} \sigma \eta_0^2 (\xi^2 + \eta^2) - \varepsilon^\alpha \eta_0^2 \left(\xi \frac{\partial}{\partial \xi_*} - \eta \frac{\partial}{\partial \eta_*} \right) \right.$$
$$\left. - \varepsilon^\alpha \left(\frac{1}{\xi} \frac{\partial}{\partial \xi_*} + \frac{1}{\eta} \frac{\partial}{\partial \eta_*} \right) \right\} \hat{C} . \tag{8.77}$$

The boundary conditions (8.15)–(8.20) are transformed to the following forms:

1. As $\eta_* \to \infty$,

$$\hat{T} \to 0, \quad \hat{C} \to 0 . \tag{8.78}$$

2. As $\eta_* \to -\infty$,

$$\hat{T}_{\mathrm{S}} \to 0 . \tag{8.79}$$

3. At the interface $\eta_* = 0$,
   (i)

$$\hat{T} = \hat{T}_{\mathrm{S}} + \lambda \hat{h} + \text{(higher-order terms)} , \tag{8.80}$$

   (ii)

$$\hat{T}_{\mathrm{S}} = \frac{\varepsilon^{2-2\alpha}}{S(\xi)} \frac{\partial^2 \hat{h}}{\partial \hat{\xi}_*^2} - M(\hat{C} - \Delta C \hat{h}) + \text{(higher-order terms)} , \tag{8.81}$$

   (iii)

$$\varepsilon^{1-\alpha} \frac{\partial}{\partial \eta_*} \left(\hat{T} - \hat{T}_{\mathrm{S}} \right) + \lambda S^2 \sigma S^2(\xi) \hat{h} + \varepsilon^{1-\alpha} \lambda \xi \frac{\partial \hat{h}}{\partial \xi_*}$$
$$= \text{(higher-order terms)} , \tag{8.82}$$

   (iv)

$$\varepsilon^{1-\alpha} \frac{\partial \hat{C}}{\partial \eta_*} + \Delta C \left\{ \sigma S^2(\xi) \hat{h} + \varepsilon^{1-\alpha} \xi \frac{\partial \hat{h}}{\partial \xi_*} \right\}$$
$$= \text{(higher-order terms)} . \tag{8.83}$$

Assume that the inner expansion is in the following form:

$$\hat{T}(\xi_*, \eta_*, t) = \left\{ \nu_0(\varepsilon)\hat{T}_0(\xi_*, \eta_*) + \nu_1(\varepsilon)\hat{T}_1(\xi_*, \eta_*) + \dots \right\} e^{\frac{\sigma t}{\varepsilon \eta_0^2}}$$

$$\hat{T}_S(\xi_*, \eta_*, t) = \left\{ \nu_0(\varepsilon)\hat{T}_{S0}(\xi_*, \eta_*) + \nu_1(\varepsilon)\hat{T}_{S1}(\xi_*, \eta_*) + \dots \right\} e^{\frac{\sigma t}{\varepsilon \eta_0^2}} \quad (8.84)$$

$$\hat{h} = \left\{ \nu_0(\varepsilon)\hat{h}_0 + \nu_1(\varepsilon)\hat{h}_1 + \cdots \right\} e^{\frac{\sigma t}{\varepsilon \eta_0^2}}.$$

Letting $\varepsilon \to 0$ in the leading order approximation, the above inner equations can be simplified to the third-order, complex ODE for the unsteady interface perturbations:

$$\mathrm{i}\frac{\varepsilon^{3-3\alpha}}{S\Lambda_1}\frac{\partial^3 \hat{h}_0}{\partial \xi_*^3} + \varepsilon^{1-\alpha}(\mathrm{i}+\xi)\frac{\partial \hat{h}_0}{\partial \xi_*} + \sigma_0 S^2 \hat{h}_0 = (\text{higher-order terms})\,. \quad (8.85)$$

Again, we use the transformation

$$\tilde{h}_0 = W_0(\xi) \exp\left\{ \frac{\mathrm{i}}{\varepsilon} \int_{\xi_c}^{\xi} k_c(\xi_1)\, \mathrm{d}\xi_1 \right\}, \quad (8.86)$$

where the reference wave number function $k_c(\xi)$ is chosen as in Chap. 6, and, in the outer region we have

$$H_1 = W_0^{(+)}(\xi) \exp\left\{ \frac{\mathrm{i}}{\varepsilon} \int_{\xi_c}^{\xi} k_c(\xi_1)\, \mathrm{d}\xi_1 \right\}$$

$$H_3 = W_0^{(-)}(\xi) \exp\left\{ \frac{\mathrm{i}}{\varepsilon} \int_{\xi_c}^{\xi} k_c(\xi_1)\, \mathrm{d}\xi_1 \right\}. \quad (8.87)$$

The general outer solutions for W_0 may be written in the form:

$$W_0 = D_1 W_0^{(+)} + D_3 W_0^{(-)}\,. \quad (8.88)$$

The remaining derivation will be completely the same as that for the pure melt system. In the vicinity of the turning point ξ_c, the inner equation is reduced to the Airy equation with the index $\alpha = \frac{1}{2}$ and the matching condition gives the connection formula

$$\frac{D_1}{D_3} = \mathrm{e}^{\mathrm{i}\frac{\pi}{2}} \quad (8.89)$$

and

$$b_1 = \frac{\mathrm{i}}{4}, \quad R_1(\xi_c) = 0\,. \quad (8.90)$$

8.6 Global Modes and the Quantization Condition

Now, we need to apply the smooth tip condition to derive the quantization condition for the eigenvalues σ. For the 3D case, the MVE solution has a singularity at the tip $\xi = 0$. Applying the tip condition means matching the outer solution with the tip inner solution, which is the same as we derived for the system of pure melt. Combining the connection condition (8.89) with the matching condition (6.160) in the tip region, we obtain the quantization condition, which is formally the same as that for the pure melt system:

$$\frac{1}{\varepsilon} \int_0^{\xi_c} \left(k_0^{(1)} - k_0^{(3)} \right) \mathrm{d}\xi = \left(2n + 1 + \frac{1}{2} + \frac{\theta_0}{2} \right) \pi - \frac{i}{2} \ln \alpha_0 \qquad (8.91)$$

$$\left(n = 0, \pm 1, \pm 2, \pm 3, \cdots \right),$$

where

$$\alpha_0 \, e^{i\theta_0 \pi} = \frac{k_0^{(1)}(0)}{k_0^{(3)}(0)}. \qquad (8.92)$$

The only difference between the case of a binary mixture and the case of a pure melt is that the wave number functions $k_0^{(1)}, k_0^{(3)}$ are now solved from the dispersion formula (8.45). Consequently, the eigenvalues σ_0 for the present system depend on the parameters ε and $\lambda, \kappa, M, \Delta C$, and T_∞.

Furthermore, from the matching condition (8.90) we obtain

$$\sigma_1 = \frac{1}{1 + \xi_c^2} \left\{ \frac{\eta_0^2}{2} \left(1 + \frac{\lambda^2 - \lambda}{2\Lambda_1} \right) \left(1 - \frac{\hat{k}_0^2}{\Lambda_1 S} \right)^2 - (1 + \eta_0^2) + \frac{\eta_0^2 \lambda}{2\Lambda_1} \left(1 - \frac{\lambda}{2} \right) \right.$$

$$\left. + \frac{M\Delta C}{2\Lambda_1} \left[1 - \eta_0^2(1 - 2\kappa) \right] \left(1 - \frac{\hat{k}_0^2}{\Lambda_1 S} \right) \right\}_{\xi = \xi_c}. \qquad (8.93)$$

Noting that when $\xi = \xi_c$,

$$1 - \frac{\hat{k}_0^2}{\Lambda_1 S} = \frac{2}{3} + \frac{i\xi_c}{3}, \qquad (8.94)$$

we have

$$\sigma_1 = \frac{1}{1 + \xi_c^2} \left\{ \frac{\eta_0^2}{2} \left(1 + \frac{\lambda^2 - \lambda}{2\Lambda_1} \right) \left(\frac{2}{3} + \frac{i\xi_c}{3} \right)^2 - (1 + \eta_0^2) + \frac{\eta_0^2 \lambda}{2\Lambda_1} \left(1 - \frac{\lambda}{2} \right) \right.$$

$$\left. + \frac{M \Delta C}{2\Lambda_1} \left[1 - \eta_0^2(1 - 2\kappa) \right] \left(\frac{2}{3} + \frac{i\xi_c}{3} \right) \right\}. \qquad (8.95)$$

Several special cases can be considered:

1. $M = 0, \lambda = 1$: this is the case for the system of pure melt. Formula (8.93) is reduced to

$$\sigma_1 = -\frac{1}{1+\xi_c^2}\left[1 + \frac{\eta_0^2}{2} - \frac{\eta_0^2}{2}\left(\frac{2}{3} + \frac{i\xi_c}{3}\right)^2\right]. \qquad (8.96)$$

We regain result (6.141).

2. $\frac{\eta_0^2}{2} \ll 1$, $\lambda \ll 1$, but $M = O(1)$: this is the case of small undercooling. Formula (8.93) is reduced to

$$\sigma_1 = -\frac{1}{1+\xi_c^2}\left(\frac{2}{3} - \frac{i\xi_c}{6}\right). \qquad (8.97)$$

Given the parameters $\varepsilon, \lambda, M, \kappa$, and T_∞, the system has a discrete set of eigenvalues $\sigma_n \sim \{\sigma_0 + \varepsilon\sigma_1 + \cdots\}_n$ $(n = 0, 1, 2, \cdots)$, and a unique global neutrally stable solution with $\sigma_R = 0$. The critical number ε_* is now a function of $\{T_\infty, \lambda, \kappa,$ and $M\}$.

The numerical computations for the selected neutrally stable modes have been carried out for various parameters. To present these results, It is appropriate to introduce some new definitions:

(i)   the effective undercooling temperature ΔT, which is defined as

$$\Delta T = -T_\infty - M ; \qquad (8.98)$$

(ii)   the dimensionless tip velocity U_{tip}, which is defined as

$$U_{\text{tip}} = \frac{(U)_{\text{D}}}{\kappa_{\text{T}}/\ell_c} = \lambda\varepsilon_*^2\eta_0^4 ; \qquad (8.99)$$

(iii) the dimensionless tip radius R_{tip}, which is defined as

$$R_{\text{tip}} = \frac{\ell_t}{\ell_c} = \frac{1}{\varepsilon_*^2\eta_0^2} ; \qquad (8.100)$$

(iv) the dimensionless frequency of oscillation of the interface Ω_*, which is defined as

$$\Omega_* = \frac{\omega_*}{\varepsilon_*\eta_0^2} . \qquad (8.101)$$

It should be remarked that, for the special case $M = 0$, the problem in hand reduces to the problem of dendrite growth from pure melt. However, the quantities ε, η_0^2 used here for a binary system are different from what we used in Chap. 6 for the case of pure melt, since the length scales are chosen differently in the two cases. To distinguish them from each other, let us denote the parameters $\varepsilon, \varepsilon_*, \eta_0^2$, and other parameters used for the case

$M = 0$, by adding the subscript 'melt'. Thus, in the case $M = 0$, we have defined

$$(\varepsilon)_{\text{melt}} = \frac{\ell_{\text{T}}/\ell_{\text{c}}}{(\eta_0^2)_{\text{melt}}} , \tag{8.102}$$

$$(U_{\text{tip}})_{\text{melt}} = \frac{(U)_{\text{D}}}{\ell_{\text{c}}/\kappa_{\text{T}}} = (\varepsilon_*^2)_{\text{melt}}(\eta_0^4)_{\text{melt}} , \tag{8.103}$$

$$(R_{\text{tip}})_{\text{melt}} = \frac{\ell_{\text{t}}}{\ell_{\text{c}}} = \frac{1}{(\varepsilon_*^2)_{\text{melt}}(\eta_0^2)_{\text{melt}}} , \tag{8.104}$$

and the frequency of oscillation of the interface

$$(\Omega_*)_{\text{melt}} = \frac{(\omega_*)_{\text{melt}}}{(\varepsilon_*)_{\text{melt}}(\eta_0^2)_{\text{melt}}} . \tag{8.105}$$

However, from (8.12), we can derive

$$(\eta_0^2)_{\text{melt}} = \lambda \lim_{M \to 0} \eta_0^2 . \tag{8.106}$$

Thus we obtain the following connection between the quantities used for the two cases: fixing T_∞, λ and κ, and letting $M \to 0$, we find that

$$\frac{\varepsilon_*^2}{\lambda} \to (\varepsilon_*^2)_{\text{melt}} \tag{8.107}$$

$$\frac{\Omega_*}{\lambda} \to (\Omega_*)_{\text{melt}} \tag{8.108}$$

$$U_{\text{tip}} \to (U_{\text{tip}})_{\text{melt}} \tag{8.109}$$

$$R_{\text{tip}} \to (R_{\text{tip}})_{\text{melt}} . \tag{8.110}$$

Relationships (8.109) and (8.110) imply that for any fixed ΔT, as $M \to 0$ the limit value of U_{tip} and R_{tip} will be irrelevant to the parameters λ and κ. Our numerical results satisfy this property.

The zeroth-order approximation, $\varepsilon_*^{(0)}$, of the critical number ε is calculated through the zeroth-order approximation of the eigenvalue σ, namely by the equation $\text{Re}\{\sigma_0(\varepsilon)\} = 0$, whereas the first-order approximation, $\varepsilon_*^{(1)}$ is calculated through the first-order approximation of σ, namely by the equation $\text{Re}\{\sigma_0(\varepsilon) + \varepsilon\sigma_1(\varepsilon)\} = 0$. The numerical computations for both $\varepsilon_0^{(0)}$ and $\varepsilon_0^{(1)}$ have been carried out for a large range of growth conditions.

In Fig. 8.1, we show $\varepsilon_*^{(0)}$ and $\varepsilon_*^{(1)}$ versus the morphological parameter M for the case $\Delta T = 0.01$. It is noticed that the numerical values of $\varepsilon_*^{(1)}$ have significant modifications compared to the values of $\varepsilon_*^{(0)}$

A remark should be made here that the asymptotic expansion solution obtained in this chapter, $\sigma \sim \sigma_0 + \varepsilon\sigma_1 + \cdots$, is derived in the limit $\varepsilon \to 0$, for fixed $M, \lambda, \kappa = O(1)$. It may be not uniformly valid in the entire space

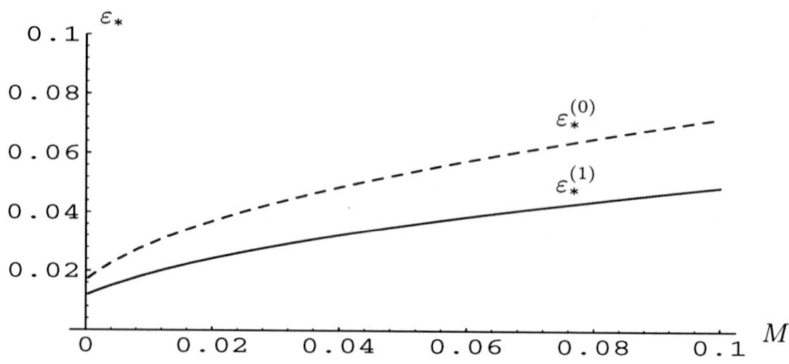

Fig. 8.1. The variation of the critical number ε_* with M for $\Delta T = 0.01$, $\lambda = 0.01114$, and $\kappa = 0.103$. The solid line represents the first-order approximation $\varepsilon_*^{(1)}$, while the dashed line represents the zeroth-order approximation $\varepsilon_*^{(0)}$

of the parameters λ, M, κ. In other words, as $\lambda = O(\varepsilon)$, or $M = O(\varepsilon)$, the solution may have a different form of asymptotic expansion. As a result, for some range of these parameters, the first-order results of our asymptotic solution, $\sigma \approx \sigma_0 + \varepsilon\sigma_1$ may not give a better numerical approximation than the zeroth-order results, $\sigma \approx \sigma_0$. This may be the case when it turns out that $\varepsilon\sigma_1 \gg \sigma_0$. Such cases have been seen in Chap. 2 when we study unidirectional solidification systems. For instance, from Fig 2.3, it is seen that as the wave number $k > 1$, the first-order approximate solution is not close to the exact solution, even qualitatively.

Thus, it is clear that for the selected neutrally stable mode solutions one cannot anticipate that, in terms of above obtained the asymptotic solution, the first-order approximation for the critical number, $\varepsilon_*^{(1)}$ will always be better than the zeroth-order approximation $\varepsilon_*^{(0)}$. Instead, it may be expected that when the difference between the first-order and zeroth-order results, $\Delta\varepsilon_* = \varepsilon_*^{(1)} - \varepsilon_*^{(0)}$, is too large for some range of the parameters, for instance, $M = O(\varepsilon)$ and $\lambda = O(\varepsilon)$, the accuracy of the first-order results $\varepsilon_* \approx \varepsilon_*^{(1)}$, may be even worse than the zeroth-order approximation, $\varepsilon_* \approx \varepsilon_*^{(0)}$. The exact number ε_* may fall between these two. In order to improve the accuracy of the zeroth-order approximation solution in that range of parameters, one needs either to incorporate the higher-order approximations from the current form of asymptotic solution, or to derive a different form of asymptotic solution.

In the following, we shall mainly give the numerical results of the zeroth-order approximation solutions.

The variation of the ratio of the critical numbers, $\varepsilon_*(M)/\varepsilon_*(0)$ with the parameter M, for various values of the effective undercooling parameter ΔT, the parameters κ and λ is shown in Figs. 8.2–8.4, respectively. Note that when $M = 0$, the zeroth-order critical number $\varepsilon_*^{(0)}(0) = (\varepsilon_*^{(0)})_{\text{melt}}\sqrt{\lambda} = 0.1590\sqrt{\lambda}$.

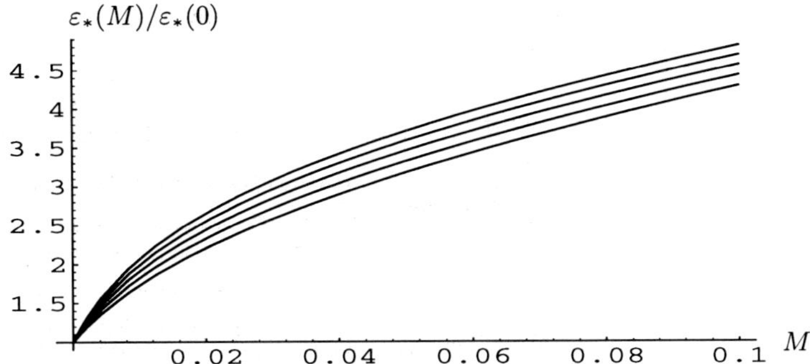

Fig. 8.2. The variation of the ratio $\varepsilon_*(M)/\varepsilon_*(0)$ with M for $\lambda = 0.01114$, $\kappa = 0.103$, and the five values $\Delta T = 0.01, 0.02, 0.03, 0.04, 0.05$ from bottom to top

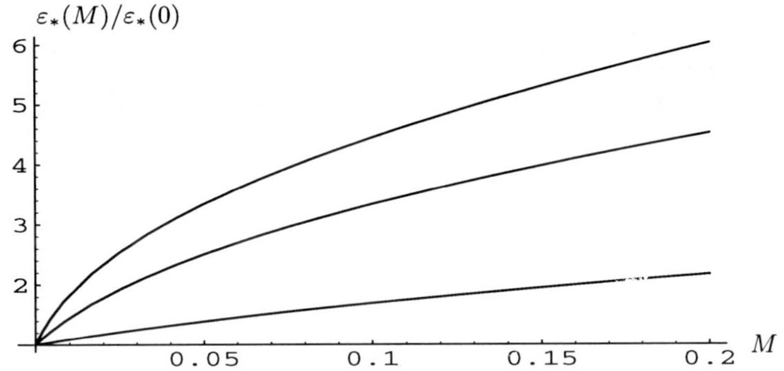

Fig. 8.3. The variation of the ratio $\varepsilon_*(M)/\varepsilon_*(0)$ with M for $\lambda = 0.01114$, $\Delta T = 0.021$, and the three values $\kappa = 0.1, 0.5, 0.9$ from top to bottom

The variation of the oscillation frequency of the interface, Ω_* with the parameter M, under different parameters ΔT, κ, and λ, is shown in Figs. 8.5–8.7, respectively. It is seen that the frequency Ω_* increases with M and ΔT, while it decreases as κ increases.

With the numerical results for ε_*, one can calculate the tip velocity U_{tip} and the tip radius R_{tip} of the dendrite. The variation of the tip velocity U_{tip} with the parameter M, for different values of the parameters ΔT, κ, and λ, is shown in Figs. 8.8–8.10, respectively.

The variation of the tip radius R_{tip} with the parameter M, for different values of the parameters ΔT, κ, and λ, is shown in Figs. 8.11–8.13, respectively.

It is seen that with other parameters fixed, as M increases the tip velocity U_{tip} has a maximum value, while R_{tip} has a minimum value.

To describe the interface pattern, in Fig. 8.14 we show the tips of dendrites for $\lambda = 0.01114, M = 0.05, \Delta T = 0.021$ with different values $\kappa = 0.1, 0.5, 0.9$. It is seen that the dendrite's shape is quite sensitive to the variation of κ .

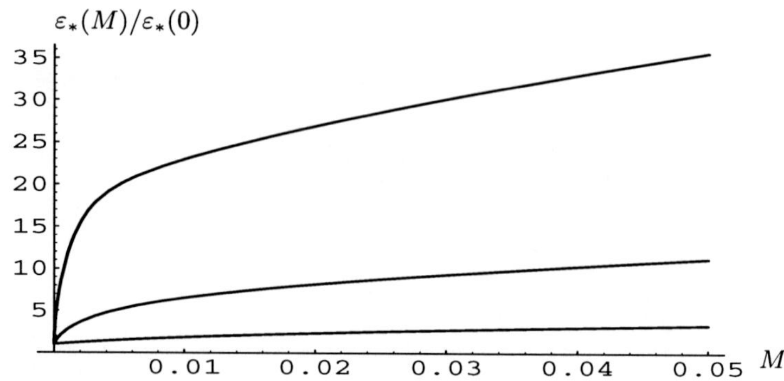

Fig. 8.4. The variation of the ratio $\varepsilon_*(M)/\varepsilon_*(0)$ with M for $\kappa = 0.103$, $\Delta T = 0.021$, and the three values $\lambda = 0.011, 0.001, 0.0001$ from top to bottom

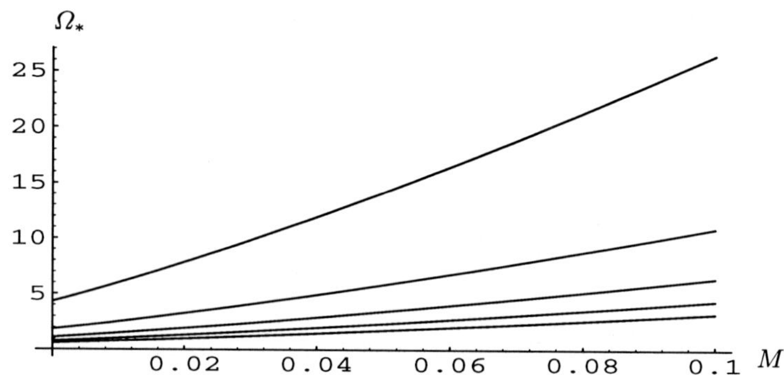

Fig. 8.5. The variation of Ω_* with M for $\lambda = 0.01114$, $\kappa = 0.103$, and the five values $\Delta T = 0.01, 0.02, 0.03, 0.04, 0.05$ from top to bottom

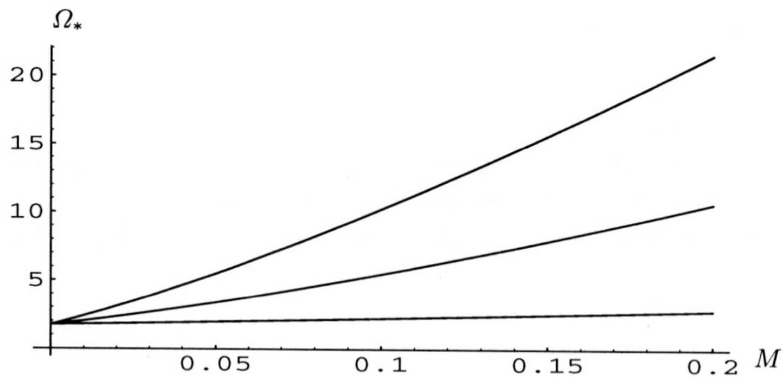

Fig. 8.6. The variation of Ω_* with M for $\lambda = 0.01114$, $\Delta T = 0.021$, and the three values $\kappa = 0.1, 0.5, 0.9$ from top to bottom

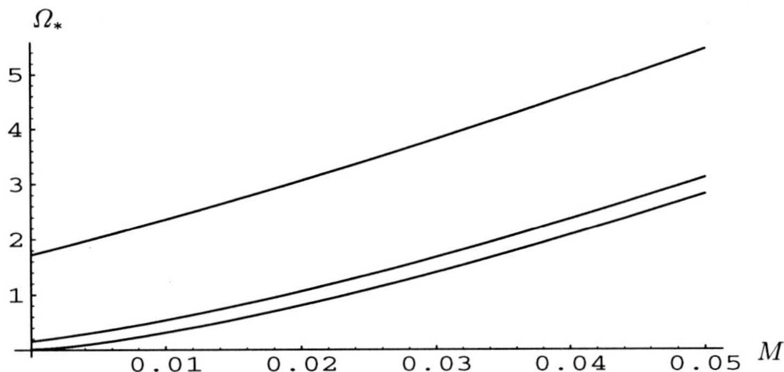

Fig. 8.7. The variation of Ω_* with M for $\kappa = 0.103$, $\Delta T = 0.021$, and the three values $\lambda = 0.011, 0.001, 0.0001$ from top to bottom

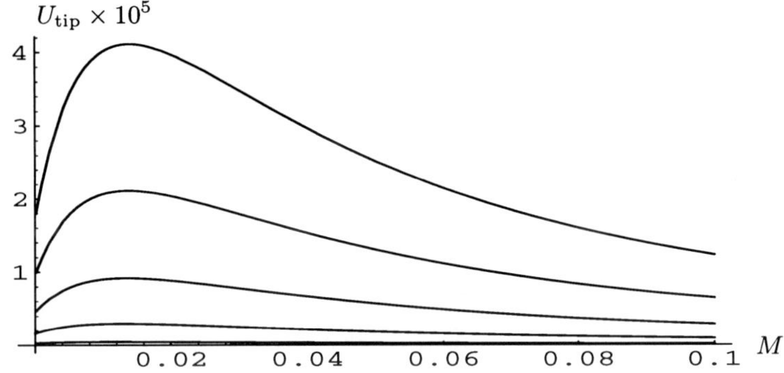

Fig. 8.8. The variation of U_{tip} with M for $\lambda = 0.01114$, $\kappa = 0.103$, and the five values $\Delta T = 0.01, 0.02, 0.03, 0.04, 0.05$ from bottom to top

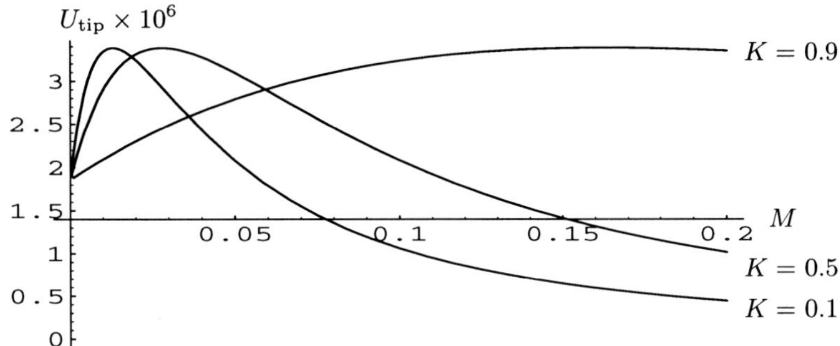

Fig. 8.9. The variation of U_{tip} with M for $\lambda = 0.01114$, $\Delta T = 0.021$, and the three values $\kappa = 0.1, 0.5, 0.9$ from bottom to top

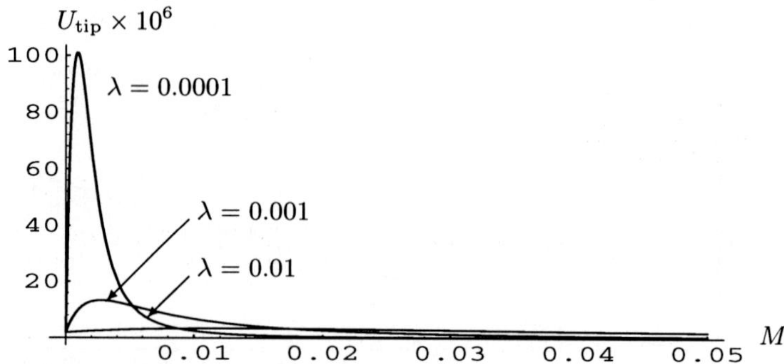

Fig. 8.10. The variation of U_{tip} with M for $\kappa = 0.103$, $\Delta T = 0.021$, and the three values $\lambda = 0.011, 0.001, 0.0001$ from bottom to top

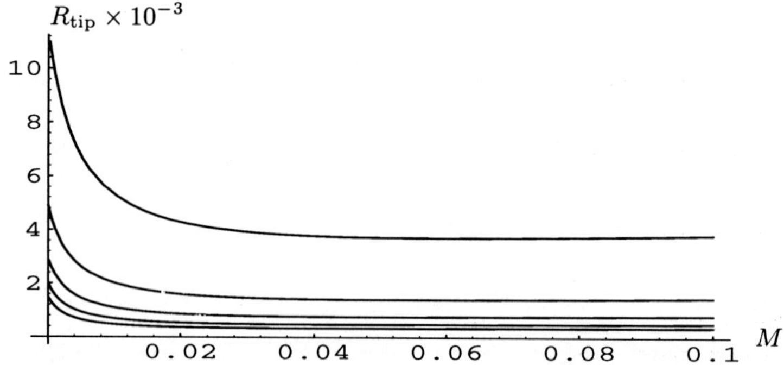

Fig. 8.11. The variation of R_{tip} with M for $\lambda = 0.01114$, $\kappa = 0.103$, and the five values $\Delta T = 0.01, 0.02, 0.03, 0.04, 0.05$ from top to bottom

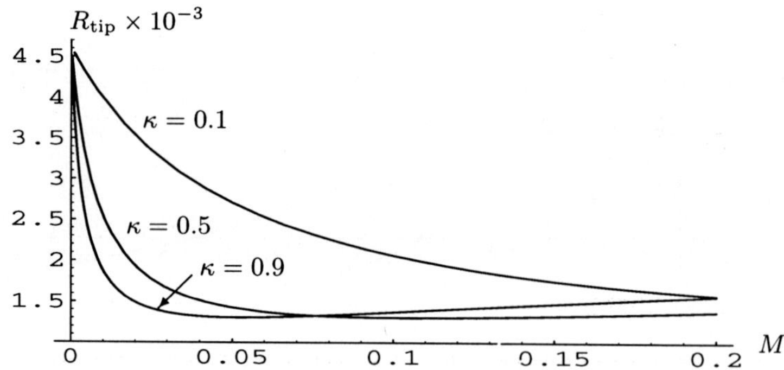

Fig. 8.12. The variation of R_{tip} with M for $\lambda = 0.01114$, $\Delta T = 0.021$, and the three values $\kappa = 0.1, 0.5, 0.9$ from top to bottom

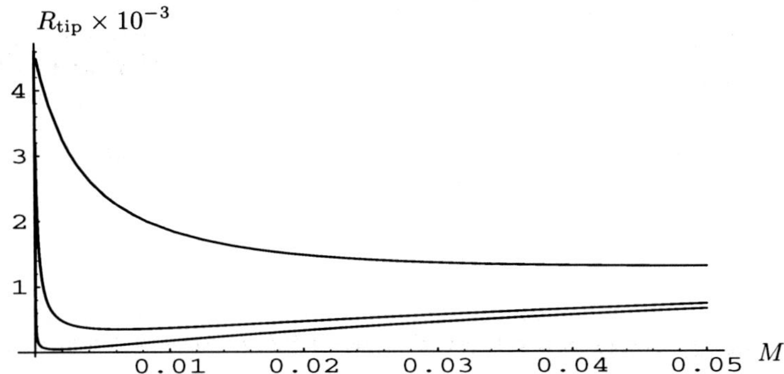

Fig. 8.13. The variation of R_{tip} with M for $\kappa = 0.103$, $\Delta T = 0.021$, and the three values $\lambda = 0.011, 0.001, 0.0001$ from top to bottom

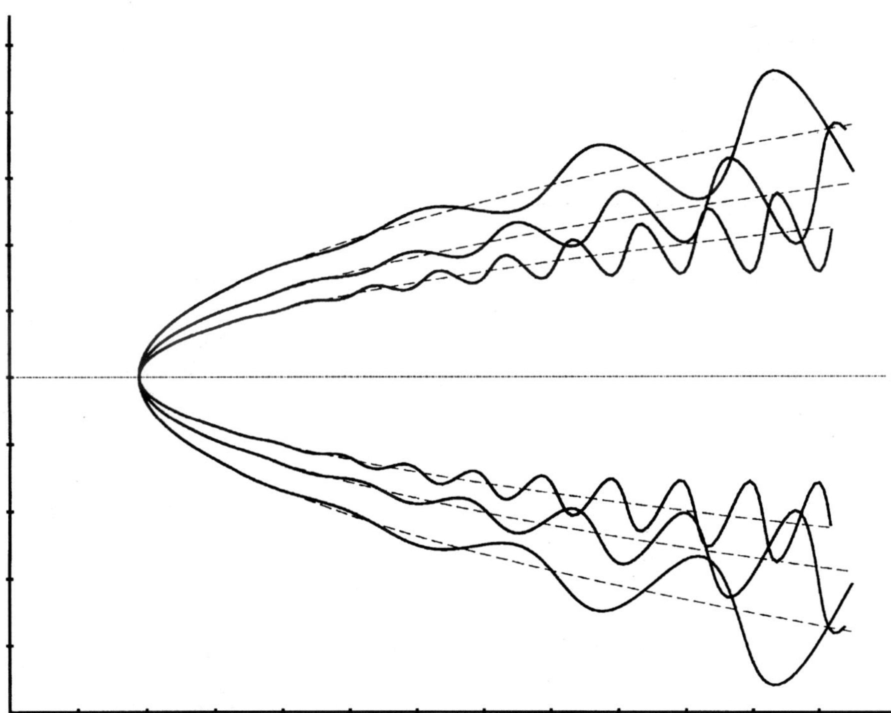

Fig. 8.14. The tips of dendrites with the three values $\kappa = 0.1, 0.5, 0.9$ from inside to outside

For a long time, researchers in materials science have phenomenologically considered that a dendrite consists of two different parts: the steady and smooth tip part, and the unsteady, side-branching part. But nobody seems to know where these two parts are separated. Now, from the IFW theory, one sees that the dendrite tip part is not really steady, nor smooth at all. Every point on the entire interface oscillates with the same frequency Ω_* but with different amplitude. Hence, in this respect, the tip part of the dendrite is not different from any other part. The tip appears 'steady' and 'smooth' only because the amplitude of oscillation in the tip part is much smaller than that in the other part. Nevertheless, from the wave diagram of the GTW mode, it is reasonable to consider that the superficial 'smooth' tip part is separated from the side-branching part at the intersection point ξ_c'. With this viewpoint, the microstructure of the dendrite is then quantitatively characterized by the location of ξ_c', the first several wavelengths of the outgoing traveling wave propagating along the Ivantsov paraboloid in the side-branching part, and the frequency Ω_*. This detailed information can be drawn from the GTW mode solutions. According to linear theory, the amplitude of this outgoing wave is determined only up to an arbitrary constant multiplier.

8.7 Comparisons of Theoretical Results with Experimental Data

So far the most systematic experiments on dendrite growth from binary mixtures have been conducted by Lipton, Glicksman and Kurz for the system SCN–acetone and SCN–argon in 1984 and 1987 [8.3]–[8.5]. The data in these papers do not completely agree with one another. The more complete and accurate data is for SCN-acetone and this was given by Chopra, Glicksman, and Singh in 1988 [8.6].

In the following, we shall compare the theory with this experimental data.

The properties of SCN–acetone mixtures are listed in Table 8.1. Based on this data, we calculate that for this system $\ell_c = 2.7726 \times 10^{-9}$(m), the unit of temperature $[T] = \frac{\Delta H}{(c_p \rho)} = 23.8761 \, \mathrm{K}$, the unit of the velocity $[U] = \frac{\kappa_T}{\ell_c} =$

Table 8.1. The thermodynamic properties of SCN–acetone mixtures.

$\lambda (= \kappa_\mathrm{D}/\kappa_\mathrm{T})$	0.01114
κ_D	$1.27 \times 10^{-9} \mathrm{m^2 s^{-1}}$
ΔH	$46.26 \times 10^3 \mathrm{J\ kg^{-1}}$
c_p	$1937.5 \mathrm{\ J\ kg^{-1} K^{-1}}$
κ_T	$1.14 \times 10^{-7} \mathrm{m^2 s^{-1}}$
$\gamma T_\mathrm{M0}/\Delta H$	$6.63 \times 10^{-8} \mathrm{K\ m}$
m	$-2.16 \mathrm{\ K\ mol\%^{-1}}$
κ	$0.103 \mathrm{\ mol\%\ mol\%^{-1}}$

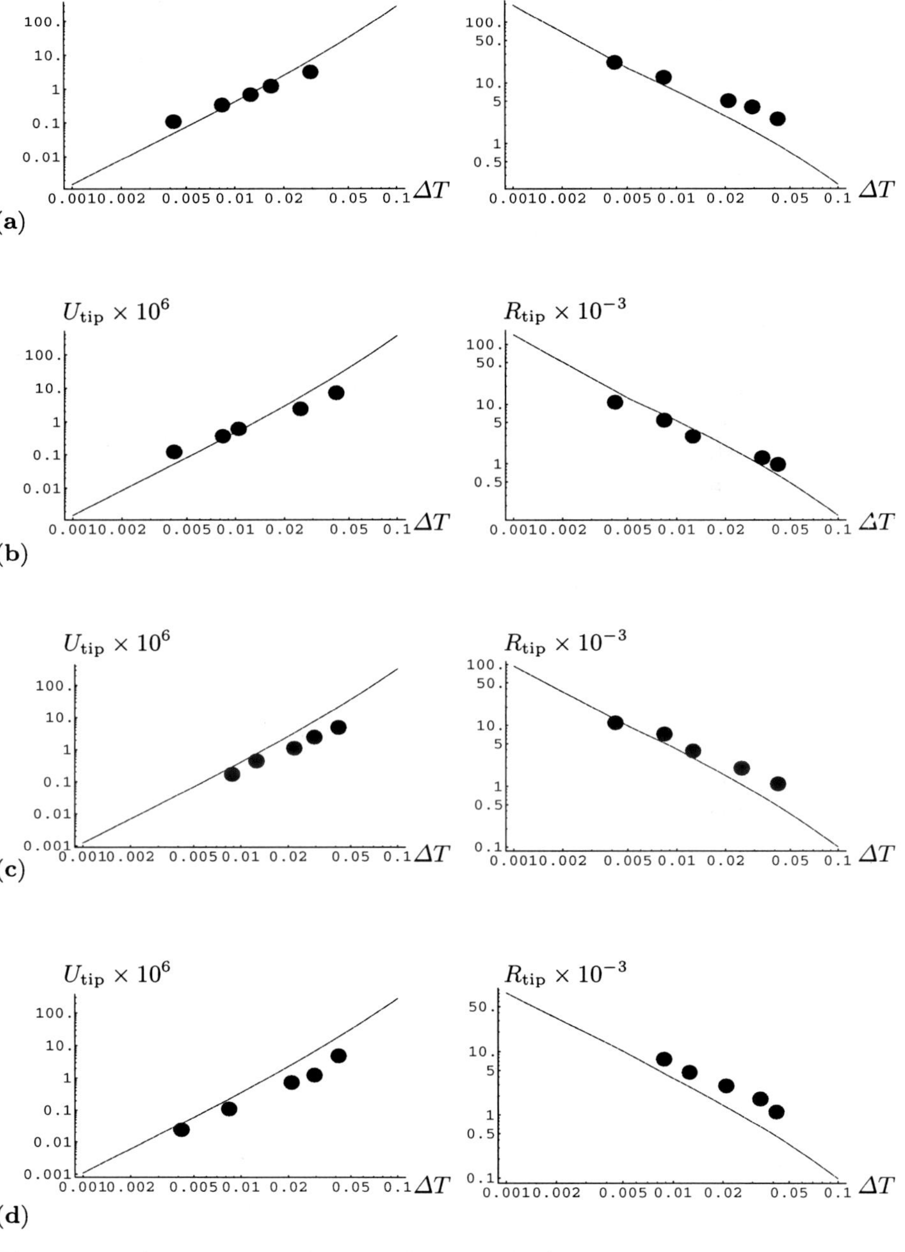

Fig. 8.15. The variations of U_{tip} and R_{tip} with ΔT for $\kappa = 0.103, \lambda = 0.01114$, and **(a)** $M = 0.00407$; **(b)** $M = 0.00995$; **(c)** $M = 0.02713$; **(d)** $M = 0.03608$

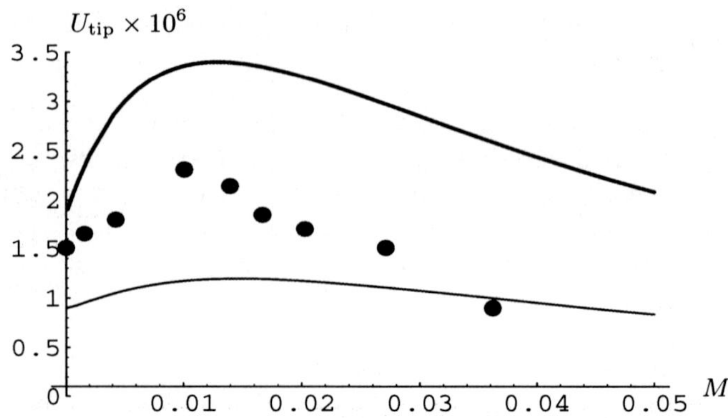

Fig. 8.16. The variations of U_{tip} with M for the case $\Delta T = 0.021, \kappa = 0.103, \lambda = 0.01114$. The bold line shows the zeroth-order solution, while the thin line is the first-order solution

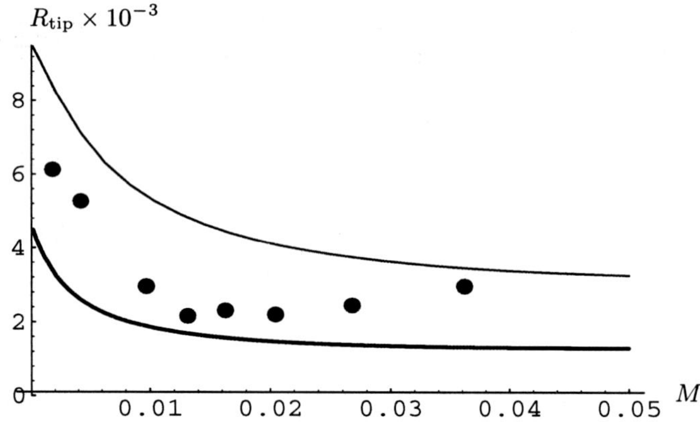

Fig. 8.17. The variations of R_{tip} with M for the case $\Delta T = 0.021, \kappa = 0.103, \lambda = 0.01114$. The bold line shows the zeroth-order solution, while the thin line is the first-order solution

$41.1173\,(\mathrm{m/s})$, the concentration of solute at the far-field is $C_\infty = 11.0583\,M$, where M is the morphological parameter. In terms of this data, we convert the experimental data given in [8.6] into our dimensionless quantities $U_{\mathrm{tip}}, R_{\mathrm{tip}}$, and M.

Comparisons between the zeroth-order approximation solutions and the experimental data for the variation of U_{tip} and R_{tip} versus ΔT for $\kappa = 0.103, \lambda = 0.01114$, and the four values $M = 0.00407, 0.00995, 0.02713, 0.03608$ are shown in Figs. 8.15(a)–(d). Noting that there is no adjustable parameter in the theory, the agreement between the theoretical results and the experimental data is reasonably good. The discrepancy between the experimental data and the theoretical curves might indicate the effects of some physical parameters neglected in our analysis, and/or the effect of higher-order contributions and nonlinearity.

Comparisons of theoretical results and experimental data for U_{tip} and R_{tip} versus M for $\Delta T = 0.021, \kappa = 0.103, \lambda = 0.01114$ are shown in Figs. 8.16 and 8.17, respectively. In these figures, the zeroth-order solution is shown by the bold lines, while the first-order solution is shown by the thin lines. The experimental data fall between the two lines. Again, it is seen that the agreement between the theoretical results and the experimental data is reasonably good.

From these figures, it is noted that when $M > 0.03$, the first-order solution agrees with the experimental data very well, while it does not agree so well in the range of $M < 0.03$. This is quite understandable. For the cases under discussion, the numerical value of the critical number is around $\varepsilon_* \approx 0.01 - 0.03$. In the range of $M \approx 0.04 - 0.05$, we have $M > \varepsilon_*$. Therefore, the asymptotic expansion that we derived may still be applicable. As a result, the first-order solution may be more accurate and shows a better agreement with the experimental data than the zeroth-order solution. However, $M < \varepsilon_*$, when $M < 0.03$. In this range, compared with ε, one can no longer consider $M = O(1)$. It would be more accurate to consider $M = O(\varepsilon)$. As a consequence, the asymptotic expansion that we derived will no longer be applicable. For a more accurate approximate solution in this range, one should derive a different asymptotic expansion solution with the assumption $M = O(\varepsilon)$.

References

8.1   J. J. Xu, "Global Instability and Pattern Formation in Dendritic Solidification of Dilute Binary Alloy System", Canad. Appl. Math. Quar. **1**, No. 2, pp. 255–292, (1993).

8.2   J. J. Xu and Z. X. Pan, "Interfacial Wave Theory of Dendritic Growth From a Binary Mixture: A Comparison with Experiments", J. Crystal Growth, No. 129, pp. 666–682, (1993).

8.3   M. Copra, PhD. Thesis, Rensselaer Polytechnic Institute, Troy, NY (1983).

8.4   J. Lipton, M.E. Glicksman and W. Kurz, "Dendritic Growth into Undercooled Alloy Melts", Mater. Sci. Eng. **65**, pp. 57–63, (1984).

8.5   J. Lipton, M. E. Glicksman and W. Kurz, "Equiaxed Dendrite Growth in Alloys at Small Supercooling", Metall. Trans. A **18**, pp. 341–345, (1987).

8.6   M. Copra, M. E. Glicksman and N. B. Singh, "Dendritic Solidification in Binary Alloys", Metall. Trans. A **19**, pp. 3087–3096, (1988).

8.7   A. Karman and J. S. Langer, "Inpurity Effects in Dendritic Solidification", Phys. Rev. A **30**, No. 6, p. 3147–3155, (1984).

9. Viscous Fingering in a Hele–Shaw Cell

In this chapter, we turn to the study of another interfacial phenomenon: the formation of viscous fingers in a Hele–Shaw cell. This phenomenon occurs in an entirely different physical system to that of dendrite growth, but it raises similar issues and can thus be treated using the same approach established in the previous chapters.

9.1 Introduction

The formation of viscous fingers in porous media has, for almost half a century, been a phenomenon of considerable interest within a broad field of science and technology, including secondary oil recovery, fixed bed regeneration in chemical processing, and underground water engineering [9.1]–[9.24]. This phenomenon can be demonstrated in systems such as those shown in Figs 9.1 and 9.2 [9.20]. It was recognized early on that the formation of these fingers is associated with some kind of instability mechanism for the interface, which is caused either by the different viscosities and densities of fluids, or by the surface tension at the interface.

The first scientific investigation and experimental observation of the fingering phenomenon in a vertical packed column driven by gravity and viscosity was made by Hill in 1952 [9.2]. Hill studied the displacement of sugar liquors by water from columns of granular bone charcoal, and carried out a stability analysis for the one-dimensional flat interface.

Fingering phenomena can also be driven by the surface tension on the interface and the difference in the viscosity of two immiscible fluids. Evidently, most fingerings which occur in a horizontal process belong to this category. These phenomena were first studied by Chouke et al. [9.3] and Saffman and Taylor [9.4] in the late 1950s. These investigators performed a rigorous linear stability analysis for a flat interface. This instability mechanism for a flat interface has become known as the 'Saffman–Taylor instability' but more precisely, it should be called the 'Chouke–Saffman–Taylor instability. To simulate the viscous fingering phenomenon in porous media, Saffman and Taylor (1958) first used a Hele–Shaw cell as a simplified tool and made systematic experimental observations.

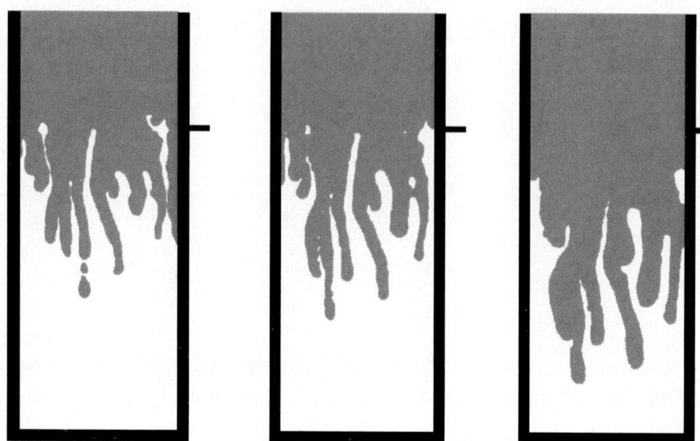

Fig. 9.1. Viscous fingering formation in a vertical packed column

The device used by Saffman and Taylor is sketched in Fig. 9.3. The cell has a very thin gap b and a width $2W$. To begin with, the cell is filled with a viscous fluid. Then a less viscous fluid is injected into the cell in order to displace the more viscous fluid. In the early stage of evolution, the flat interface between the two fluids will immediately develop into fingers with different sizes, due to the Chouke–Saffman–Taylor instability. However, at a later stage of evolution, experiments show that when the injection pressure is sufficiently small, or the surface tension fairly large, these small figures gradually disappear and the cell is eventually dominated by a single, smooth, steady finger which occupies about one-half the cell width, as shown in Fig 9.4 (a) [9.19]. This kind of interfacial pattern evolution was observed in detail by Saffman and Taylor in 1958.

When the injection pressure increases, or the surface tension decreases, the system may permit another type of stationary interfacial pattern: namely, oscillatory, dendrite-like fingers which have a much narrower width. These oscillatory fingers were first discovered by Couder et al. in 1986 [9.7], and later by Kopf-Sill and Homsy in 1987 [9.8].

The oscillatory fingers of Couder et al.'s type are easy to produce, as they put a tiny bubble or a wire which may be considered as a rigid bubble at the finger's tip. On the other hand, the oscillatory fingers of the Kopf-Sill and Homsy type, with no bubble or wire at the tip, are very difficult to observe. When the injection pressure becomes even larger, or the surface tension even smaller, the oscillatory fingers of the Couder et al. type are not sustained. These fingers will split and spread. Thus the viscous fingers appear as a chaotic pattern.

The smooth finger solution with zero surface tension was first found by Zhuravlev in 1956. This solution contains an undetermined parameter which

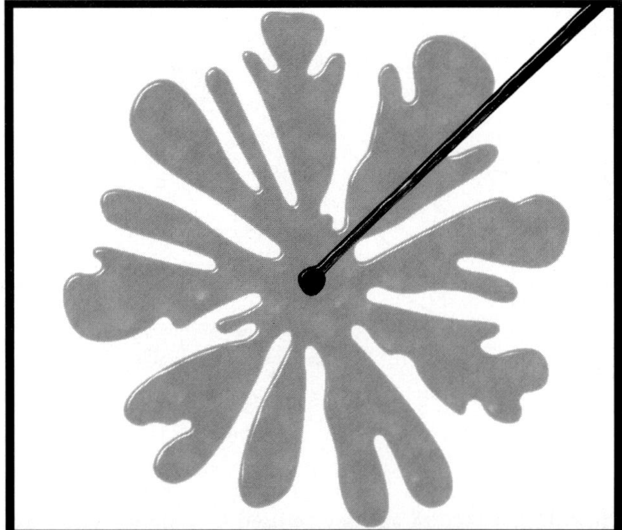

(a)

(b)

Fig. 9.2. Viscous fingering formation in some horizontal devices: (**a**) a liquid with lower viscosity is injected into the cell from the center of the top plate; (**b**) a liquid with lower viscosity is injected into the cell from the lower-left corner of the top plate and the fluid is drawn out from the upper-right corner

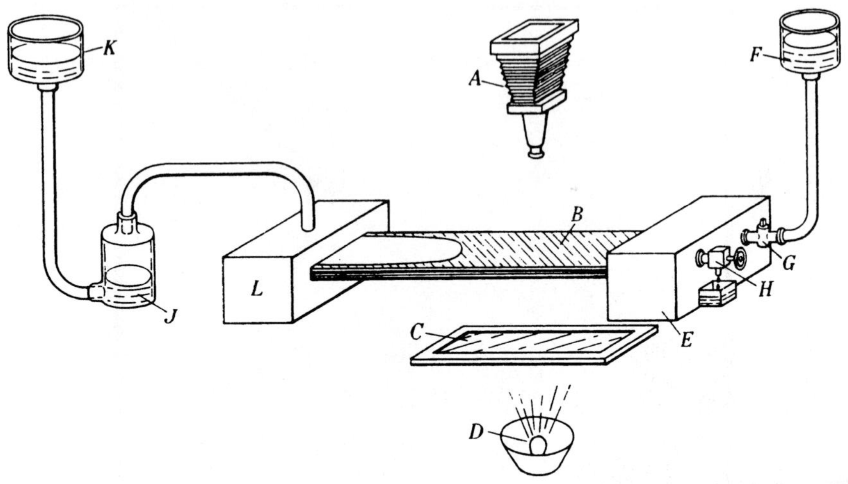

Fig. 9.3. A sketch of the device used by Saffman and Taylor

represents the tip velocity or the asymptotic width of the finger. This solution was also derived independently by Saffman and Taylor in 1958.

The most significant contribution made by Saffman and Taylor was their identification of the selection problem. On the basis of their experimental observations, Saffman and Taylor discovered that the tip velocity was actually uniquely determined by the operating conditions. They also initiated a linear stability analysis for the steady finger solution with zero surface tension and found that the solution was unstable. They did not find the steady smooth finger solutions with nonzero surface tension, nor perform a stability analysis for such a general case. As such, the selection problem remained unsolved.

After Saffman and Taylor, many other researchers devoted themselves to this subject. The central issues were how to specify the basic state solution when the surface tension is not zero, and how to perform a stability analysis for the basic state solution and determine the instability mechanisms induced by surface tension.

The smooth finger solutions with zero surface tension obtained by Zhuravlev, Saffman and Taylor describe a steady, smooth finger with an infinitely long root, and so they may be called 'the classic steady finger solutions'. For the case of symmetric finger formation with nonzero surface tension, the system involves two parameters: the surface tension parameter ε, which is proportional to the surface tension at the finger interface, and the tip velocity parameter U (or the effective finger width parameter λ, which is defined as the reciprocal of the tip velocity, i.e., $\lambda = \frac{1}{U}$). For this case, it is a nontrivial matter to predict whether or not the system still allows a classic steady finger solution.

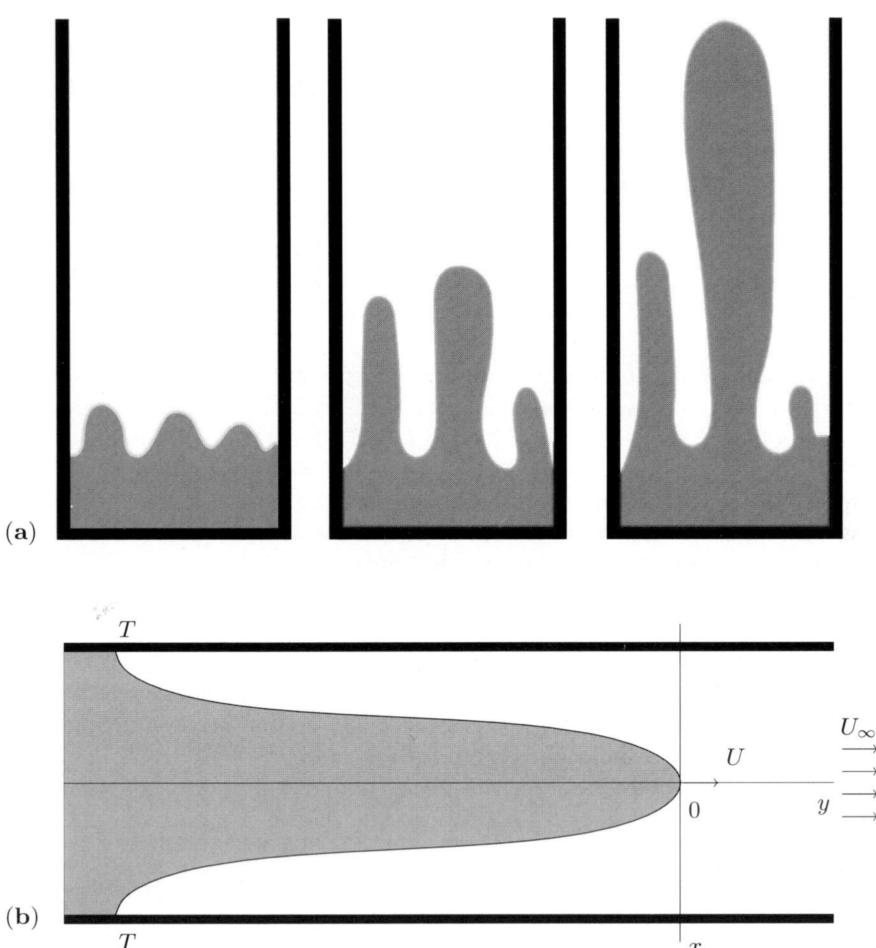

Fig. 9.4. A typical process of smooth finger formation in a Hele–Shaw cell: (**a**) finger evolution at an early stage of the process; (**b**) sketch of a single dominant smooth finger formed in the final stage of the process

During the past few decades, great effort has been made by a number of investigators to look for the classic steady finger solution to the problem with nonzero surface tension. It was finally found numerically by Romero (1981) [9.10], and later confirmed by Vanden-Broeck (1983) [9.11], that for the surface tension parameter $\varepsilon \neq 0$ and for any given operating condition, the system only allows a discrete set of classic steady finger solutions, with width parameters λ_i $(i = 0, 1, 2, \ldots)$. For other width parameters λ, fingers with an infinitely long smooth root will have a cusp at the tip. As the surface tension parameter $\varepsilon \to 0$, this cusp will be transcendentally small. Based on these results Vanden-Broeck was able to plot a set of curves $\{C_i\}(i = 0, 1, 2, 3, \ldots)$

in the (λ, ε) parameter plane. The system does not permit a classic finger solution for points not on these curves. Among the curves $\{C_i\}$, the lowest branch $\{C_0\}$ was found numerically in 1980 by McLean and Saffman. The results of Vanden-Broeck were soon verified by a number of investigators in terms of the microscopic solvability approach (see, for instance, [9.13]).

A number of researchers, such as Tanveer (1987) [9.12] and Bensimon, Pelce, and Schraiman (1986) [9.9] have attempted to study the stability around these classic steady finger solutions. Tanveer concluded that the solution with $n = 0$ is neutrally stable for all surface tension parameters $0 < \varepsilon \ll 1$. The same statement was made by Bensimon (1986) and other investigators.

In the meantime, numerical solutions for the initial value problem were performed by a number of investigators such as DeGregoria and Schwartz (1985, 1986) [9.17], Kessler and Levine (1986) [9.14] and Bensimon (1987) [9.6]. The numerical results of DeGregoria and Schwartz showed that the slowly evolving fingers at the later stage of evolution spontaneously split when the surface tension is sufficiently small. As a consequence, the classic steady finger solutions will all be unstable when the surface tension parameter of fingering is less than a critical number.

The numerical results of DeGregoria and Schwartz apparently contradict the linear stability analysis by Tanveer and others. To reconcile the disagreement, Bensimon (1986) speculated, on the basis of his numerical experiments, that the observed instability in the numerical simulation is a nonlinear mechanism where the threshold amplitude for destabilization decreases with the surface tension parameter of fingering. Bensimon did not perform a nonlinear bifurcation analysis to support his conjecture. So the contradiction between the numerical evidence and linear stability analysis by Tanveer and others had not been reconciled until the global trapped wave (GTW) instability for Hele–Shaw flow was discovered in 1991 [9.22].

The theoretical difficulty encountered in the fingering problem, due to the nonexistence of a classic steady finger solution with arbitrary tip velocity, is quite similar to what we have seen for the problem of dendritic growth. For points off the curves $\{C_i\}$ in the (ε, λ) parameter plane, stability analysis for the classic steady state is meaningless, since these steady state solutions do not exist. Consequently, one cannot divide the entire parameter plane into a stable region and an unstable region.

In order to overcome the difficulty, one needs to consider a broader class of basic state solutions, such as the 'nearly' steady needle solution at the later stage of evolution $(t \gg 1)$ and study their global stability properties. The results show that the fingering system also possesses two types of instability mechanism:

1. the low-frequency (LF) instability mechanism which is caused by perturbations with low frequencies and is associated with the real spectrum of eigenvalues. The neutral modes of the LF mechanism coincide with

the classic steady state solutions obtained by Vanden-Broeck and others based on the microscopic solvability condition (MSC) theory.

2. the global trapped wave (GTW) instability mechanism, which is induced by perturbations with a high frequency and is connected with a complex spectrum of eigenvalues.

When the surface tension parameter of fingering ε is smaller than a critical number, the system is dominated by the GTW mechanism, and the steady finger solutions will be linearly unstable. Hence, Tanveer's conclusion is incorrect as $\varepsilon \to 0$.

In this chapter we shall restrict ourselves to the study of viscous finger formation in a rectangular Hele–Shaw cell and explore the finger selection mechanism in the later stages of evolution.

9.2 Mathematical Formulation of the Problem

Consider the evolution of a finger developing in a Hele–Shaw cell in the positive y direction of a moving coordinate system $\{x, y\}$, which is fixed at the finger's tip. The flow velocity in the up-stream far field is set to a constant U_∞ but the tip velocity, in general, will change with time (see Fig. 9.4 (b)).

For simplicity, we shall assume that the dynamic viscosity of the more viscous fluid is μ, while the viscosity of the less viscous fluid is zero; all physical quantities have been averaged along the thickness of the cell, so that the problem is treated as being two-dimensional. We use one-half of the cell width W as the length unit and use the flow velocity in the up-stream, U_∞, as the velocity unit; the product $U_\infty W$ is used as the unit of the potential function $\phi(x, y, t)$ of the absolute velocity field $\mathbf{u}(x, y, t)$. In terms of these units, $y_\mathrm{s}(x, t)$ represents the interface's shape while the dimensionless tip velocity is denoted by $U(t)$. The effective width of the finger is defined as $\lambda(t) = 1/U$. Moreover, we assume that for any finite time $t > 0$, the finger has a finite length with a triple point T $(x = 1, y = y_\mathrm{T})$ down-stream. This triple point is also a stagnation point of the flow field.

The general unsteady flow in a Hele–Shaw cell, in its dimensionless form, is therefore described by Laplace's equation

$$\nabla^2 \phi = \frac{\partial^2 \phi}{\partial x^2} + \frac{\partial^2 \phi}{\partial y^2} = 0 \tag{9.1}$$

with the following boundary and regularity conditions:

1. Up-stream far-field condition: as $y \to \infty$,

$$\mathbf{u} = \mathbf{e}_y : \quad \frac{\partial \phi}{\partial y} = 1, \quad \frac{\partial \phi}{\partial x} = 0. \tag{9.2}$$

2. The slipping condition on the sidewalls: at $x = \pm 1$,

$$\frac{\partial \phi}{\partial x} = 0 \,. \tag{9.3}$$

3. At the interface: $\tilde{S}(x, y, t) = y - y_{\mathrm{s}}(x, t) = 0$,

   (i) the dynamic condition: according to Darcy's law, the fluid velocity $\mathbf{u}$ is proportional to the gradient of the pressure. Namely, in the dimensional form, we have

$$\mathbf{u} = -\frac{b^2}{12\mu}\nabla p \,, \tag{9.4}$$

   where b is the thickness of the cell. Thus, the potential field ϕ is proportional to the pressure field. So again in the dimensional form, we have

$$\phi = -\frac{b^2 p}{12\mu} \,. \tag{9.5}$$

   Moreover, one can derive from thermodynamics that, due to the presence of the surface tension, the pressure will have a jump across the interface, which is proportional to the curvature of the interface, $\mathcal{K}$ and the surface tension coefficient, γ. Assuming that the pressure p_0 of the less viscous fluid inside the finger is constant and equal to zero, in the dimensionless form it is found that, at the interface,

$$\phi = \varepsilon^2 \mathcal{K}\{y_{\mathrm{s}}(x, t)\} \,; \tag{9.6}$$

   (ii) the kinematic condition: assuming the two fluids are immiscible, the interface is a contact surface of two fluids with no transverse mass flux. Thus, we have

$$\frac{\partial \phi}{\partial n} = V_n = U(t)(\mathbf{e}_y \cdot \mathbf{n}) - \frac{\frac{\partial \tilde{S}}{\partial t}}{|\nabla \tilde{S}|} \,. \tag{9.7}$$

   Here, $\{\mathbf{e}_x, \mathbf{e}_y\}$ are the unit coordinate vectors along the x- and y-axes, respectively; $\mathbf{n}$ is the outward normal vector of the interface and the surface tension parameter of fingering is defined as

$$\varepsilon^2 = \frac{b^2 \gamma}{12\mu U_\infty W^2} \,. \tag{9.8}$$

4. The down-stream triple point condition: at $x = \pm 1, y = y_{\mathrm{T}}(t)$,

$$\mathbf{u} = 0, \quad \text{or} \quad \frac{\partial \phi}{\partial x} = \frac{\partial \phi}{\partial y} = 0 \,. \tag{9.9}$$

   Since the triple point T is a stagnation point, from kinematic considerations it follows that

$$y = y_{\mathrm{T}}(t) = -L - \int_0^t U \mathrm{d}t. \tag{9.10}$$

5. The tip smoothness conditions.

In order to fully determine a time evolving solution one also needs to specify the initial conditions. The above PDE system is quite similar to that of two-dimensional dendrite growth described in Chap. 7. The dynamic condition (9.6) has the same form as the Gibbs–Thomson condition. Hence, it should not be surprising that the solutions for these two systems may display very similar mathematical structures. Note also that the mathematical formulation given above is different from 'the classic Saffman–Taylor problem'. The problem formulated by Saffman and Taylor is a special case of the above system with the triple point T set at infinity. Clearly, the system given here is more realistic and allows a wider class of solutions than the classic Saffman–Taylor problem.

9.3 The Smooth Finger Solution with Zero Surface Tension

For the case of zero surface tension ($\varepsilon = 0$), the system allows a special class of steady finger solution with a constant tip velocity $U_0 = \frac{1}{\lambda_0}$ and the triple point at infinity. For this case, the parameter $0 < \lambda_0 < 1$ also describes the asymptotic width of the finger. Conventionally, this special steady solution is called the Saffman–Taylor solution. However, two years prior to Saffman and Taylor, Zhuravlev published this solution [9.1]. It might therefore be more appropriate to call this solution the Zhuravlev–Saffman–Taylor (ZST) solution. The ZST solution plays the same role for the system of viscous fingering as the Ivantsov solution for dendrite growth. In our coordinate system, the ZST solution is subject to the reduced system

$$\nabla^2 \phi_* = 0 \,, \tag{9.11}$$

with the boundary conditions:

1. As $y \to \infty$, $\ |x| < 1$,

$$\frac{\partial \phi_*}{\partial y} = 1 \,, \qquad \frac{\partial \phi_*}{\partial x} = 0 \,. \tag{9.12}$$

2. As $y \to -\infty$, $\ \lambda_0 < |x| < 1$,

$$\frac{\partial \phi_*}{\partial x} = \frac{\partial \phi_*}{\partial y} = 0 \,. \tag{9.13}$$

3. At $x = \pm 1$,

$$\frac{\partial \phi_*}{\partial x} = 0 \,. \tag{9.14}$$

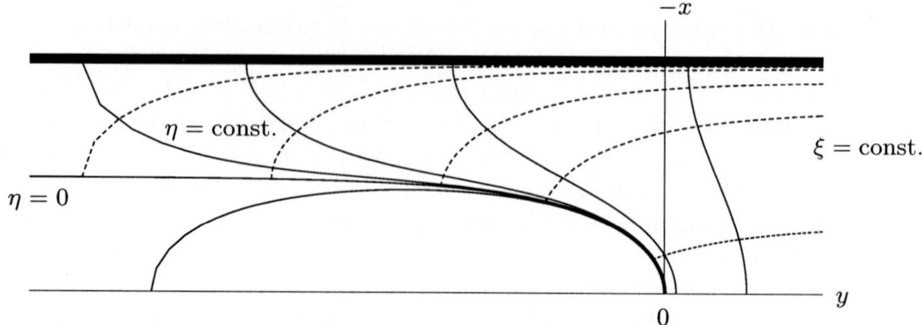

Fig. 9.5. Sketch of the orthogonal curvilinear coordinate system (ξ, η) based on the ZST zero surface tension steady state solutions

4. At the interface, $y = y_*(x)$,

$$\phi_* = 0 \,, \tag{9.15}$$

$$\frac{\partial \phi_*}{\partial y} - y'_* \frac{\partial \phi_*}{\partial x} = \frac{1}{\lambda_0}. \tag{9.16}$$

Hereby, it is assumed that $0 < \lambda_0 = \frac{1}{U_0} < 1$.

One can denote the stream function by $\psi_*(x, y)$ and define the following complex variables:

$$\begin{cases} Z = x + iy \\ W = \phi_* + i\psi_* \,. \end{cases} \tag{9.17}$$

On the interface, from (9.16), one has

$$\frac{\mathrm{d}}{\mathrm{d}x}\left\{\psi_*(x, y_*(x))\right\} = \frac{\partial \psi_*}{\partial x} + y'_* \frac{\partial \psi}{\partial y} = -\left(\frac{\partial \phi_*}{\partial y} - y'_* \frac{\partial \phi}{\partial x}\right) = -\frac{1}{\lambda_0} \,.$$

Therefore, on the interface $\phi_* = 0$, $\psi_* = -\frac{x}{\lambda_0}$ holds. By separation of variables or conformal mapping, one can derive the solution $\{\phi_*(x, y), \psi_*(x, y)\}$ for the above system (9.1)–(9.9).

By defining

$$\begin{cases} \xi = -\psi_* \\ \eta = \phi_* \\ \zeta = \xi + i\eta \,, \end{cases} \tag{9.18}$$

the Zhuravlev–Saffman–Taylor solution can be expressed in terms of the mapping function

$$Z = x + iy = Z(\zeta) = \lambda_0 \zeta + i\frac{2(1 - \lambda_0)}{\pi} \ln \cos\left(\frac{\pi\zeta}{2}\right). \tag{9.19}$$

The variables $\{\xi = \xi(x,y); \eta = \eta(x,y)\}$ constitute a new orthogonal curvilinear coordinate system on the (x,y)-plane as shown in Fig. 9.5:

$$\begin{cases} x = X(\xi, \eta) \\ y = Y(\xi, \eta). \end{cases} \tag{9.20}$$

The Lamé coefficients $\mathcal{G}_1$ and $\mathcal{G}_2$ along the two coordinate curves under the transformation (9.20) are:

$$\mathcal{G}_1 = \mathcal{G}_2 = \mathcal{G} = |Z'(\zeta)| = \sqrt{X_\xi^2 + Y_\xi^2}. \tag{9.21}$$

From the ZST solution, we have

$$\begin{cases} Z'(\zeta) = \lambda_0 - \mathrm{i}(1 - \lambda_0)\tan\left(\frac{\pi\zeta}{2}\right), \qquad Z'(0) = \lambda_0 = \frac{1}{U_0} \\ Z''(\zeta) = -\mathrm{i}\frac{(1-\lambda_0)}{2}\sec^2\left(\frac{\pi\zeta}{2}\right). \end{cases} \tag{9.22}$$

Since on the interface, we have

$$\eta = 0 \quad \text{and} \quad \zeta = \xi = \frac{x}{\lambda_0}, \tag{9.23}$$

the interface shape function is

$$y = \frac{2(1 - \lambda_0)}{\pi} \ln\cos\left(\frac{\pi x}{2\lambda_0}\right). \tag{9.24}$$

Furthermore, on the interface we have

$$X_\xi(\xi, 0) = Y_\eta(\xi, 0) = \lambda_0$$

$$X_\eta(\xi, 0) = -Y_\xi(\xi, 0) = (1 - \lambda_0)\tan\left(\frac{\pi\xi}{2}\right)$$

$$X_{\xi\xi}(\xi, 0) = -X_{\eta\eta}(\xi, 0) = Y_{\xi\eta}(\xi, 0) = 0 \tag{9.25}$$

$$Y_{\xi\xi}(\xi, 0) = -Y_{\eta\eta}(\xi, 0) = -X_{\xi\eta}(\xi, 0) = -\frac{\pi(1 - \lambda_0)}{2}\sec^2\left(\frac{\pi\xi}{2}\right)$$

$$\mathcal{G}_0 = \mathcal{G}(\xi, 0) = \sqrt{\lambda_0^2 + (1 - \lambda_0)^2\tan^2\left(\frac{\pi\xi}{2}\right)}.$$

For any given constant parameter $0 < \lambda_0 < 1$, the ZST solution is an analytic function in the domain $-\infty < y < \infty$, $-1 < x < 1$. The interface of the solution has an infinitely long root, smoothly extending to $y = -\infty$, and has no singularity at the tip.

Saffman obtained a set of analytical solutions for the initial value problem with zero surface tension and a special class of initial conditions. Those time evolving fingers have finite roots and their tip velocities vary with time. Saffman showed that as $t \to \infty$ these solutions converge to the ZST steady solutions [9.5].

9.4 Formulation of the General Problem in Curvilinear Coordinates (ξ, η) and the Basic State Solutions

To study the general unsteady flow in a Hele–Shaw cell with $\varepsilon \neq 0$, we adopt a curvilinear coordinate system (ξ, η). This curvilinear system plays the same role as the paraboloidal coordinate system does for the problem of dendrite growth. In this coordinate system, the ZST solution has the simple form

$$\phi_* = \eta, \quad \eta_* = 0. \tag{9.26}$$

We denote the potential function of a general unsteady flow by $\phi = \phi(\xi, \eta, t)$ and the interface shape by $\eta = \eta_\mathrm{s}(\xi, t)$. The system (9.1)–(9.9) is then rewritten as

$$\frac{\partial^2 \phi}{\partial \xi^2} + \frac{\partial^2 \phi}{\partial \eta^2} = 0 \tag{9.27}$$

with the following boundary conditions:

1. As $\eta \to \infty$,

$$\frac{\partial \phi}{\partial \eta} = 1. \tag{9.28}$$

2. At $\xi = \pm 1$,

$$\frac{\partial \phi}{\partial \xi} = 0. \tag{9.29}$$

3. At $\eta = \eta_\mathrm{s}(\xi, t)$,

$$\phi = \varepsilon^2 \mathcal{K}\Big\{ \eta_\mathrm{s}(\xi, t) \Big\} \tag{9.30}$$

and

$$\frac{\partial \phi}{\partial \eta} - \eta_\mathrm{s}' \frac{\partial \phi}{\partial \xi} = U_0 \big(Y_\eta - \eta_\mathrm{s}' Y_\xi \big) + \mathcal{G}^2 \Big(\frac{\partial \eta_\mathrm{s}}{\partial t} \Big). \tag{9.31}$$

Here, the prime represents the derivative with respect to ξ, $U_0 = \frac{1}{\lambda_0}$ is the tip velocity of the ZST solution observed in the laboratory frame, and the curvature operator is

$$\mathcal{K}\Big[\eta_\mathrm{s}(\xi, t) \Big] = -\frac{1}{\mathcal{G}(\xi, \eta_\mathrm{s})} \Bigg[\frac{\eta_\mathrm{s}''}{\big(1 + \eta_\mathrm{s}'^2 \big)^{3/2}} + \frac{\Pi_0(\xi, \eta_\mathrm{s})}{\mathcal{G}^2(\xi, \eta_\mathrm{s}) \big(1 + \eta_\mathrm{s}'^2 \big)^{1/2}}$$

$$- \frac{\Pi_1(\xi, \eta_\mathrm{s})}{\mathcal{G}^2(\xi, \eta_\mathrm{s}) \big(1 + \eta_\mathrm{s}'^2 \big)^{1/2}} \eta_\mathrm{s}' \Bigg], \tag{9.32}$$

where

$$\Pi_0(\xi, \eta) = \left(Y_{\xi\xi}X_\xi - X_{\xi\xi}Y_\xi\right)$$
$$\Pi_1(\xi, \eta) = \left(Y_{\xi\xi}Y_\xi + X_{\xi\xi}X_\xi\right) \tag{9.33}$$

4. At the triple point T, $\xi = \pm 1, \eta = \eta_T(t) \approx e^{\frac{\pi y_T}{1-\lambda_0}}$,

$$\frac{\partial \phi}{\partial \eta} = \frac{\partial \phi}{\partial \xi} = 0. \tag{9.34}$$

This system is highly nonlinear due to the interface conditions (9.30). As before, we consider the 'nearly' steady finger solutions when $t \gg 1$. Thus we employ the slow time variable defined as

$$\tau = \varepsilon(t - t_0) \quad (t_0 \gg 1), \tag{9.35}$$

and assume that the solutions $\{\phi_B, \eta_B\}$ are functions of the variables (ξ, η, τ). In terms of this slow variable τ, the tip velocity $U(t)$ is written

$$U(t) = U(\tau, \varepsilon), \tag{9.36}$$

and the location of the triple point T is expressed in the (x, y)-coordinate system as

$$x_T = \pm 1; \quad y_T(\tau) = -\frac{1}{\varepsilon}\left(\bar{L} + \int_0^\tau U(\tau, \varepsilon)d\tau\right) = \frac{\bar{y}_T(\tau, \varepsilon)}{\varepsilon}, \tag{9.37}$$

where $\bar{L} = O(1)$. In the (ξ, η)-coordinate system, the triple point is expressed as

$$\xi = \xi_T = \pm 1; \quad \eta = \bar{\eta}_T(\tau) \approx e^{\frac{\pi \bar{y}_T}{\varepsilon(1-\lambda_0)}} \ll 1. \tag{9.38}$$

The down-stream triple point condition (9.9) and the kinematic interface condition (9.7) can therefore be transformed into the following forms:

$$\frac{\partial \phi}{\partial \xi} = \frac{\partial \phi}{\partial \eta} = 0 \tag{9.39}$$

at $\xi = \pm 1, \eta = \bar{\eta}_T(\tau)$, and

$$\frac{\partial \phi_B}{\partial \eta} - \eta_B'\frac{\partial \phi_B}{\partial \xi} = U_0\left(Y_\eta - \eta_s'Y_\xi\right) + \varepsilon\mathcal{G}^2\frac{\partial \eta_B}{\partial \tau} \tag{9.40}$$

at $\eta = \eta_s(\xi, \tau)$.

It is quite evident that the ZST solution is the solution for the above system at $\varepsilon = 0$. Finding the solutions $q_B =: \{\phi_B, \eta_B\}$ for $0 < \varepsilon \ll 1$ is a

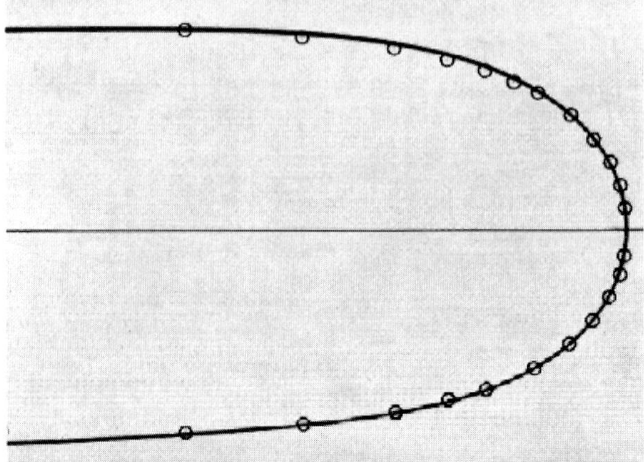

Fig. 9.6. Comparison of a realistic smooth finger with the ZST solution with a properly selected width λ_0. The solid line is the experimental curve, while the dots are calculated from the ZST solution

singular perturbation problem. Letting $\varepsilon \to 0$, one can make the following regular perturbation expansion (RPE):

$$\phi_{\mathrm{B}}(\xi,\eta,\tau,\varepsilon) \sim \phi_* + \varepsilon^2 \phi_1(\xi,\eta) + \varepsilon^2 \phi_2(\xi,\eta) + \cdots, \tag{9.41}$$

$$\eta_{\mathrm{B}}(\xi,\tau,\varepsilon) \sim \eta_* + \varepsilon^2 \eta_1(\xi) + \varepsilon^2 \eta_2(\xi) + \cdots, \tag{9.42}$$

or

$$q_{\mathrm{B}}(\xi,\eta,\tau,\varepsilon) \sim q_* + \varepsilon^2 q_1(\xi,\eta) + \varepsilon^2 q_2(\xi,\eta) + \cdots, \tag{9.43}$$

where the first term q_* is the ZST solution. By substituting the expansion into the system (9.27)–(9.40), one can find terms ϕ_n, η_n. These solutions are all time independent and, at $\xi = \pm 1$, the solutions η_n all vanish, i.e.,

$$\eta_n(\pm 1) = 0 \qquad (n = 1, 2, 3, \cdots). \tag{9.44}$$

Therefore, the solution q_{B} is expected to have the structure

$$q_{\mathrm{B}} = q_* + (\mathcal{R}_N) + (\mathcal{S}_N), \tag{9.45}$$

where $\mathcal{R}_N$ is the partial sum of first N terms in the RPE and the remaining part $\mathcal{S}_N$ may depend on the slow time variable τ. Given τ, for any large number N, $\mathcal{S}_N \ll \varepsilon^{2N}$ as $\varepsilon \to 0$. Therefore this time-dependent part is smaller beyond all orders. In this sense, the solutions q_{B} are called the 'nearly' steady solutions, which we shall define as the basic state solutions for stability analysis.

The exact form of these basic state solutions is unknown, but this is not important. Our goal is to study the stability of these solutions. Here what is most important to us is that these basic state solutions exist, and that for any fixed $0 < \tau < \infty$, in the outer region away from the triple point, they are close to the ZST solutions, namely

$$
\begin{cases}
\phi_{\mathrm{B}}(\xi, \eta, \tau, \varepsilon) = \phi_*(\eta) + O(\varepsilon^2) \\[2mm]
\eta_{\mathrm{B}}(\xi, \tau, \varepsilon) = O(\varepsilon^2) \\[2mm]
\bar{U}(\tau, \varepsilon) = U_0 + O(\varepsilon^2) \\[2mm]
\lambda(\tau, \varepsilon) = \lambda_0 + O(\varepsilon^2) \,.
\end{cases}
\tag{9.46}
$$

The argument that the basic state solution can be approximated by a ZST solution with a high accuracy is well supported by the experimental evidence. For instance, the classic experiments conducted by Saffman and Taylor have shown that the fingers are very close to a ZST solution, once the width λ_0 is properly selected (see Fig 9.6 and refer to [9.4]). The higher order small error terms in (9.46) only have higher order small effects on the calculations of the eigenvalues and eigenmodes which will be carried out in the following sections.

9.5 The Linear Perturbed System and the Outer Solutions

To develop a linear stability theory we consider perturbations around the basic state induced by infinitesimal initial perturbations characterized by the parameter $\delta \ll 1$. The unsteady solution is written as

$$
\phi(\xi, \eta, t) = \phi_{\mathrm{B}} + \tilde{\phi}(\xi, \eta, t)
$$
$$
\eta_{\mathrm{s}}(\xi, t) = \eta_{\mathrm{B}} + \tilde{h}(\xi, t) \,.
\tag{9.47}
$$

In terms of the parameter $\delta \ll 1$, we linearize the nonlinear interface conditions around the basic state. The homogeneous linear perturbed system is

$$
\nabla^2 \tilde{\phi} = \frac{\partial^2 \tilde{\phi}}{\partial \xi^2} + \frac{\partial^2 \tilde{\phi}}{\partial \eta^2} = 0
\tag{9.48}
$$

with the following boundary conditions:

1. As $\eta \to \infty$,

$$
\frac{\partial \tilde{\phi}}{\partial \eta} = \frac{\partial \tilde{\phi}}{\partial \xi} = 0 \,.
\tag{9.49}
$$

2. At $\xi = \pm 1$,

$$
\frac{\partial \tilde{\phi}}{\partial \xi} = 0 \,.
\tag{9.50}
$$

3. On the interface, $\eta = \eta_{\mathrm{B}}$,

$$\tilde{\phi} + \tilde{h} = -\frac{\varepsilon^2}{\hat{\mathcal{G}}(\xi)}\left[\frac{\partial^2 \tilde{h}}{\partial \xi^2} - \frac{\hat{\Pi}_1(\xi)}{\hat{\mathcal{G}}^2(\xi)}\frac{\partial \tilde{h}}{\partial \xi}\right.$$

$$\left. + \left(\frac{\partial \Pi_0(\xi,\eta_B)}{\partial \eta} - \frac{2\hat{\Pi}_0(\xi)\mathcal{G}_\eta(\xi,\eta_B)}{\hat{\mathcal{G}}(\xi)}\right)\frac{\tilde{h}}{\hat{\mathcal{G}}^2(\xi)}\right], \quad (9.51)$$

$$\frac{\partial \tilde{\phi}}{\partial \eta} + U_0 \hat{Y}_\xi(\xi)\frac{\partial \tilde{h}}{\partial \xi} - \hat{\mathcal{G}}^2(\xi)\frac{\partial \tilde{h}}{\partial t} - U_0 \hat{Y}_{\eta\eta}(\xi)\tilde{h} = 0 , \quad (9.52)$$

where

$$\begin{cases} \hat{\mathcal{G}}(\xi) = \mathcal{G}(\xi,\eta_B), \quad \hat{Y}_\xi(\xi) = Y_\xi(\xi,\eta_B), \quad \hat{Y}_{\eta\eta}(\xi) = Y_{\eta\eta}(\xi,\eta_B) \\ \hat{\Pi}_0(\xi) = \Pi_0(\xi,\eta_B), \quad \hat{\Pi}_1(\xi) = \Pi_1(\xi,\eta_B). \end{cases} \quad (9.53)$$

The left-hand side of (9.53) actually depends on the variables ξ, τ, and ε but, for the sake of simplicity, we only explicitly show the variable ξ.

4. At the triple point, $\xi = \pm 1$, $\eta = \eta_{\mathrm{T}}$,

$$\tilde{h} = \frac{\partial \tilde{\phi}}{\partial \xi} = \frac{\partial \tilde{\phi}}{\partial \eta} = 0 . \quad (9.54)$$

5. At the tip, $\xi = 0$,

$$\frac{\partial}{\partial \xi}\left\{\tilde{\phi},\ \tilde{h}\right\} = 0 \quad \text{for a symmetrical mode (S-mode);} \quad (9.55)$$

and

$$\left\{\tilde{\phi},\tilde{h}\right\} = 0 \quad \text{for an anti-symmetrical mode (A-mode).} \quad (9.56)$$

The above system contains the two parameters ε and λ_0. It leads to a linear eigenvalue problem, as one looks for the following type of solutions:

$$\tilde{\phi} = \hat{\phi}(\xi,\eta,\tau,\varepsilon)e^{\sigma t}$$
$$\tilde{h} = \hat{h}(\xi,\tau,\varepsilon)e^{\sigma t}. \quad (9.57)$$

The eigenvalues $\sigma = \sigma_{\mathrm{R}} - i\omega$ must be a function of (λ_0,ε).

One only needs to solve the problem in the half plane $\xi \geq 0$. The solution in the other half plane $\xi < 0$ can then be derived from the symmetry or antisymmetry. As before, we shall solve this eigenvalue problem in two steps. As the first step, we solve the system (9.48)–(9.54) for any given parameters $(\sigma,\lambda_0,\varepsilon)$. For this purpose, we shall look for the uniformly valid asymptotic solution to (9.48)–(9.54) with fixed parameters (σ,λ_0) in the limit $\varepsilon \to 0$. The solutions obtained satisfy all the boundary conditions except the tip condition (9.55) or (9.56). Then, as the second step, we apply the tip conditions. This determines the eigenvalue $\sigma = \sigma(\lambda_0,\varepsilon)$.

To obtain asymptotic solutions, we introduce a set of stretched fast variables $\{\xi_+, \eta_+, t_+\}$ similar to those used for dendrite growth:

$$\xi_+ = \frac{1}{\varepsilon} \int_{\xi_0}^{\xi} k(\xi', \eta, \tau, \varepsilon) \mathrm{d}\xi'$$

$$\eta_+ = \frac{1}{\varepsilon} \int_0^{\eta} g(\xi, \eta', \tau, \varepsilon) \mathrm{d}\eta' \tag{9.58}$$

$$t_+ = \frac{t}{\varepsilon}.$$

Furthermore, we assume that

$$\begin{cases} \dfrac{\partial k}{\partial \eta} = m \dfrac{\partial k}{\partial \xi} \\[2mm] \dfrac{\partial g}{\partial \eta} = s \dfrac{\partial g}{\partial \xi}. \end{cases} \tag{9.59}$$

In terms of these multiple variables $(\xi, \eta, \tau, \xi_+, \eta_+, t_+)$, we make the following multiple variables expansion for the perturbed state:

$$\tilde{\phi} = \left\{ \tilde{\phi}_0(\xi, \eta, \tau, \xi_+, \eta_+) + \varepsilon \tilde{\phi}_1(\xi, \eta, \tau, \xi_+, \eta_+) + \cdots \right\} e^{\sigma t_+}$$

$$\tilde{h} = \left\{ \tilde{h}_0(\xi, \tau, \xi_+) + \varepsilon \tilde{h}_1(\xi, \tau, \xi_+) + \cdots \right\} e^{\sigma t_+}$$

$$k(\xi, \tau, \varepsilon) = k_0 + \varepsilon k_1 + \varepsilon^2 k_2 + \cdots$$

$$g(\xi, \tau, \varepsilon) = k_0 + \varepsilon k_1 + \varepsilon^2 g_2 + \cdots \tag{9.60}$$

$$m(\xi, \eta, \varepsilon) = m_0 + \varepsilon m_1 + \varepsilon^2 m_2 + \cdots$$

$$s(\xi, \eta, \varepsilon) = m_0 + \varepsilon s_1 + \varepsilon^2 s_2 + \cdots$$

$$\sigma(\tau, \varepsilon) = \sigma_0 + \varepsilon \sigma_1 + \varepsilon^2 \sigma_2 \cdots.$$

Hereby, one can assume that m_0 is a constant.

At the interface $\eta = 0$, $\eta_+ = 0$, we shall write

$$\xi_+ = \hat{\xi}_+ = \frac{1}{\varepsilon} \int_{\xi_0}^{\xi} \hat{k}(\xi', \tau, \varepsilon) \mathrm{d}\xi'$$

$$k(\xi, 0, \tau, \varepsilon) = \hat{k}(\xi, \tau, \varepsilon). \tag{9.61}$$

Clearly, as $\varepsilon \to 0$, we have

$$\hat{k}(\xi, \tau, \varepsilon) = \hat{k}_0 + \varepsilon \hat{k}_1 + \varepsilon^2 \hat{k}_2 + \cdots, \tag{9.62}$$

$$\hat{k}_i(\xi) = k_i(\xi, 0) \quad (i = 0, 1, 2, \cdots). \tag{9.63}$$

One can convert the linear system (9.48)–(9.52) into a system with multiple variables and successively derive each order of approximation, as we did for the problem of dendrite growth in the previous sections.

For the readers' convenience, we write down the converted multiple variables system below:

$$k^2 \frac{\partial^2 \tilde{\phi}}{\partial \xi_+^2} + g^2 \frac{\partial^2 \tilde{\phi}}{\partial \eta_+^2} = -2\varepsilon \left(k \frac{\partial^2 \tilde{\phi}}{\partial \xi \partial \xi_+} + g \frac{\partial^2 \tilde{\phi}}{\partial \eta \partial \eta_+} \right)$$

$$-\varepsilon \left(\frac{\partial k}{\partial \xi} \frac{\partial \tilde{\phi}}{\partial \xi_+} + \frac{\partial g}{\partial \eta} \frac{\partial \tilde{\phi}}{\partial \eta_+} \right) - \varepsilon^2 \left(\frac{\partial^2 \tilde{\phi}}{\partial \xi^2} + \frac{\partial^2 \tilde{\phi}}{\partial \eta^2} \right) \qquad (9.64)$$

$$(0 \leq \xi < 1, \ 0 \leq \eta, \xi_+, \eta_+ < \infty)$$

with the boundary conditions:

1. In the up-stream far-field, as $\eta_+ \to \infty$ and $\eta \to \infty$,

$$\frac{\partial \tilde{\phi}}{\partial \eta_+} = O(\varepsilon), \qquad \frac{\partial \tilde{\phi}}{\partial \xi_+} = O(\varepsilon) . \qquad (9.65)$$

2. At the sidewall, $\xi = 1$ and $\xi_+ \to \infty$,

$$\frac{\partial \tilde{\phi}}{\partial \xi_+} = O(\varepsilon) . \qquad (9.66)$$

3. On the interface, $\eta_+ = \eta = 0$,

$$\tilde{\phi} + \tilde{h} = -\frac{1}{\hat{\mathcal{G}}(\xi)} \left\{ k^2 \frac{\partial^2 \tilde{h}}{\partial \xi_+^2} + 2\varepsilon k \frac{\partial^2 \tilde{\phi}}{\partial \xi \partial \xi_+} + \varepsilon \frac{\partial k}{\partial \xi} \frac{\partial \tilde{\phi}}{\partial \xi_+} + \varepsilon^2 \frac{\partial^2 \tilde{h}}{\partial \xi^2} \right.$$

$$\left. - \varepsilon \frac{\hat{\Pi}_1(\xi)}{\hat{\mathcal{G}}^2(\xi)} \left(k \frac{\partial \tilde{h}}{\partial \xi_+} + \frac{\partial \tilde{h}}{\partial \xi} \right) + \varepsilon^2 \left(\frac{\partial \hat{\Pi}_0(\xi)}{\partial \eta} - \frac{2\hat{\Pi}_0(\xi)\hat{\mathcal{G}}_\eta(\xi)}{\hat{\mathcal{G}}(\xi)} \right) \frac{\tilde{h}}{\hat{\mathcal{G}}^2(\xi)} \right\},$$

$$+ \text{ (higher order terms)} , \qquad (9.67)$$

$$g \frac{\partial \tilde{\phi}}{\partial \eta_+} + \varepsilon \frac{\partial \tilde{\phi}}{\partial \eta} + U_0 \hat{Y}_\xi(\xi) \left(k \frac{\partial \tilde{h}}{\partial \xi_+} + \varepsilon \frac{\partial \tilde{h}}{\partial \xi} \right) - \hat{\mathcal{G}}^2(\xi) \frac{\partial \tilde{h}}{\partial t_+}$$

$$-\varepsilon U_0 \hat{Y}_{\eta\eta}(\xi)\tilde{h} = \text{(higher order terms)} . \qquad (9.68)$$

4. At the triple point, $\xi = 1$, $\xi_+ \to \infty$ and $\eta \to 0$,

$$\tilde{h} = 0, \qquad \frac{\partial \tilde{\phi}}{\partial \xi_+} = O(\varepsilon), \qquad \frac{\partial \tilde{\phi}}{\partial \eta_+} = O(\varepsilon) . \qquad (9.69)$$

5. At the tip, $\xi = \xi_+ = 0$,

$$\frac{\partial}{\partial \xi_+} \left\{ \tilde{\phi}, \ \tilde{h} \right\} = 0 \qquad \text{for a symmetrical mode (S-mode);} \qquad (9.70)$$

and

$$\left\{ \tilde{\phi}, \tilde{h} \right\} = 0 \qquad \text{for an anti-symmetrical mode (A-mode).} \qquad (9.71)$$

Later, for convenience, we shall employ a new variable ρ defined by

$$\rho = -U_0 \hat{Y}_\xi(\xi), \tag{9.72}$$

rather than using ξ. The variable ρ is connected with the coordinate y along the interface of the basic state. It is also related to the arclength $\ell(\xi)$ measured along the interface of the basic state starting from the tip and is determined by

$$\ell = \int_0^\xi \hat{\mathcal{G}}(\xi_1) \mathrm{d}\xi_1 . \tag{9.73}$$

One can derive

$$\xi = \int_0^\rho \frac{\mathrm{d}\rho_1}{G(\rho_1)} \tag{9.74}$$

and

$$\ell = \int_0^\rho F(\rho_1)\mathrm{d}\rho_1 , \tag{9.75}$$

where

$$G(\rho) = -U_0 \hat{Y}_{\xi\xi}(\xi), \quad F(\rho) = -\frac{\hat{\mathcal{G}}(\xi)}{U_0 \hat{Y}_{\xi\xi}(\xi)} . \tag{9.76}$$

As ξ goes from 0 to 1, ℓ goes from 0 to L_T, whereas ρ goes from 0 to ρ_T. Note that $L_T = \int_0^{\rho_T} F(\rho)\mathrm{d}\rho < \infty$ is the total arclength of the finger interface of the basic state.

The coefficients of the above linear perturbed equation, $\hat{Y}_\xi(\xi,\tau,\varepsilon)$, $\hat{Y}_{\xi\xi}(\xi,\tau,\varepsilon)$, and $\hat{\mathcal{G}}(\xi,\tau,\varepsilon)$, are determined by the basic state. Due to (9.46), as $\varepsilon \to 0$

$$\hat{Y}_\xi(\xi,\tau,\varepsilon) = Y_\xi(\xi,0) + O(\varepsilon^2)$$

$$\hat{\mathcal{G}}(\xi,\tau,\varepsilon) = \mathcal{G}(\xi,0) + O(\varepsilon^2)$$

$$\hat{\Pi}_0(\xi,\tau,\varepsilon) = \Pi_0(\xi,0) + O(\varepsilon^2) \tag{9.77}$$

$$\hat{\Pi}_1(\xi,\tau,\varepsilon) = \Pi_1(\xi,0) + O(\varepsilon^2) .$$

In the zeroth-order approximation, we obtain

$$\left(\frac{\partial^2}{\partial \xi_+^2} + \frac{\partial^2}{\partial \eta_+^2} \right) \tilde{\phi}_0 = 0 \tag{9.78}$$

or

$$\mathrm{i}\frac{\partial \tilde{\phi}_0}{\partial \xi_+} = \frac{\partial \tilde{\phi}_0}{\partial \eta_+} \tag{9.79}$$

$$(0 \le \xi \le 1, \ 0 \le \eta, \xi_+, \eta_+ < \infty)$$

with the boundary conditions:

1. As $\eta_+ \to \infty$,

$$\frac{\partial \tilde{\phi}_0}{\partial \eta_+} = \frac{\partial \tilde{\phi}_0}{\partial \xi_+} = 0 \,. \tag{9.80}$$

2. At $\xi = 1$ and $\xi_+ \to \infty$,

$$\frac{\partial \tilde{\phi}_0}{\partial \xi_+} = 0 \,. \tag{9.81}$$

3. At $\eta = \eta_+ = 0$,

$$\tilde{\phi}_0 + \tilde{h}_0 = -\frac{\hat{k}_0^2}{\mathcal{G}_0} \frac{\partial^2 \tilde{h}_0}{\partial \hat{\xi}_+^2} \,, \tag{9.82}$$

$$\hat{k}_0 \frac{\partial \tilde{\phi}_0}{\partial \eta_+} + U_0 \hat{k}_0 Y_{\xi,0}(\xi) \frac{\partial \tilde{h}_0}{\partial \hat{\xi}_+} - \sigma_0 \mathcal{G}_0{}^2(\xi) \tilde{h}_0 = 0 \,, \tag{9.83}$$

where

$$Y_{\xi,0}(\xi) = Y_\xi(\xi,0), \quad \mathcal{G}_0(\xi) = \mathcal{G}(\xi,0) \,.$$

4. At $\xi = 1, \eta = 0$, and $\xi_+ \to \infty$,

$$\left\{ \tilde{\phi}_0; \ \tilde{h}_0 \right\} = 0 \,. \tag{9.84}$$

We consider the mode solutions for the above system:

$$\tilde{\phi}_0 = A_0(\xi,\eta) \ \exp\left\{ i\xi_+ - \eta_+ \right\}$$

$$\tilde{\psi}_0 = B_0(\xi,\eta) \ \exp\left\{ i\xi_+ - \eta_+ \right\} \tag{9.85}$$

$$\tilde{h}_0 = \hat{D}_0 \ \exp\left\{ i\hat{\xi}_+ \right\} \,.$$

The coefficient $\hat{D}_0$ in the zeroth-order approximation is a constant. To satisfy the interface conditions (9.82) and (9.83) at $\eta = 0$, the wavenumber function $\hat{k}_0(\xi)$ is subject to the local dispersion relation

$$\mathcal{G}_0^2 \sigma_0 = i\hat{k}_0 U_0 Y_{\xi,0}(\xi) + \hat{k}_0 \left(1 - \frac{\hat{k}_0^2}{\mathcal{G}_0} \right). \tag{9.86}$$

We now change from the variable ξ to ρ. To leading order

$$\rho = -\frac{Y_{\xi,0}(\xi)}{\lambda_0} = \left(\frac{1 - \lambda_0}{\lambda_0} \right) \tan\left(\frac{\pi\xi}{2} \right). \tag{9.87}$$

Accordingly,

$$\mathcal{G}_0(\xi) = \lambda_0 S(\rho)$$

$$G(\rho) = \frac{\pi}{2} \frac{\lambda_0}{1 - \lambda_0} \left[\rho^2 + \left(\frac{1 - \lambda_0}{\lambda_0} \right)^2 \right] \tag{9.88}$$

$$F(\rho) = \frac{\lambda_0 S(\rho)}{G(\rho)} \,,$$

where

$$S(\rho) = \sqrt{1 + \rho^2} \,. \tag{9.89}$$

Moreover, in terms of ρ, the normal mode solution is expressed in the form

$$\tilde{h}_0 = \tilde{h}_0(\rho) = \hat{D}_0 \exp\left\{ \frac{\mathrm{i}}{\varepsilon} \int_{\rho_0}^{\rho} \tilde{k}_0 \mathrm{d}\rho \right\}, \tag{9.90}$$

where $\hat{k}_0(\xi) = \tilde{k}_0(\rho)G(\rho)$ and $\rho_0 = \rho(\xi_0)$. The corresponding local dispersion relation is then transformed to

$$\sigma_0 = \Sigma_*(\rho, \tilde{k}_0) = \frac{G(\rho)\tilde{k}_0}{\lambda_0^2 S^2(\rho)} \left\{ (1 - \mathrm{i}\rho) - \frac{G^2(\rho)\tilde{k}_0^2}{\lambda_0 S(\rho)} \right\}. \tag{9.91}$$

The local dispersion relation can be written in the form

$$\sigma_e = \Sigma(\rho, k_e) = \frac{k_e}{S^2(\rho)} \left(1 - \mathrm{i}\rho - \frac{k_e^2}{S(\rho)} \right), \tag{9.92}$$

provided one sets

$$k_e = \frac{G(\rho)}{\lambda_0^{\frac{1}{2}}} \tilde{k}_0, \quad \text{and} \quad \sigma_e = \lambda_0^{\frac{3}{2}} \sigma_0 \,. \tag{9.93}$$

Notice that this form is exactly as that for dendrite growth.

For any given constant σ_0, one can find three roots of (9.92) (see Fig. 9.7), namely

$$\left\{ \begin{array}{ll} k_e^{(1)}(\rho) = M(\rho) \cos\left\{ \frac{1}{3} \cos^{-1}\left(\frac{\sigma_e}{N(\rho)} \right) \right\} & \text{(short-wavelength branch)} \\[2ex] k_e^{(2)}(\rho) = M(\rho) \cos\left\{ \frac{1}{3} \cos^{-1}\left(\frac{\sigma_e}{N(\rho)} \right) + \frac{2\pi}{3} \right\} & \tag{9.94} \\[2ex] k_e^{(3)}(\rho) = M(\rho) \cos\left\{ \frac{1}{3} \cos^{-1}\left(\frac{\sigma_e}{N(\rho)} \right) + \frac{4\pi}{3} \right\} & \text{(long-wavelength branch)} \,, \end{array} \right.$$

where $M(\rho) = \sqrt{\frac{4S(\rho)}{3}} \left(1 - \mathrm{i}\rho \right)^{\frac{1}{2}}$, and $N(\rho) = -\frac{M(\rho)}{3S^2(\rho)} \left(1 - \mathrm{i}\rho \right)$.

In order for the potential $\hat{\phi}_0$ to satisfy the boundary condition (9.80), $\mathrm{Re}\{\tilde{k}_0\} > 0$ is needed. Consequently, only $\{k_e^{(1)}(\rho), k_e^{(3)}(\rho)\}$ in (9.94) are meaningful. Thus, the general solution in the outer region is:

$$\tilde{h} = D_1 H_1(\rho) + D_3 H_3(\rho)$$

$$= D_1 \exp\left\{ \frac{\mathrm{i}}{\varepsilon} \int_{\rho_0}^{\rho} \left(\tilde{k}_0^{(1)} + \varepsilon\tilde{k}_1^{(1)} + \cdots \right) \mathrm{d}\rho_1 \right\}$$

$$+ D_3 \exp\left\{ \frac{\mathrm{i}}{\varepsilon} \int_{\rho_0}^{\rho} \left(\tilde{k}_0^{(3)} + \varepsilon\tilde{k}_1^{(3)} + \cdots \right) \mathrm{d}\rho_1 \right\}, \tag{9.95}$$

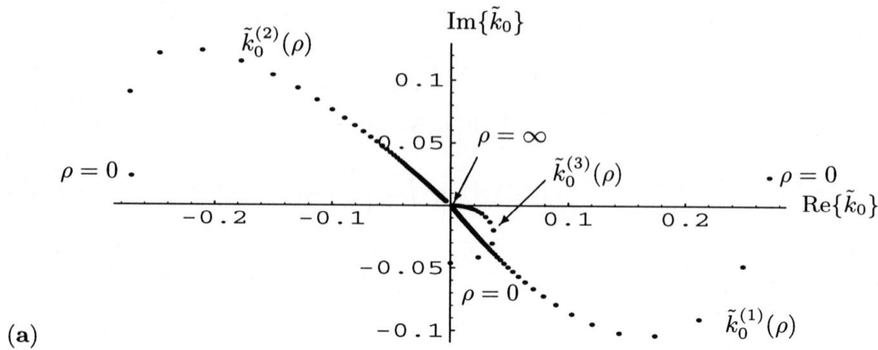

(a)

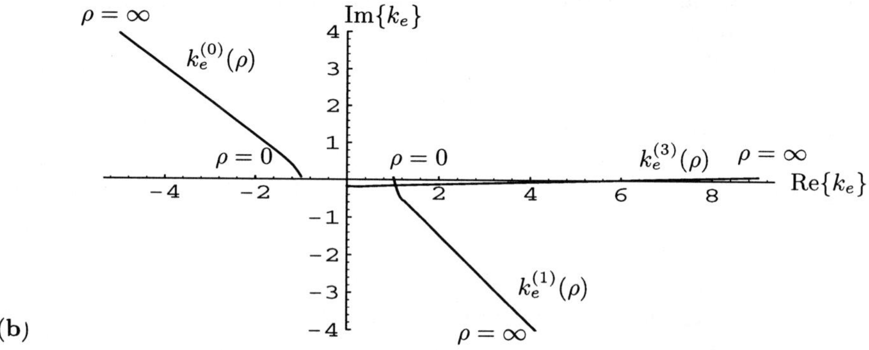

(b)

Fig. 9.7. The distribution of curves of the wavenumber functions for a typical case, $\sigma_0 = (0, -0.7)$, $\lambda_0 = 0.4$: (a) the variation of $\left\{ \tilde{k}_0^{(1)}, \tilde{k}_0^{(2)}, \tilde{k}_0^{(3)} \right\}$ with ρ in the complex $\tilde{k}_0$-plane; (b) the variation of $\left\{ k_e^{(1)}, k_e^{(2)}, k_e^{(3)} \right\}$ with ρ in the complex k_e-plane

where the constants D_1 and D_3 are to be determined. As we have seen in the case of dendrite growth, the MVE solution (9.95) is not uniformly valid in the entire complex ρ (or ξ)-plane. There is a singularity for this solution at a point ρ_c (or ξ_c). This singularity is not yet evident in the leading order approximation (9.95), but it can be clearly seen in higher order approximations. We shall not show the results of the first-order approximation solution. The derivation is very similar to that in Chap. 6. The singular point ρ_c satisfies the equation

$$\frac{\partial \Sigma_*(\rho, \tilde{k}_0)}{\partial \tilde{k}_0} = 0 \quad \text{or} \quad \left(\tilde{k}_0^{(1)}(\xi) - \tilde{k}_0^{(3)}(\xi) \right) = 0 \,. \tag{9.96}$$

Combining (9.96) and (9.91), one finds that ρ_c is also the root of

$$\sigma_0 = e^{-i\frac{3\pi}{4}} \sqrt{\frac{4}{27\lambda_0^3}} \left(\frac{\rho_c + i}{\rho_c - i}\right)^{\frac{3}{4}}. \tag{9.97}$$

As before, we choose ρ_c (corresponding to ξ_c) as the lower limit of the integrals in (9.95).

For the finger system, we still have $\mathrm{Im}\{\tilde{k}_0^{(1)}\} < 0$ and $\mathrm{Im}\{\tilde{k}_0^{(3)}\} > 0$, as $\rho \gg 1$. So, as $\rho \to \infty$, H_1 increases exponentially, whereas H_3 decreases exponentially. To satisfy the root condition (9.84) in the far fields, the solution must be approximated by the subdominant function H_3. Namely,

$$\tilde{h}_0 \sim D H_3, \tag{9.98}$$

where the constant D is proportional to the characteristic amplitude of the initial perturbation δ.

As we demonstrated in Chap. 6, the asymptotic form (9.98) cannot be applied to the entire complex ρ-plane, due to the Stokes phenomenon. The structure of the Stokes lines for the finger system will be shown later. It will be seen that, similar to dendrite growth, one anti-Stokes line (A_2) in the complex ρ (or ξ)-plane crosses the real axis at the point ρ'_c, and divides the whole complex plane into sector (S_1) and sector (S_2). For a uniformly valid solution, the coefficients $\{D_1, D_3\}$ in (9.95) will, in general, be different in different sectors. We denote the coefficients of the solution (9.95), in (S_1), by $\{D_1, D'_3\}$, and in (S_2) by $\{D'_1 = 0, D'_3 = D\}$. To determine the connection condition for these two pairs of constants, one needs to derive the inner solution in the inner region near the singular point ρ_c and match the inner solution with the outer region.

9.6 The Inner Equation near the Singular Point ξ_c

In order to obtain the asymptotic expansion in the inner region $|\xi - \xi_c| \ll 1$, $|\eta| \ll 1$, we introduce the inner variables

$$\xi_* = \frac{\xi - \xi_c}{\varepsilon^\alpha},$$
$$\eta_* = \frac{\eta}{\varepsilon^\alpha}, \tag{9.99}$$

and denote the inner solution by

$$\tilde{h}(\xi, t) = \varepsilon^\alpha \hat{h}(\xi_*, t)$$
$$\tilde{\phi}(\xi, \eta, t) = \varepsilon^\alpha \hat{\phi}(\xi_*, \eta_*, t) \tag{9.100}$$
$$\tilde{\psi}(\xi, \eta, t) = \varepsilon^\alpha \hat{\psi}(\xi_*, \eta_*, t)$$

where α is to be determined. We make the following inner expansion:

$$\hat{\phi}(\xi_*, \eta_*, t) = \left[\nu_0(\varepsilon)\hat{\phi}_0(\xi_*, \eta_*) + \nu_1(\varepsilon)\hat{\phi}_1(\xi_*, \eta_*) + \ldots\right]e^{\frac{\sigma t}{\varepsilon}}$$

$$\hat{\psi}(\xi_*, \eta_*, t) = \left[\nu_0(\varepsilon)\hat{\psi}_0(\xi_*, \eta_*) + \nu_1(\varepsilon)\hat{\psi}_1(\xi_*, \eta_*) + \ldots\right]e^{\frac{\sigma t}{\varepsilon}} \qquad (9.101)$$

$$\hat{h} = \left[\nu_0(\varepsilon)\hat{h}_0 + \nu_1(\varepsilon)\hat{h}_1 + \cdots\right]e^{\frac{\sigma t}{\varepsilon}}\,.$$

In terms of the inner variables, (9.99) is expressed in the form:

$$\left(\frac{\partial^2}{\partial\xi_*^2} + \frac{\partial^2}{\partial\eta_*^2}\right)\hat{\phi} = 0\,. \qquad (9.102)$$

The boundary conditions (9.49), (9.51), and (9.52) can be written as:

1. As $\eta_* \to \infty$,

$$\hat{\phi} \to 0\,. \qquad (9.103)$$

2. At $\eta_* = 0$,

$$\hat{\phi} + \hat{h} = -\frac{\varepsilon^{2-2\alpha}}{\mathcal{G}_0}\frac{\partial^2\tilde{h}}{\partial\xi_*^2} + \text{(higher order terms)}, \qquad (9.104)$$

$$\varepsilon^{1-\alpha}\frac{\partial\hat{\phi}}{\partial\eta_*} - \sigma_0\mathcal{G}_0^2(\xi)\hat{h} + \varepsilon^{1-\alpha}U_0 Y_{\xi,0}(\xi)\frac{\partial\hat{h}}{\partial\xi_*}$$

$$= \text{(higher order terms)}. \qquad (9.105)$$

Letting $\varepsilon \to 0$, the above inner equations are reduced to the following third-order, complex, ordinary differential equation for the interface perturbation $\hat{h}$:

$$-\mathrm{i}\frac{\varepsilon^{3-3\alpha}}{\mathcal{G}_0}\frac{\mathrm{d}^3\hat{h}}{\mathrm{d}\xi_*^3} + \varepsilon^{1-\alpha}\left\{\mathrm{i} - U_0 Y_{\xi,0}(\xi)\right\}\frac{\mathrm{d}\hat{h}}{\mathrm{d}\xi_*} + \sigma_0\mathcal{G}_0^2\hat{h}$$

$$= \text{(higher order terms)}. \qquad (9.106)$$

With the new inner variable $\rho_* = (\rho - \rho_c)/\varepsilon^\alpha$, (9.106) can be written as

$$\mathrm{i}\frac{\varepsilon^{3-3\alpha}G^3}{\lambda_0 S}\frac{\mathrm{d}^3\hat{h}}{\mathrm{d}\rho_*^3} + \varepsilon^{1-\alpha}G(\rho)(\rho + \mathrm{i})\frac{\mathrm{d}\hat{h}}{\mathrm{d}\rho_*} + \lambda_0^2 S^2\sigma_0\hat{h}$$

$$= \text{(higher order terms)}. \qquad (9.107)$$

As before, we use the transformation:

$$\tilde{h} = W(\xi)\,\exp\left\{\frac{\mathrm{i}}{\varepsilon}\int_{\rho_c}^{\rho} \tilde{k}_c(\rho_1)\,\mathrm{d}\rho_1\right\}, \qquad (9.108)$$

where the reference wavenumber function $\tilde{k}_c(\rho)$ is chosen in the same way as in Chap. 6, i.e.,

$$\left(\frac{\partial S^2 \Sigma_*}{\partial \tilde{k}_0}\right)_{\tilde{k}_0 = \tilde{k}_c} = \frac{3G^2 \tilde{k}_c^2}{\lambda_0 S} - (1 - i\rho) = 0\,, \tag{9.109}$$

or

$$\tilde{k}_c(\rho) = \frac{\sqrt{\frac{\lambda_0 S}{3}\,(1 - i\rho)}}{G(\rho)} = \frac{e^{-\frac{i\pi}{4}} \lambda_0^{\frac{1}{2}} (\rho - i)^{\frac{1}{4}} (\rho + i)^{\frac{3}{4}}}{3^{\frac{1}{2}} G(\rho)} \tag{9.110}$$

$$(\mathrm{Re}\{\tilde{k}_c\} > 0)\,.$$

From this definition, it is known that

$$\mathrm{Re}\{\tilde{k}_0^{(3)}\} < \mathrm{Re}\{\tilde{k}_c\} < \mathrm{Re}\{\tilde{k}_0^{(1)}\}\,. \tag{9.111}$$

Thus in the outer region we have

$$\begin{aligned}
H_1 &= W_0^{(+)}(\rho) \exp\left\{\frac{i}{\varepsilon} \int_{\rho_c}^{\rho} \tilde{k}_c(\rho_1)\, d\rho_1\right\} \\
H_3 &= W_0^{(-)}(\rho) \exp\left\{\frac{i}{\varepsilon} \int_{\rho_c}^{\rho} \tilde{k}_c(\rho_1)\, d\rho_1\right\}\,.
\end{aligned} \tag{9.112}$$

One can draw the same diagram of the relationship between the H waves and W waves as that for dendrite growth:

$$H_1 \text{ wave} \qquad\qquad \Longleftrightarrow \qquad W^{(+)}$$

(short-wavelength branch)   (outgoing wave);

$$H_3 \text{ wave} \qquad\qquad \Longleftrightarrow \qquad W^{(-)}$$

(long-wavelength branch)   (incoming wave).

In accordance with the above, in the inner region, we set

$$\hat{h} = \hat{W}(\rho_*) \, \exp\left\{\frac{i}{\varepsilon} \int_{\rho_c}^{\rho} \tilde{k}_c(\rho_1)\, d\rho_1\right\}\,. \tag{9.113}$$

In terms of the transformation (9.108), equation (9.107) is transformed to

$$i\varepsilon^{3-3\alpha} \Omega_3 \frac{d^3 \hat{W}}{d\rho_*^3} - \varepsilon^{2-2\alpha} \Omega_2 \frac{d^2 \hat{W}}{d\rho_*^2} - \lambda_0^2 S^2 \left\{\sigma_0 - \Sigma_c(\rho)\right\} \hat{W}$$

$$= \text{(higher order terms)}\,, \tag{9.114}$$

where

$$\Omega_3 = -\frac{G^3}{\lambda_0 S}$$

$$\Omega_2 = -\frac{3G^3 \tilde{k}_c}{\lambda_0 S} = -\mathrm{e}^{-\mathrm{i}\frac{\pi}{4}} \frac{3^{\frac{1}{2}} G^2}{\lambda_0^{\frac{1}{2}}} \left(\frac{\rho + \mathrm{i}}{\rho - \mathrm{i}}\right)^{\frac{1}{4}}$$

$$\Sigma_c(\rho) = \Sigma_*(\tilde{k}_c, \rho) = \mathrm{e}^{-\mathrm{i}\frac{3\pi}{4}} \frac{2}{3^{\frac{3}{2}} \lambda_0^{\frac{3}{2}}} \left(\frac{\rho + \mathrm{i}}{\rho - \mathrm{i}}\right)^{\frac{3}{4}} .$$

(9.115)

Clearly,

$$\sigma_0 - \Sigma_c(\rho_c) = 0 .$$

(9.116)

Therefore ρ_c is a simple turning point of (9.114).

Note that with the use of the outer variable ρ, equation (9.114) may be approximately applied to the whole outer region and yield the same dispersion relationship as (9.91). Its three fundamental solutions, in the WKB form, are $\{W_0^{(+)}(\rho), W_0^{(-)}(\rho), W_0^{(2)}(\rho)\}$, where the solution $W_0^{(2)}(\rho)$, corresponding to the H_2-wave, is not meaningful. In order to eliminate the solution $W_0^{(2)}$ from the general solution, one may reduce equation (9.114) into the following second order ODE by dropping its third order derivative term:

$$\varepsilon^{2-2\alpha} \Omega_2 \frac{\mathrm{d}^2 \hat{W}}{\mathrm{d}\rho_*^2} + \lambda_0^2 S^2 \{\sigma_0 - \Sigma_c(\rho)\} \hat{W} = 0 .$$

(9.117)

Two fundamental solutions of this reduced equation will not exactly equal to the outer solutions $\{W_0^{(+)}(\rho), W_0^{(-)}(\rho)\}$. But, in the intermediate region between the inner region and the outer region, the fundamental solutions of this reduced equation do approximately equal to these outer solutions.

Furthermore, we need to expand all coefficient functions of (9.117) in a Taylor series near $\rho = \rho_c$. Note that

$$\frac{\lambda_0^2 S^2(\rho)}{\Omega_2(\rho)} \Sigma_c'(\rho_c) = \frac{\lambda_0}{3G^2(\rho)} \frac{(\rho + \mathrm{i})^{\frac{3}{4}}}{(\rho_c + \mathrm{i})^{\frac{1}{4}}} \frac{(\rho - \mathrm{i})^{5/4}}{(\rho_c - \mathrm{i})^{7/4}}$$

$$G(\rho) = \frac{\pi}{2a}(\rho - \mathrm{i}a)(\rho + \mathrm{i}a)$$

(9.118)

$$a = \frac{1 - \lambda_0}{\lambda_0} .$$

Besides ρ_c, the inner equation has two other turning points, $\rho = \pm\mathrm{i}$, and two polar points, $\rho = \pm\mathrm{i}a$. The relative positions of these singular points with respect to ρ_c are related to the parameters σ_0, λ_0. For a single-valued, analytical solution, we set the branch cut along the Stokes line (L_1) (see Fig 9.8). Thus, we must choose the singular point ρ_c with $\mathrm{Re}\{\rho_c\} > 0$ and $\mathrm{Im}\{\rho_c\} < 0$. The singular points $\rho = \mathrm{i}, \mathrm{i}a$ are away from the turning point

ρ_c, so their influence on the inner solution is negligible. However, as $\sigma_0 \to 0$, $\rho_c \to -i$, while as $\lambda_0 \to \frac{1}{2}$, $-ia \to -i$. Hence, the singular points $\rho = -i, -ia$ may enter the inner region of ρ_c, and consequently influence the behavior of the inner solution. To discuss these cases, we set

$$\sigma_0 = \hat{\sigma}_0 \varepsilon^{\nu_0}, \qquad a = 1 + \hat{a}\varepsilon^{\mu_0}, \quad \text{or} \quad \lambda_0 = \frac{1}{2} - \frac{\hat{a}}{4}\varepsilon^{\mu_0}, \qquad (9.119)$$

and rewrite the inner equation (9.114) as

$$-\varepsilon^{2-2\alpha}\frac{d^2\hat{W}}{d\rho_*^2} + \frac{A_0^2}{(\rho_c - i)^{\frac{1}{2}}}\frac{(\rho - \rho_c)}{(\rho - ia)(\rho + ia)}\frac{(\rho + i)^{\frac{3}{4}}}{(\rho_c + i)^{\frac{1}{4}}}\hat{W}$$

$$= \text{(higher order terms)} , \qquad (9.120)$$

where

$$A_0 = \frac{2a\lambda_0^{\frac{1}{2}}}{3^{\frac{1}{2}}\pi} . \qquad (9.121)$$

Two cases are found to be significant, as detailed in the following two sections.

9.6.1 Case I: $|\sigma_0| = O(1)$

It follows that $|\rho_c + i| \gg O(\varepsilon^\alpha)$ and $|\rho_c + ia| \gg O(\varepsilon^\alpha)$.

In this case, all the singular points $\rho = \pm i$ and $\rho = \pm ia$ are away from ρ_c. By balancing the leading terms of (9.120) we derive $\alpha = \frac{2}{3}$. We must have

$$|\sigma_0| \gg O(\varepsilon^{\frac{1}{2}}) \quad \text{or} \quad 0 \leq \nu_0 < \frac{1}{2} . \qquad (9.122)$$

To leading order, the inner equation is reduced to Airy's equation

$$\frac{d^2\hat{W}_0}{d\hat{\rho}_*^2} + \hat{\rho}_*\hat{W}_0 = 0 , \qquad (9.123)$$

where

$$\hat{\rho}_* = \frac{B}{\varepsilon^{\frac{2}{3}}}(\rho - \rho_c) ,$$

$$B = \hat{A}_1^{\frac{2}{3}} \qquad (9.124)$$

$$\hat{A}_1 = \frac{iA_0}{(\rho_c - ia)^{\frac{1}{2}}(\rho_c + ia)^{\frac{1}{2}}}\left(\frac{\rho_c + i}{\rho_c - i}\right)^{\frac{1}{4}} ,$$

and

$$0 < \arg(\hat{A}_1) \leq \frac{3\pi}{4}, \quad \text{and} \quad 0 < \arg(B) \leq \frac{\pi}{2} . \qquad (9.125)$$

The structure of the Stokes lines is sketched in Fig. 9.8(a).

9.6.2 Case II: $|\sigma_0| \ll 1$

We further assume $|\rho_c + i| = O(\varepsilon^\alpha)$ and $|\rho_c + ia| = O(\varepsilon^\alpha)$.

In this case, the turning point ρ_c will be close to the singular points $\rho = -i, -ia$. We derive

$$\rho_c = -i + i\hat{\delta}\varepsilon^{\frac{4\nu_0}{3}}, \quad \hat{\delta} = \frac{9}{2^{\frac{1}{3}}}\lambda_0^2\hat{\sigma}_0^{\frac{4}{3}}, \tag{9.126}$$

and thus

$$\begin{cases} \rho + ia = (\rho - \rho_c) + (\rho_c + ia) = \varepsilon^\alpha\rho_* + i\hat{\delta}\varepsilon^{\frac{4\nu_0}{3}} + i\hat{a}\varepsilon^{\mu_0} \\[2mm] \rho + i = (\rho - \rho_c) + (\rho_c + i) \ = \varepsilon^\alpha\rho_* + i\hat{\delta}\varepsilon^{\frac{4\nu_0}{3}}. \end{cases} \tag{9.127}$$

Furthermore, $\alpha = \frac{4\nu_0}{3} = \mu_0$. The left-hand side of (9.120) is dominated by the second and third terms, namely $(1) \ll (2) \approx (3)$. By balancing the leading order terms, it is found that

$$\alpha = \frac{4}{3}, \quad \nu_0 = 1, \quad \text{and} \quad \mu_0 = \frac{4}{3}. \tag{9.128}$$

Therefore we have $(\lambda_0 - \frac{1}{2})^{\frac{3}{4}} = O(\sigma_0)$. This relationship between the quantities σ_0 and $(\lambda_0 - \frac{1}{2})$ is consistent with the quantization conditions that we derive later.

In the far-field as $|\hat{\rho}_*| \to \infty$, the leading order approximation of the inner equation is

$$\frac{d\hat{W}_0}{d\hat{\rho}_*^2} + \hat{\rho}_*^{-\frac{1}{4}}\hat{W}_0 = 0, \tag{9.129}$$

where

$$\hat{\rho}_* = \frac{B}{\varepsilon^{\frac{4}{3}}}(\rho - \rho_c)$$

$$B = \frac{\hat{A}_3^{8/7}}{(i\hat{\delta})^{1/7}} \tag{9.130}$$

$$\hat{A}_3 = \frac{iA_0}{(\rho_c - ia)(\rho_c - i)^{\frac{1}{4}}}$$

The structure of the Stokes lines when $\frac{1}{2} \le \lambda_0 < 1$ (or $0 < a < 1$) is sketched in Fig. 9.8(b).

We see that for both cases I and II, the inner equation in the far-field of the inner region can be written in the following unified form:

$$\frac{d^2\hat{W}_0}{d\hat{\rho}_*^2} + \hat{\rho}_*^{p_0}\hat{W}_0 = 0, \tag{9.131}$$

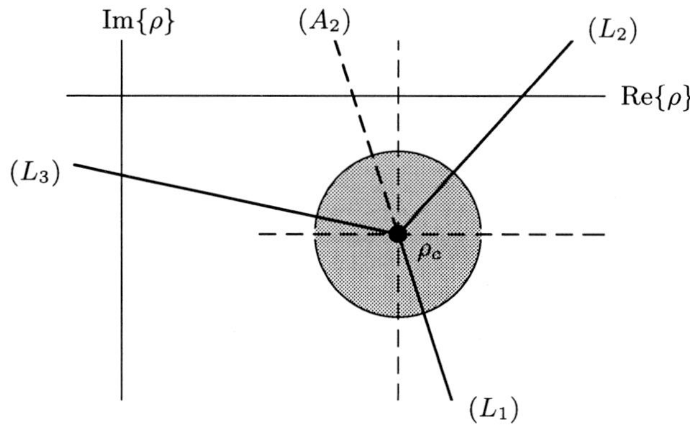

Fig. 9.8. A sketch of the Stokes lines (L_1), (L_2), (L_3) and the anti-Stokes line (A_2): **(a)** for case I; **(b)** for case II

where

$$\hat{\rho}_* = \frac{B}{\varepsilon^\alpha}(\rho - \rho_c) \gg 1. \tag{9.132}$$

For the different cases, the constants B, p_0, and the exponent α are different. Moreover, the anti-Stokes line, which is tangent to the direction $\arg(\zeta) = \frac{3\pi}{2}$ always intersects the real ρ-axis at a point $\rho'_c > 0$.

The general solution of (9.131) is

$$\hat{W}_0 = D_1 \hat{\rho}_*^{\frac{1}{2}} H_\nu^{(1)}(\zeta) + D_2 \hat{\rho}_*^{\frac{1}{2}} H_\nu^{(2)}(\zeta) \qquad (\zeta = 2\nu \hat{\rho}_*^{\frac{1}{2\nu}}), \tag{9.133}$$

where $\nu = \frac{1}{p_0+2}$ and $H_\nu^{(1)}(\zeta)$ and $H_\nu^{(2)}(\zeta)$ are the νth-order Hankel functions of the first and second kind, respectively. For future use, we summarize the results below.

- Case I: $|\sigma_0| \geq O(\varepsilon^{\frac{1}{2}})$,

$$\alpha = \frac{2}{3}, \quad p_0 = 1, \quad \text{and} \quad \nu = \frac{1}{3};$$

- Case II: $|\sigma_0| \leq O(\varepsilon^{\frac{3}{5}})$, $(\lambda_0 - \frac{1}{2})^{\frac{3}{4}} = O(\sigma_0)$,

$$\alpha = \frac{4}{3}, \quad p_0 = -\frac{1}{4}, \quad \text{and} \quad \nu = \frac{4}{7}.$$

We now turn to match the inner solution (9.133) with the outer solution (9.95) in the intermediate regions in each sector. Matching the inner solution with the outer solution in sector (S_2) gives $D_1 = 0$ and $D_3 = D_3' \times$ constant $\neq 0$. Furthermore, in sector (S_1), the matching condition gives the connection formula

$$\frac{D_1}{D_3} = \mathrm{i}2 \cos(\nu\pi). \tag{9.134}$$

So far, we have derived the asymptotic solution in the limit $\varepsilon \to 0$, for any given parameters σ_0, λ_0. As the second step, we shall require the solution to satisfy the tip condition. Thus the parameter σ_0 must be chosen as a proper function of λ_0 and ε. As in the case of dendrite growth, we find two different types of spectra, namely

1. the complex spectrum $\sigma_0 = \sigma_{\mathrm{R}} - \mathrm{i}\omega$ $(\omega > 0)$;
2. the real spectrum (within the accuracy of the asymptotic solution).

These spectra correspond to two different instability mechanisms:

1. the global trapped wave (GTW) instability which is induced by perturbations with a high frequency $(|\omega| = O(1))$.
2. the low-frequency (LF) instability mechanism which is induced by perturbations with low frequency $(\omega \leq O(\varepsilon))$.

9.7 Eigenvalues Spectra and Instability Mechanisms

The derivations that follow are almost completely the same as those for 2D dendrite growth with anisotropy.

9.7.1 The Spectrum of Complex Eigenvalues and GTW Instability

Consider $\sigma_0 = \sigma_R - i\omega$ with $\omega > 0$. With the complex eigenvalue σ_0, the physical solution in the outer region is

$$\mathrm{Re}\left\{\tilde{h}_0(\xi, t)\right\} = \mathrm{Re}\left\{H(\rho)e^{\frac{\sigma_0 t}{\varepsilon}}\right\}, \tag{9.135}$$

where

$$H(\rho) = D_1 H_1 + D_3 H_3. \tag{9.136}$$

We define

$$\begin{cases} d_1 = D_1 e^{-i\chi_1} \\ d_3 = D_3 e^{-i\chi_3}, \end{cases} \tag{9.137}$$

where

$$\begin{aligned} \chi_1 &= \frac{1}{\varepsilon}\int_0^{\rho_c} \tilde{k}_0^{(1)}\mathrm{d}\rho \\ \chi_3 &= \frac{1}{\varepsilon}\int_0^{\rho_c} \tilde{k}_0^{(3)}\mathrm{d}\rho. \end{aligned} \tag{9.138}$$

Thus, it follows from the connection formula (9.134) that

$$\frac{d_1}{d_3} = i2\cos(\nu\pi)e^{-i\chi}, \tag{9.139}$$

where

$$\chi = \frac{1}{\varepsilon}\int_0^{\rho_c}\left(\tilde{k}_0^{(1)} - \tilde{k}_0^{(3)}\right)\mathrm{d}\rho. \tag{9.140}$$

To satisfy the smooth tip conditions, the coefficients d_1 and d_3 must obey

(i) for the symmetrical S-modes,

$$\frac{d_3}{d_1} = -\frac{\tilde{k}_0^{(1)}(0)}{\tilde{k}_0^{(3)}(0)}; \tag{9.141}$$

(ii) for the anti-symmetrical A-modes,

$$d_1 = -d_3. \tag{9.142}$$

Combining (9.139) with (9.141) or (9.142), one obtains the following quantization conditions:

$$\frac{1}{\varepsilon} \int_0^{\rho_c} \left(\tilde{k}_0^{(1)} - \tilde{k}_0^{(3)} \right) \mathrm{d}\rho \; = \; \left(2n + 1 + \frac{1}{2} + \theta_0 \right) \pi$$

$$-\mathrm{i} \ln \alpha_0 \,, \quad \left(n = 0, \pm 1, \pm 2, \cdots \right) \tag{9.143}$$

where

$$\begin{cases} \alpha_0 \, \mathrm{e}^{\mathrm{i}\theta_0 \pi} \; = \; \dfrac{\tilde{k}_0^{(1)}(0)}{\tilde{k}_0^{(3)}(0)} & \text{(for the S-modes);} \\[2ex] \alpha_0 = 1, \; \theta_0 = 0 & \text{(for the A-modes) .} \end{cases} \tag{9.144}$$

The system has complex eigenvalues with $|\sigma_0| = O(1)$, corresponding to case I discussed in the previous section. This spectrum contains two discrete sets of complex eigenvalues for S-modes and A-modes respectively,

$$\sigma_{0n} \quad (n = 0, \pm 1, \pm 2, \cdots),$$

which are functions of ε and λ_0. Given λ_0, the system allows a set of neutrally stable modes ($\sigma_{\mathrm{R}} = 0$) when $\varepsilon = \varepsilon_{*n}$ $(n = 0, 1, 2, \cdots)$ with $\varepsilon_* = \varepsilon_{*0} > \varepsilon_{*1} > \varepsilon_{*2} > \cdots$. Or, given ε, the system allows a set of global neutrally stable modes with $\lambda_0 = \lambda_{0n}$ $(n = 0, 1, 2, \cdots)$ and $\lambda_{00} < \lambda_{01} < \cdots$. These eigenmodes are all traveling waves propagating along the interface. The neutral curves for the S-mode and A-mode are shown in Fig. 9.9. It is found that the neutral A-modes are more stable than the neutral S-modes. Furthermore, in Fig. 9.10, we show the variation with ε of the frequency ω_0 of the corresponding neutral GTW modes.

As in Chap. 7, we can prove that the system does not allow other branches in the complex spectrum corresponding to case II. We can first simplify the quantization condition (9.143) with the assumption $|\sigma_0| \ll 1$. From the local dispersion formula (9.92),

$$\begin{cases} k_e^{(3)}(\rho, \sigma_e) = (1 + \mathrm{i}\rho)\sigma_e + \cdots \\[2ex] k_e^{(1)}(\rho, \sigma_e) = \mathrm{e}^{-\mathrm{i}\pi/4}(\rho + \mathrm{i})^{\frac{3}{4}}(\rho - \mathrm{i})^{\frac{1}{4}} - \frac{1}{2}(1 + \mathrm{i}\rho)\sigma_e + \cdots \end{cases} \tag{9.145}$$

as $\sigma_0 \to 0$. By definition

$$\begin{cases} \sigma_e = \lambda_0^{\frac{3}{2}} \sigma_0 \\[2ex] \tilde{k}_0 = \dfrac{\lambda_0^{\frac{1}{2}}}{G(\rho)} k_e = \dfrac{2(1-\lambda_0)}{\pi \lambda_0^{\frac{1}{2}}} \dfrac{k_e}{(\rho + \mathrm{i}a)(\rho - \mathrm{i}a)} \,, \end{cases} \tag{9.146}$$

then one has

$$\left(\tilde{k}_0^{(1)} - \tilde{k}_0^{(3)} \right) = \frac{2(1 - \lambda_0)}{\pi \lambda_0^{\frac{1}{2}}}$$

$$\times \left\{ \frac{(1 - \mathrm{i}\rho)^{\frac{3}{4}}(1 + \mathrm{i}\rho)^{\frac{1}{4}}}{(\rho - \mathrm{i}a)(\rho + \mathrm{i}a)} - \frac{3\sigma_0}{2} \lambda_0^{\frac{3}{2}} \frac{(1 + \mathrm{i}\rho)}{(\rho - \mathrm{i}a)(\rho + \mathrm{i}a)} \right\} + O(\sigma_0^2). \tag{9.147}$$

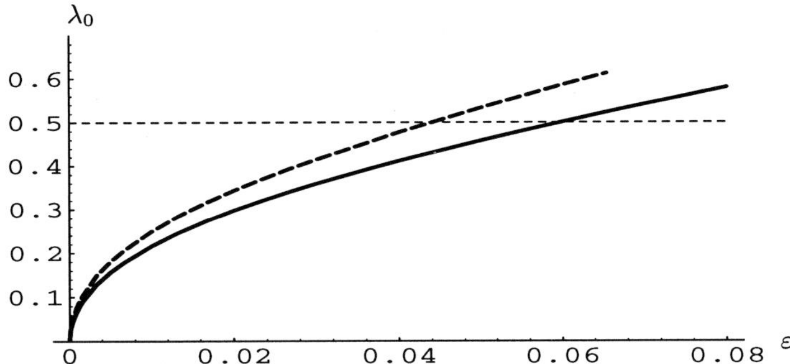

Fig. 9.9. The neutral curves of GTW modes in the (λ_0, ε)-plane. The dashed line represents the S-mode, while the solid line is for the A-mode

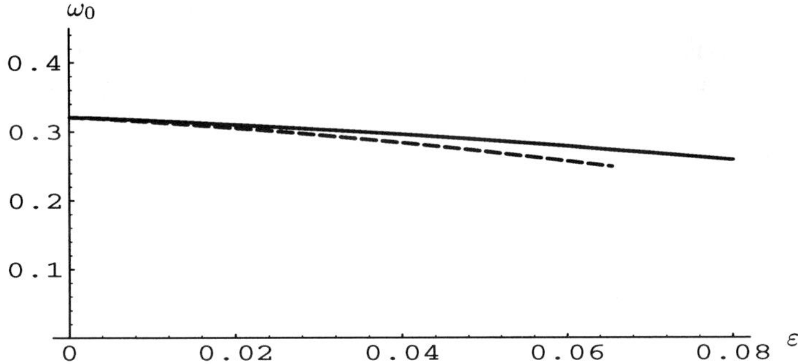

Fig. 9.10. Variation of the frequency, ω_0, of the neutral GTW modes with ε. The dashed line represents the S-mode, while the solid line is for the A-mode

To continue the calculation for χ, one must consider two distinct ranges of λ_0.

(1) $\frac{1}{2} \leq \lambda_0 < 1$ **or** $0 \leq a < 1$**. For this case,**

$$\int_0^{\rho_c} \left(\tilde{k}_0^{(1)} - \tilde{k}_0^{(3)}\right) d\rho = \left(\int_0^{-i(a-0)}\right) + \left(\int_{-i(a+0)}^{-i}\right) + \left(\int_{(C_r)}\right) + \left(\int_{-i}^{\rho_c}\right)$$

$$= -i\frac{2(1-\lambda_0)}{\pi\lambda_0^{\frac{1}{2}}} \int_0^1 \frac{(1-x)^{\frac{3}{4}}(1+x)^{\frac{1}{4}}}{(a-x)(a+x)}dx$$

$$+ i\sigma_0\frac{3\lambda_0(1-\lambda_0)}{\pi} \int_0^1 \frac{(1+x)}{(a-x)(a+x)}dx$$

$$+ \frac{(2\lambda_0 - 1)^{\frac{3}{4}}}{\lambda_0^{\frac{1}{2}}} - \frac{3\lambda_0}{2}\sigma_0 + i\sigma_0\frac{4 \cdot 3^{\frac{3}{2}}\lambda_0^3(1-\lambda_0)}{2\lambda_0 - 1} + O(\sigma_0^2), \tag{9.148}$$

where the integral path (C_r) is a semicircle, connecting the points $\rho = -i(a - 0)$ and $\rho = -i(a + 0)$ clockwisely. It follows that

$$\chi = \frac{1}{\varepsilon} \int_0^{\rho_c} \left(\tilde{k}_0^{(1)} - \tilde{k}_0^{(3)} \right) d\rho \approx \frac{1}{\varepsilon} \left[\frac{(2\lambda_0 - 1)^{\frac{3}{4}}}{\lambda_0^{\frac{1}{2}}} - i\hat{B}_1 \frac{1 - \lambda_0}{\lambda_0^{\frac{1}{2}}} \right]$$

$$- \frac{\sigma_0}{\varepsilon} \left[\frac{3\lambda_0}{2} - i\lambda_0(1 - \lambda_0)\hat{B}_2 \right] , \tag{9.149}$$

where

$$\hat{B}_1 = \frac{2}{\pi} \int_0^1 \frac{(1 - x)^{\frac{3}{4}}(1 + x)^{\frac{1}{4}}}{(a - x)(a + x)} dx , \tag{9.150}$$

$$\hat{B}_2 = \frac{3}{\pi} \int_0^1 \frac{(1 - x)}{(a - x)(a + x)} dx + \frac{4 \cdot 3^{\frac{3}{2}} \lambda_0^2}{2\lambda_0 - 1} . \tag{9.151}$$

(2) $0 < \lambda_0 < \frac{1}{2}$ or $a > 1$. For this case,

$$\int_0^{\rho_c} \left(\tilde{k}_0^{(1)} - \tilde{k}_0^{(3)} \right) d\rho = -i\frac{2(1 - \lambda_0)}{\pi\lambda_0^{\frac{1}{2}}} \int_0^1 \frac{(1 - x)^{\frac{3}{4}}(1 + x)^{\frac{1}{4}}}{(a - x)(a + x)} dx$$

$$+ i\sigma_0 \frac{3\lambda_0(1 - \lambda_0)}{\pi} \int_0^1 \frac{(1 - x)}{(a - x)(a + x)} dx + O(\sigma_0^2) , \tag{9.152}$$

so,

$$\chi \approx \frac{i}{\varepsilon} \left\{ \sigma_0\lambda_0(1 - \lambda_0)\hat{B}_2 - \hat{B}_1 \frac{1 - \lambda_0}{\lambda_0^{\frac{1}{2}}} \right\} . \tag{9.153}$$

By substituting χ into the quantization condition (9.143) we find that the eigenvalues are $\sigma_0 = O(1)$ for both cases. This contradicts our original assumption that $|\sigma_0| \ll 1$. Therefore we conclude that the system only allows the complex spectrum with $|\sigma_0| = O(1)$.

9.7.2 The Spectrum of Real Eigenvalues and the LF Instability

We now consider the spectrum of real eigenvalues σ_0. The physical solution in the outer region for this case is

$$\mathrm{Re}\{\tilde{h}(\xi, t)\} = \mathrm{Re}\{H(\xi)\}e^{\frac{\sigma_0 t}{\varepsilon}} . \tag{9.154}$$

The tip smoothness condition for symmetric modes is

$$\mathrm{Re}\{H(\xi)\}'(0) = \mathrm{Re}\{H'(0)\} = \mathrm{Re}\left\{ \frac{id_1}{\varepsilon} \tilde{k}_0^{(1)}(0) + \frac{id_3}{\varepsilon} \tilde{k}_0^{(3)}(0) \right\} = 0 . \tag{9.155}$$

Without loss of generality, we assume d_1 to be a positive real number and write

$$d_1 > 0, \quad \text{and} \quad d_3 = |d_3|e^{i\chi_0\pi}. \tag{9.156}$$

Since for a real eigenvalue σ_0, $k_0^{(1)}(0)$ and $k_0^{(3)}(0)$ are both real. The tip condition (9.55) for the S-mode becomes $d_{3I} = 0$. We therefore have

$$\frac{d_1}{d_3} = \left|\frac{d_1}{d_3}\right| e^{-i\chi_0\pi} = i2\cos(\nu\pi)e^{-i\chi} \quad (\chi_0 = 0,1). \tag{9.157}$$

This leads to the quantization condition:

$$\mathrm{Re}\left\{\frac{1}{\varepsilon}\int_0^{\rho_c}\left(\tilde{k}_0^{(1)} - \tilde{k}_0^{(3)}\right)d\rho\right\} = \left(2n + \frac{1}{2} + \chi_0\right)\pi \tag{9.158}$$

$$(n = 0, \pm 1, \pm 2, \cdots),$$

$$\left|\frac{d_1}{d_3}\right| = 2\cos(\nu\pi)e^{\mathrm{Im}\{\chi\}}. \tag{9.159}$$

For the anti-symmetric modes, the tip condition is

$$\mathrm{Re}\{H(0)\} = \mathrm{Re}\{d_1 + d_3\} = 0, \tag{9.160}$$

which leads to

$$\left|\frac{d_1}{d_3}\right| = -\cos(\chi_0\pi) \le 1. \tag{9.161}$$

We now have

$$\frac{d_1}{d_3} = \left|\frac{d_1}{d_3}\right| e^{-i\chi_0\pi} = -\cos(\chi_0\pi)e^{-i\chi_0\pi} = i2\cos(\nu\pi)e^{-i\chi}, \tag{9.162}$$

and this leads to the quantization condition:

$$\mathrm{Re}\left\{\frac{1}{\varepsilon}\int_0^{\rho_c}\left(\tilde{k}_0^{(1)} - \tilde{k}_0^{(3)}\right)d\rho\right\} = \left(2n + \frac{1}{2} + \chi_0\right)\pi \tag{9.163}$$

$$(n = 0, \pm 1, \pm 2, \cdots),$$

$$\left|\frac{d_1}{d_3}\right| = -\cos(\chi_0\pi) = 2\cos(\nu\pi)e^{\mathrm{Im}\{\chi\}}. \tag{9.164}$$

The above two quantization conditions can be written in the same form as follows:

$$\mathrm{Re}\left\{\frac{1}{\varepsilon}\int_0^{\rho_c}\left(\tilde{k}_0^{(1)}-\tilde{k}_0^{(3)}\right)\mathrm{d}\rho\right\}=\left(2n+\frac{1}{2}+\chi_0\right)\pi \qquad (9.165)$$

$$(n=0,\pm1,\pm2,\cdots),$$

$$\left|\frac{d_1}{d_3}\right|=2\cos(\nu\pi)\mathrm{e}^{\mathrm{Im}\{\chi\}}. \qquad (9.166)$$

and

$$\chi_0=\begin{cases}0\text{ or }1 & \text{(for S-modes)}\\[2mm]1+\dfrac{\cos^{-1}\left|\dfrac{d_1}{d_3}\right|}{\pi} & \text{(for A-modes)}.\end{cases} \qquad (9.167)$$

The quantization condition (9.165) determines the eigenvalues σ_{0n}, while formula (9.166) determines the corresponding eigenfunction. We only found the real spectrum with $\sigma_0 \ll 1$. In this case, the quantization condition (9.165) can be simplified in terms of formula (9.149) or (9.153).

When $0 < \lambda_0 < \frac{1}{2}$ or $a > 1$, from (9.153), it follows that $\mathrm{Re}\{\chi\} = 0$ and, as a result, the system will have no eigenvalues. On the other hand, when $\frac{1}{2} \le \lambda_0 < 1$ or $0 \le a < 1$, in terms of (9.149) we obtain

$$\mathrm{Re}\{\chi\}=\frac{1}{\varepsilon}\left(\frac{(2\lambda_0-1)^{\frac{3}{4}}}{\lambda_0^{\frac{1}{2}}}-\frac{3\lambda_0}{2}\sigma_0\right) \qquad (9.168)$$

$$\mathrm{Im}\{\chi\}=\frac{1}{\varepsilon}\left(-\hat{B}_1\frac{1-\lambda_0}{\lambda_0^{\frac{1}{2}}}+\sigma_0\hat{B}_2\lambda_0(1-\lambda_0)\right). \qquad (9.169)$$

Based on these results, the following simplified quantization conditions follow:

(1) For the S-modes

$$\frac{3}{2}\lambda_0\sigma_0=\frac{(2\lambda_0-1)^{\frac{3}{4}}}{\lambda_0^{\frac{1}{2}}}-\varepsilon\left(n+\frac{1}{2}\right)\pi, \qquad \left(\frac{1}{2}\le\lambda_0<1\right). \qquad (9.170)$$

(2) For the A-modes

$$\frac{3}{2}\lambda_0\sigma_0=\frac{(2\lambda_0-1)^{\frac{3}{4}}}{\lambda_0^{\frac{1}{2}}}-\varepsilon\left(2n+\frac{1}{2}+\chi_0\right)\pi$$

$$\cos(\chi_0\pi)=-2\cos\left(\frac{4\pi}{7}\right)\mathrm{e}^{\mathrm{Im}\{\chi\}} \qquad (9.171)$$

$$\left(\frac{1}{2}\le\lambda_0<1\right).$$

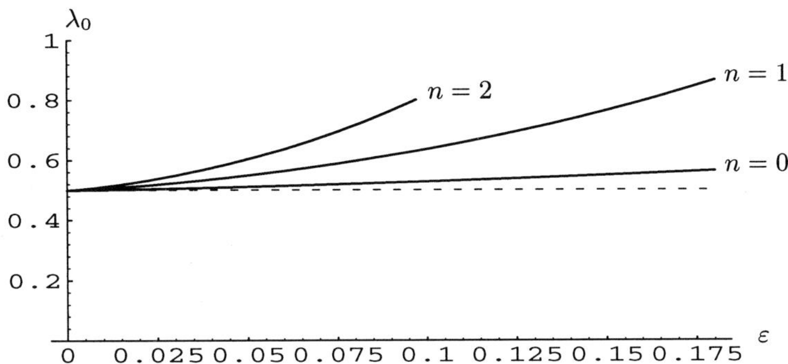

Fig. 9.11. The neutral curves of the LF S-modes in the (λ_0, ε)-plane, with $n = 0, 1, 2$ from bottom to top

It is seen from the quantization conditions (9.170) and (9.171) that one must have eigenvalues $\sigma_0 = O(\varepsilon)$ and $(\lambda_0 - \frac{1}{2})^{\frac{3}{4}} = O(\sigma_0)$. Case II, discussed in the previous section, is obviously consistent with this relationship. Hence $\nu = \frac{4}{7}$.

For any fixed ε and λ_0 , from the above quantization condition (9.170) or (9.171) one can obtain a discrete set of eigenvalues $\{\sigma_{0n}\}$ $(n = 0, 1, 2, 3, \cdots)$. Given λ_0, the system allows a discrete set of neutral stable modes $(\sigma_{0n} = 0)$ corresponding to $\varepsilon = \varepsilon'_n$ $(n = 1, 2, \cdots)$ where $\varepsilon'_0 > \varepsilon'_1 > \varepsilon'_2 > \cdots > \varepsilon'_n > \cdots$. On the other hand, given $\varepsilon > 0$ the system allows a discrete set of neutral modes with width parameters $\lambda_0 = \lambda_{0n}$, or tip velocity $U_0 = U_{0n}$. The neutral S-modes are determined by the formula:

$$\frac{(2\lambda_{0n} - 1)^{\frac{3}{4}}}{\sqrt{\lambda_{0n}}} = \varepsilon\left(n + \frac{1}{2}\right)\pi , \qquad \left(\frac{1}{2} \le \lambda_0 < 1\right). \tag{9.172}$$

The neutral curves of the S-modes in the (λ_0, ε)-plane are shown in Fig. 9.11.

The neutral A-modes are determined by

$$\frac{(2\lambda_{0n} - 1)^{\frac{3}{4}}}{\sqrt{\lambda_{0n}}} = \varepsilon\left(2n + \frac{1}{2} + \chi_0\right)\pi ,$$

$$\cos(\chi_0\pi) = -2\cos\left(\frac{4\pi}{7}\right)\exp\left\{-\frac{\hat{B}_1}{\varepsilon}\frac{(1 - \lambda_{0n})}{\lambda_{0n}^{\frac{1}{2}}}\right\} , \tag{9.173}$$

for $\frac{\pi}{2} < \chi_0 < \pi$. It is found that the neutral S-mode $(n = 0)$ is more stable than the neutral A-mode $(n = 0)$. These neutral S-modes are steady finger solutions that coincide with the classic steady finger solutions discovered numerically by Vanden-Broeck (1983) and by some other investigators analytically in terms of the microscopic solvability condition (MSC) theory. The LF instability was first discovered numerically by Kessler and Levine (1986). It was also studied by Bensimon, Pelce, and Shraiman using a different approach (1987).

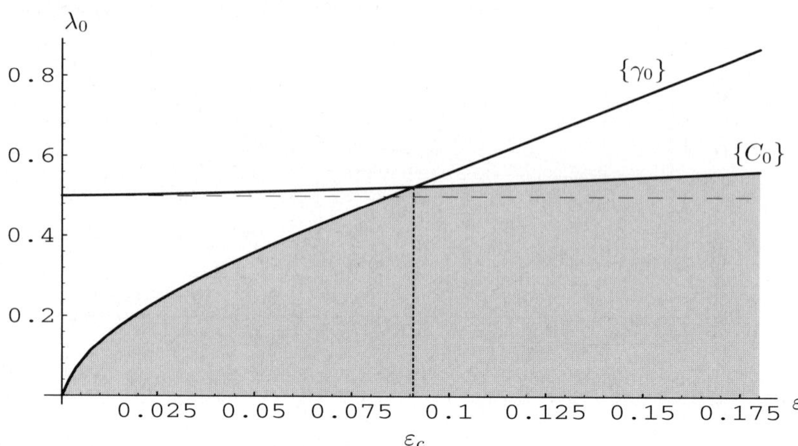

Fig. 9.12. The stability diagram of viscous fingering in the system without nose bubble. {C_0} is the neutral curve of the S-mode ($n = 0$) of the LF mechanism, while {γ_0} is the neutral curve of the A-mode ($n = 0$) of the GTW mechanism. The shaded region is stable, whereas the remaining part of plane is the unstable region

In Fig. 9.12, we plot the neutral curve {γ_0} of the A-mode ($n = 0$) of the GTW mechanism, and the neutral curve {C_0} of the S-mode ($n = 0$) of the LF mechanism, in the parameter plane (λ, ε). These two neutral curves intersect each other at a critical number $\varepsilon_c = 0.0908$. The shaded region below these two curves shown in Fig. 9.12 is stable, while the region above either of the curves is unstable region. It is evident that when the injection velocity U_∞ is very large, or the surface tension γ very small such that $\varepsilon < 0.0908$, the steady finger solutions are all linearly unstable.

9.8 Fingering Flow with a Nose Bubble

Now we turn to study the finger formation problem with a tiny nose bubble as sketched in Fig. 9.13.

9.8.1 The Basic State of Finger Formation with a Nose Bubble and Its Linear Perturbation

The presence of the bubble introduces an additional perturbation to the Zhuravlev–Saffman–Taylor solution. The bubble's effect can be simulated by modifying the interface condition at the finger tip. Assume that the effective bubble radius is r_b and the interface shape the coordinate system (x, y) is denoted by the function $y_\mathrm{s}(x, t)$. The entire interface of the finger is comprised

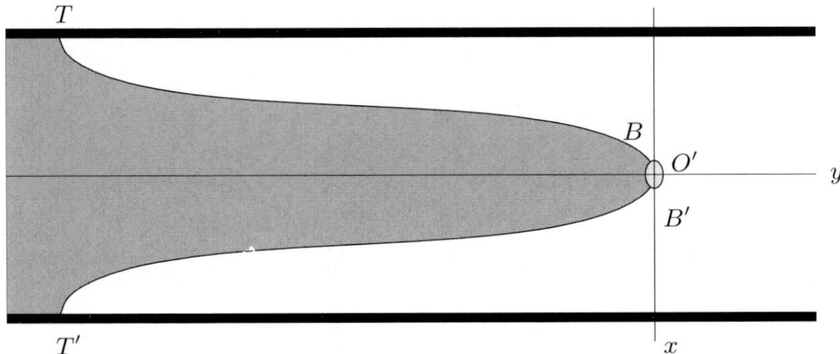

Fig. 9.13. Sketch of a finger with a nose bubble

of the curves: $T \to B$, $B \to O' \to B'$, and $B' \to T'$, which are joined at the triple points B and B'.

For simplicity we assume that the small nose bubble is solid, with shape and size unchanged during the whole process and that the slip condition $u_n = 0$ is applicable. Hence, the bubble surface $BO'B'$ coincides with the streamline $\psi = 0$. Furthermore, several kinds of boundary conditions may be imposed at the triple points B and B', where the surface of the bubble intersects with the interface. Either the triple point location on the bubble's surface or the contact angle between the bubble and the interface of the finger are prescribed. Specifically, we may consider:

1. The position of the triple point on the bubble is fixed but the contact angle is free. Namely,

$$y_\mathrm{s}(r_\mathrm{b}, t) = b.$$ (9.174)

    We call this condition *the stick condition of the bubble*.
    or

2. The contact angle of the finger surface with the tangential direction $\hat{\tau}$ along the bubble is fixed, but the position of the triple point can freely slip along the bubble. Namely,

$$\left\{ \frac{\mathrm{d}y_\mathrm{s}}{\mathrm{d}\hat{\tau}} \right\}_\mathrm{jump} (r_\mathrm{b}, t) = a.$$ (9.175)

    We call this condition *the slip condition of the bubble*.

In the above, the constants a and b are material parameters depending on the physical properties of the bubble and the fluids.

The problem can be formulated in the (ξ, η) coordinate system based on the ZST solution. The triple point location on the nose bubble $\xi_\mathrm{b}, \eta_\mathrm{b}$ can be determined by the effective size parameter r_b. Accordingly, the triple point conditions at B can be specified as follows:

(1)′ The stick bubble condition,

$$\eta_{\mathrm{s}}(\xi_{\mathrm{b}}, t) = \tilde{b}.$$ (9.176)

(2)′ The slip bubble condition,

$$\left\{ \frac{\mathrm{d}\eta_{\mathrm{s}}}{\mathrm{d}\xi} \right\}_{\mathrm{jump}} (\xi_{\mathrm{b}}, t) = \tilde{a},$$ (9.177)

where the constants $\tilde{a}$ and $\tilde{b}$ are determined by the material parameters a and b.

The exact form of the generalized, steady state solution for the present system is even harder to obtain than the previous case without nose bubble. However, since, in practice, ε and r_{b} are both small, it is expected that these basic state solutions q_{B} can be well approximated by the ZST solution in the region away from the finger's root. This argument can be justified by the experimental evidence, as shown in Fig. 9.14 [9.7]. In the figure, the solid line shows the experimental curve of a smooth finger with a nose bubble, while the cross points represent the ZST solution with the selected tip velocity through the experimental data. One sees an excellent agreement between the experimental curve and the ZST data. This experimental curve is supposed to represent the exact solution q_{B}. It shows that the presence and the nature of the nose bubble has very little effect on the basic state in the region away from the bubble.

Next we consider infinitesimal perturbations around the basic state and perform a linear stability analysis. One can see that although the modification of the basic state induced by presence of the nose bubble is negligible, the effect of the bubble on the perturbed state is significant.

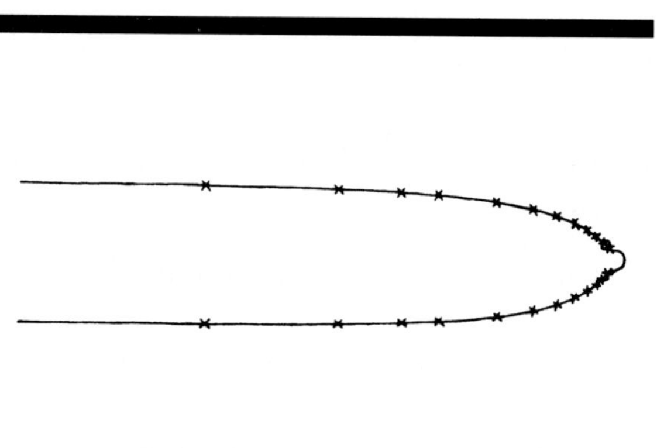

Fig. 9.14. The comparison of an experimental record of smooth finger with nose bubble with a properly selected ZST solution. The solid line is the experimental curve, while the cross points are calculated from the ZST solution

The linear perturbed system, in the leading approximation as ε and $r_b \to 0$, remains the same as that discussed in previous sections except that the tip condition is now changed due to the presence of the bubble. We shall once again apply the new variable ρ instead of ξ. At the triple point B, $\xi = \xi_b$ and $\rho = \rho_b$. We define the effective bubble size r_b by the curve length $\ell = \ell_b$ measured from $\xi = 0$ to $\xi = \xi_b$ along the ZST finger, namely

$$r_b = \ell_b \approx F(0)\rho_b = \frac{2}{\pi}\left(\frac{\lambda_0^2}{1 - \lambda_0}\right)\rho_b \,. \tag{9.178}$$

Thus, at the bubble triple point, we have

(i) the stick condition:

$$\text{as} \quad \rho \to \rho_b + 0 \,, \quad \tilde{h} = 0 \quad \text{and} \quad \frac{\partial \tilde{h}}{\partial \rho} < \infty \,, \tag{9.179}$$

or

(ii) the slip condition:

$$\text{as} \quad \rho \to \rho_b + 0 \,, \quad \tilde{h} < \infty \quad \text{and} \quad \frac{\partial \tilde{h}}{\partial \rho} = 0 \,. \tag{9.180}$$

9.8.2 The Quantization Conditions for the System with a Nose Bubble

The entire procedure for the stability analysis remains the same as in the previous sections for the system without a nose bubble. Both the GTW mechanism and LF mechanism remain valid for the present system. The respective quantization conditions in the leading order approximation are almost the same as (9.143) and (9.165) or (9.170) and (9.171). The only change is in the lower limit of the integrations due to the bubble effect.

The results are as follows:

The GTW mechanism. The branch of complex eigenvalues $\sigma_0 = O(1)$ is determined by the quantization condition

(i) for the S-modes:

$$\frac{1}{\varepsilon} \int_{\rho_b}^{\rho_c} \left(\tilde{k}_0^{(1)} - \tilde{k}_0^{(3)} \right) d\rho = \left(2n + 1 + \frac{1}{2} + \theta_0 \right)\pi - i \ln \alpha_0 \tag{9.181}$$

$$\left(n = 0, \pm 1, \pm 2, \cdots \right)$$

where

$$\alpha_0 \, e^{i\theta_0 \pi} = \frac{\tilde{k}_0^{(1)}(\rho_b)}{\tilde{k}_0^{(3)}(\rho_b)} = \frac{k_0^{(1)}(\rho_b)}{k_0^{(3)}(\rho_b)} \,, \tag{9.182}$$

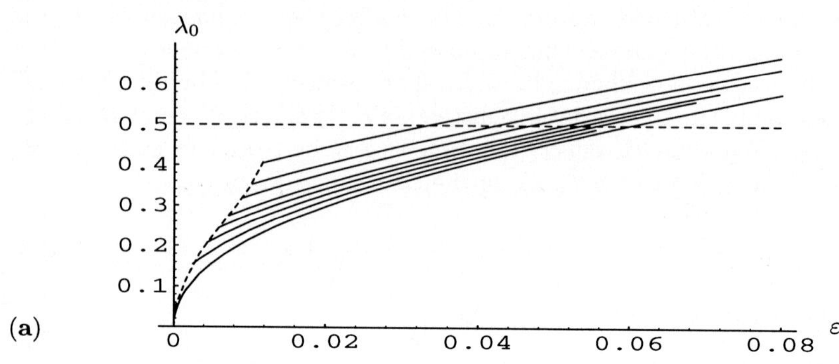

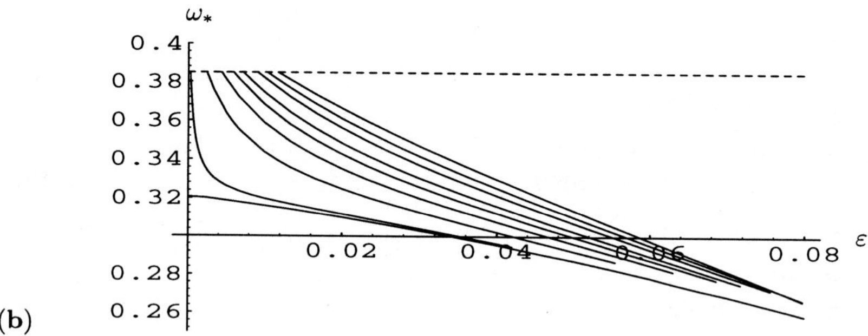

Fig. 9.15. For a variety of values of parameter ℓ_{b} (from top to bottom, $\ell_{\mathrm{b}} = 0.12$, 0.08, 0.06, 0.04, 0.03, 0.02, 0.01, 0.0): **(a)** the variation of λ_0 of the antisymmetric neutral mode $(n = 0)$ with ε; **(b)** the variation of ω_* of the antisymmetric neutral mode $(n = 0)$ with ε

(ii) for the A-mode:

$$\frac{1}{\varepsilon} \int_{\rho_{\mathrm{b}}}^{\rho_c} \left(\tilde{k}_0^{(1)} - \tilde{k}_0^{(3)} \right) \mathrm{d}\rho \; = \left(2n + 1 + \frac{1}{2} \right)\pi \qquad (9.183)$$

$$\left(n = 0, \pm 1, \pm 2, \cdots \right).$$

These quantization conditions give two discrete sets of complex eigenvalues

$$\sigma_{0n} \quad (n = 0, \pm 1, \pm 2, \cdots)$$

which are functions of ε, λ_0, and ρ_{b}.

The global neutrally stable modes $(\sigma_{\mathrm{R}} = 0)$ correspond to $\varepsilon = \varepsilon_{*0} = \varepsilon_* > \varepsilon_{*1} > \varepsilon_{*2} > \cdots$, for given $\lambda_0, \ell_{\mathrm{b}}$, or $\lambda_0 = \lambda_{00} < \lambda_{01} < \cdots$, for given $\varepsilon, \ell_{\mathrm{b}}$. In Fig. 9.15 we show the neutral curves of A-modes $(n = 0)$ in the (ε, λ_0)-plane for various bubble sizes ℓ_{b}, and the variation of ω_* of the corresponding neutral modes with ε, respectively.

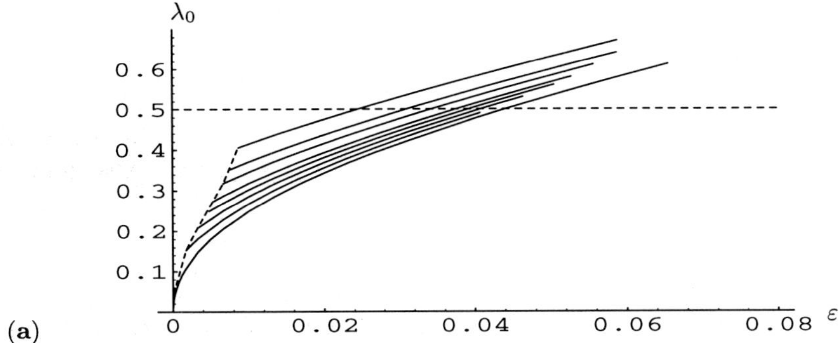

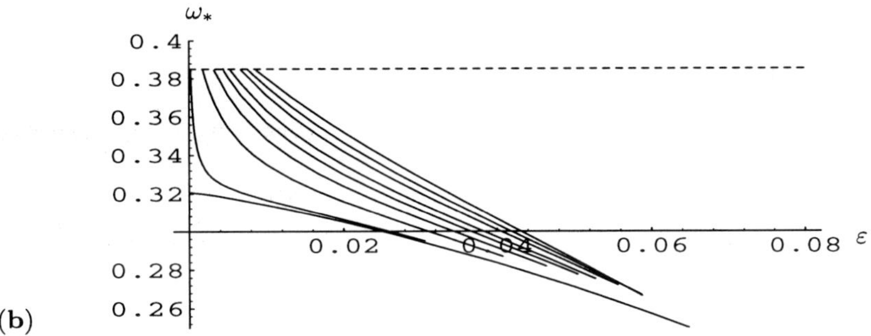

Fig. 9.16. For a variety of values of the parameter ℓ_b (from top to bottom, $\ell_b =$ 0.12, 0.08, 0.06, 0.04, 0.03, 0.02, 0.01, 0.0): (**a**) the variation of λ_0 of the symmetric neutral mode ($n = 0$) with ε; (**b**) the variation of ω_* of the symmetric neutral mode ($n = 0$) with ε

In Fig. 9.16, we show the neutral curves of S-modes with ($n = 0$) in the parameter (ε, λ_0) plane for various of bubble sizes ℓ_b, and the variations of ω_* of the corresponding neutral modes with ε, respectively.

We found that there exists a 'prohibition zone' in the parameter plane where a single-valued oscillatory neutral mode is no longer permitted. The boundary of this zone is marked by the dashed lines in Figs. 9.15 and 9.16.

The LF mechanism. The branch of real eigenvalues $|\sigma_0| \ll 1$ is determined by the quantization condition

$$\mathrm{Re}\,\{\chi\} = \left(2n + \tfrac{1}{2} + \chi_0\right)\pi\,, \qquad \left(\tfrac{1}{2} \le \lambda_0 < 1\right) \tag{9.184}$$

$$(n = 0, \pm 1, \pm 2, \cdots)\,, \tag{9.185}$$

where

$$\begin{cases} \chi_0 = 0,\ \text{or}\ 1 & \text{for S-modes,} \\[2mm] \cos(\chi_0\pi) = -2\cos(\tfrac{4\pi}{7})e^{\mathrm{Im}\{\chi\}} & \text{for A-modes,} \end{cases} \tag{9.186}$$

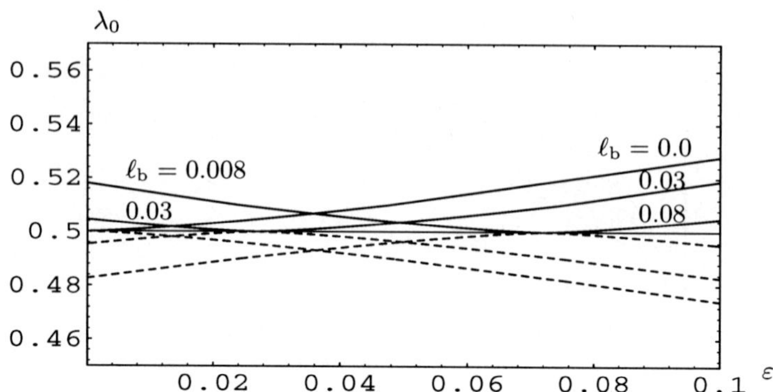

Fig. 9.17. The neutral curves of the LF mode ($n = 0$) for the systems with a nose bubble: $\ell_{\rm b} = 0.0, 0.03, 0.08$

and

$$\chi = \frac{1}{\varepsilon} \int_{\rho_{\rm b}}^{\rho_c} \left(\tilde{k}_0^{(1)} - \tilde{k}_0^{(3)} \right) d\rho . \tag{9.187}$$

Since $|\sigma_0| \ll 1$, the above quantization condition can be further simplified. For S-modes, it reduces to

$$\frac{3}{2}\lambda_0\sigma_0 = \frac{(2\lambda_0 - 1)^{\frac{3}{4}}}{\lambda_0^{\frac{1}{2}}} + \frac{\ell_{\rm b}\left(1 - \frac{2}{3}\lambda_0^{\frac{3}{2}}\sigma_0\right)}{\lambda_0^{\frac{1}{2}}} - \varepsilon\left(n + \frac{1}{2}\right)\pi \tag{9.188}$$

$$\left(\frac{1}{2} \le \lambda_0 < 1\right) .$$

For any fixed ε, we find a discrete set of neutral modes with width parameters $\lambda_0 = \lambda_{0n}$ determined by the formula

$$\frac{(2\lambda_{0n} - 1)^{\frac{3}{4}} + \ell_{\rm b}}{\sqrt{\lambda_{0n}}} = \varepsilon\left(n + \frac{1}{2}\right)\pi , \qquad \left(\frac{1}{2} \le \lambda_{0n} < 1\right) . \tag{9.189}$$

These neutral modes are steady smooth finger solutions with a tiny nose bubble. For a given operation condition with ε fixed, the width of the steady smooth finger λ_0 is reduced, due to the effect of the nose bubble. The neutral curves of the S-modes ($n = 0$) for various bubble sizes are shown in Fig. 9.17. In this figure the dashed lines should be disregarded, as they describe the branches with $\lambda_0 < 0.5$.

9.9 The Selection Criteria of Finger Solutions

Based on our understanding of instability mechanisms obtained in the preceding sections, the selection criteria for finger solutions can be derived. For this purpose we can utilize the stability diagram in Fig. 9.12.

Setting λ_0 as the effective width of the finger, we can examine the basic state solution of the system in the (λ_0, ε)-plane. Noting that, if the system has a steady limit solution as $t \to \infty$, this fixed point solution must be the classic steady finger solution with infinitely long stem and it must be a LF neutral mode, one can make the following statements:

1. If the finger system exhibits a steady pattern as $t \to \infty$, the limit solution must be on the neutral curve $\{\mathcal{C}_0\}$ and it occurs only when $\varepsilon \geq \varepsilon_c$. In other words, for a small surface tension when $\varepsilon < \varepsilon_c$, the steady needle solution can never persist due to the existence of a number of growing oscillatory GTW modes. Therefore the conclusion drawn by Tanveer and others that the steady finger solution $(n = 0)$ is linearly stable for the entire range of $0 < \varepsilon \ll 1$ is incorrect and the selection criterion given by the MSC theory is not applicable in the range $0 \leq \varepsilon < \varepsilon_c$.

2. If the finger system exhibits a time periodic, oscillatory pattern as $t \to \infty$, the limit solution, in the scope of a linear theory, must correspond to a GTW neutral mode on the neutral curve $\{\gamma_0\}$ and it occurs only when $0 \leq \varepsilon \leq \varepsilon_c$. As we have explained for the case of dendrite growth, the unsteady oscillatory pattern determined by the GTW neutral mode is self-sustaining. It does not need a continuously acting noise for its persistence. This statement also implies that for large surface tension, when $\varepsilon > \varepsilon_c$, no self-sustaining, oscillatory finger is possible.

3. If the finger system exhibits a chaotic pattern as $t \to \infty$, it must be in the unstable region.

The above statements appear to be in good agreement with both experimental observations and numerical simulations of the initial value problem. Experimental observations show that the realistic finger system approaches a steady limit within a certain range of ε, as $t \to \infty$. Therefore, from the above statement (1), one may further conclude that the neutral modes of the LF mechanism will in fact be selected for $\varepsilon > \varepsilon_c$. This conclusion has been verified by many experimental observations, as well as numerical simulations.

In what follows we attempt to directly examine the theoretical statement (2) in terms of the available experimental evidence. Although the time periodic, oscillatory finger is hard to observe in experiments with a 'pure' system, it is easily observed experimentally for systems with a nose bubble. Once a system has time-periodic limit solutions, according to statement (2), such a time-periodic, oscillatory finger solution must correspond to the GTW neutral mode and will only occur in the range $0 < \varepsilon \leq \varepsilon_c$. Thus, one may use experimental data to check whether or not this conclusion is correct.

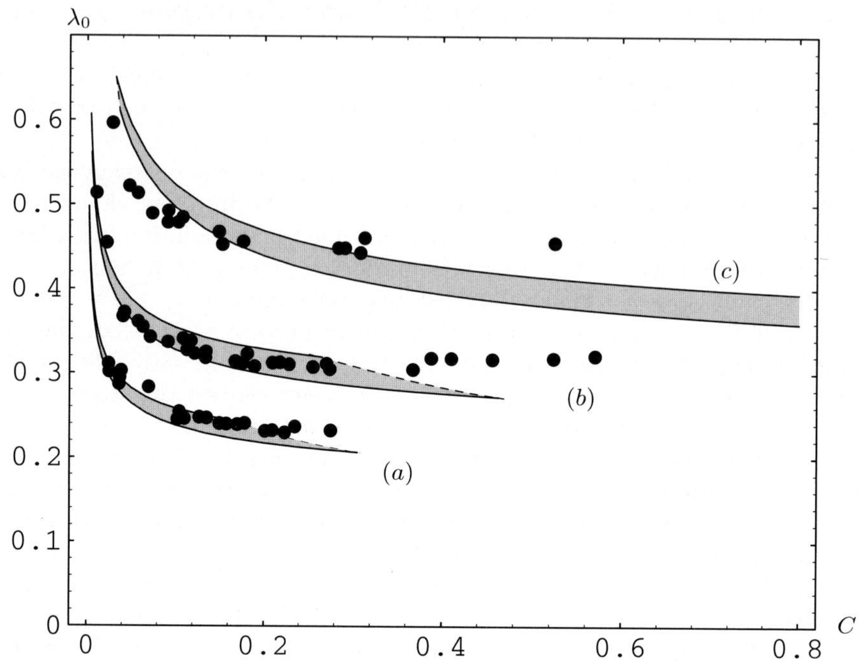

Fig. 9.18. The theoretical predictions compared with experimental data of Couder et al. (1986) for the cases: (a) $W = 12$ cm, $r_b = 0.24-0.30$ cm, $\ell_b = 0.02-0.025$; (b) $W = 6$ cm, $r_b = 0.24-0.30$ cm, $\ell_b = 0.04-0.05$; (c) $W = 2$ cm, $r_b = 0.12-0.18$ cm, $\ell_b = 0.06 - 0.09$. The shaded regions show the neutrally stable states predicted by the theory while the dots are the experimental data

The dependence of the GTW mechanism on the nose bubble size has been derived in the preceding section. In Fig. 9.18 we compare the theory with the experimental data of Couder et al. (1986).

In the experiments by Couder et al. (1986), the average width of the fingers, λ, was measured. Three sets of experimental data provided by Couder et al. correspond to three different cell widths ($W = 12$ cm, 6 cm, and 2 cm). The thickness of the cell was $b = 0.1$ cm. The bubble size for each run was not precisely recorded but the authors estimated that the range of bubble diameters was $(0.1 - 0.8$ cm$)$. Couder et al. used a stability parameter C different from the ε that we used. The relationship between the two is

$$C = \sqrt{\frac{\lambda_0}{\varepsilon}}\, \bar{b}, \tag{9.190}$$

where $\bar{b}$ is the dimensionless thickness parameter of the cell defined by $\bar{b} = b/W$ and b is the thickness of the cell. The dimensional, effective bubble size is $R_b = \ell_b W$. Thus for any given dimensional bubble size and cell width W, we can determine the bubble size parameter ℓ_b, and plot the neutral curve in the same (λ_0, C) plane as shown in Fig. 9.12. Thus, we can compare the theoretical results with the experimental data of Couder et al.

It is seen that the overall agreement between the two is surprisingly good. Specifically, the theory predicts that in case (a) ($W = 12$ cm) the bubble size in their experiments is in the range: $R_b = (0.24 - 0.30$ cm), or $\ell_b = (0.020 - 0.025)$. The experimental data for this case are fully covered by the neutral curves with the values of ℓ_b within the above range. In case (b), the bubble size parameter is in the range $\ell_b = (0.04 - 0.05)$. The quantitative agreement between the selected width λ predicted by our calculations and the measured data is still very good but a small discrepancy is evident. For case (c) ($W = 2$ cm), we assume that the dimensional bubbles size in the experiments is in the range $R_b = (0.12 - 0.18$ cm), so that ($\ell_b = 0.06 - 0.09$). The selected λ predicted by our theory is in good qualitative agreement with the experimental data but a larger numerical discrepancy appears. The discrepancy between the theory and the experiments may be attributed to the effect of three-dimensionality, since the discrepancy increases as the width of the cell decreases. Moreover, for a given nose bubble size, the existence of the prohibition zone in the (λ_0, ε)-plane predicted by the theory is also very well verified by the experimental data. The boundary of this zone is shown by the dashed lines in Fig. 9.18. It is very interesting to see that no experimental data for the values of C for case (a) were reported beyond the dashed line. This implies that, inside the prohibition zone, oscillatory fingers are indeed not observable.

References

9.1   P. A. Zhuravlev, "On the Motion of a Fluid in Channels", Zap. Leningr. Gorn. In-ta. **33**, No. 3, pp. 54–61, (1956).

9.2   S. Hill, "Channeling in Packed Columns", Chem. Eng. Sci. **1**, pp. 247–253, (1952).

9.3   R. L. Chouke, P. Van Meurs and C. Van der Pol, "The Instability of Slow Immiscible Viscous Liquid-Liquid Displacements in Permeable Media", Trans. AIME **216**, pp. 188–194, (1959).

9.4   P. G. Saffman and G. I. Taylor, "The Penetration of a Fluid into a Porous Medium or Hele–Shaw Cell Containing a More Viscous Liquid", Proc. R. Soc. London Ser. A. **245**, pp. 312–329, (1958).

9.5   P. G. Saffman, "Exact Solution for the Growth of Fingers from a Flat Interface between Two Fluids", Q. J. Mech. Appl. Math. **12**, pp. 146–150, (1959).

9.6   D. Bensimon, "Stability of Viscous Fingering", Phys. Rev. A **33**, pp. 1302–1308, (1986).

9.7   Y. Couder, N. Gerard and M. Rabaud, "Narrow Fingers in the Saffman–Taylor Instability", Phys. Rev. A **34**, pp. 5175–5178, (1986).

9.8   A. R. Kopf-Sill and G. M. Homsy, "Narrow Fingers in a Hele–Shaw Cell", Phys. Fluids **30**, No. 9, pp. 2607–2609, (1987).

9.9   D. Bensimon, P. Pelce and B. I. Shraiman, "Dynamics of Curved Fronts and Pattern Selection", J. Physique **48**, pp. 2081–2087, (1987).

9.10 L. A. Romero, PhD. Thesis, California Institute Of Technology, (1981).

9.11 Jean-Marc Vanden-Broeck, "Fingers in a Hele–Shaw Cell with Surface Tension", Phys. Fluids **26**, No. 8, pp. 2033–2034, (1983).

9.12 S. Tanveer, "Analytic Theory for the Selection of a Symmetric Saffman–Taylor Instability", Phys. Fluids **30**, No. 8, pp. 1589–1605, (1987).

9.13 D. C. Hong and J. S. Langer, "Analytic Theory of the Selection Mechanism in the Saffman–Taylor Problem", Phys. Rev. Lett. **56**, pp. 2032–2035, (1986).

9.14 D. A. Kessler and H. Levine, "Stability of Finger Patterns in Hele–Shaw Cells", Phys. Rev. A **32**, pp. 1930–1933, (1985).

9.15 D. A. Kessler and H. Levine, "Theory of the Saffman–Taylor Finger Pattern", Phys. Rev. A **33**, pp. 2621–2633, (1986).

9.16 J. Mclean and P. G. Saffman, "The Effect of Surface Tension on the Shape of Fingers in a Hele–Shaw Cell", J. Fluid Mech. **102**, pp. 455–469, (1981).

9.17 A. J. DeGregoria and L. W. Schwartz, "A Boundary Integral Method for Two-Phase Displacement in Hele–Shaw Cell", J. Fluid Mech. **164**, pp. 383–400, (1986).

9.18 E. Meiburg and G. M. Homsy, "Nonlinear Unstable Viscous Fingers in Hele–Shaw Cell: Numerical Simulation", Phys. Fluids **31**, No. 3, pp. 429–439, (1988).

9.19 P. Tabling, G. Zocchi and A. Libchaber, "An Experimental Study of the Saffman–Taylor Instability", J. Fluid Mech. **177**, pp. 67–82, (1987).

9.20 G. M. Homsy, "Viscous Fingering in Porous Media", Ann. Rev. Fluid Mech. **19**, pp. 271–311, (1987).

9.21 S. D. Howison, A. A. Lacey and J. R. Ockendon, "Hele–Shaw Free Boundary Problems with Suction", Q. J. Mech. Appl. Math. **41**, pp. 183–193, (1988).

9.22 J. J. Xu, "Global Instability of Viscous Fingering in a Hele–Shaw Cell: Formation of Oscillatory Fingers", Europ. J. Appl. Math. **2**, pp. 105–132, (1991).

9.23 J. J. Xu, "Interfacial Wave Theory for Oscillatory Finger's Formation in a Hele–Shaw Cell: a Comparison with Experiments" Europ. J. Appl. Math. **7**, pp. 169–199, (1996).

9.24 J. J. Xu, "Interfacial Instabilities and Fingering Formation in Hele–Shaw Flow", IMA J. Appl. Math. **57**, pp. 101–135, (1996).

9.25 H. Segur, S. Tanveer and H. Levine (Eds.), '*Asymptotics Beyond All Orders*', NATO ASI Series, Series B: Physics, Vol. 284, (Plenum, New York 1991).

Bibliography

1. M. Abramovitz and I. A. Stegun (Eds.), *Handbook of Mathematical Functions*, (Dover, New York 1964).
2. C. Godreche (Ed.), *Solids Far from Equilibrium*, (Cambridge University Press, Cambridge, New York 1991).
3. D. T. J. Hurle (Ed.), *Handbook of Crystal Growth, Volume 1: Fundamentals, Part A: Thermodynamics and Kinetics; Part B: Transport and Stability*, (Elsevier Science, North–Holland, Amsterdam 1993).
4. P. Hartman, (Ed.), *Crystal Growth: an Introduction*, (Elsevier Science, North–Holland, Amsterdam 1973).
5. J. Kevorkian, J. D. Cole, *Multiple Scale and Singular Perturbation Methods*, Applied Mathematical Sciences, Vol. 114, (Springer, Berlin, Heidelberg 1996).
6. C. C. Lin, *The Theory of Hydrodynamic Stability*, (Cambridge University Press, Cambridge 1955).
7. Paul Manneville, *Dissipative Structures and Weak Turbulence*, Perspectives in Physics, Ed. by Huzihiro Araki, Albert Libchaber, Giorgio Parisi, (Academic, New York 1990).
8. R. E. O'Malley, *Singular Perturbation Methods for Ordinary Differential Equations*, Applied Mathematical Sciences, Vol. 89, (Springer, Berlin, Heidelberg 1991).
9. P. Pelce, *Dynamics of Curved Front*, (Academic, New York 1988).
10. F. Rosenberger, *Fundamentals of Crystal Growth I*, (Springer, New York 1979).

Index